1

Computational Methods of Signal Recovery and Recognition

WILEY SERIES IN TELECOMMUNICATIONS

Donald L. Schilling, Editor
City College of New York

Computational Methods of Signal Recovery and Recognition

Edited by
Richard J. Mammone
Department of Electrical and Computer Engineering
Rutgers University
Piscataway, New Jersey

A Wiley-Interscience Publication
JOHN WILEY & SONS, Inc.
New York • Chichester • Brisbane • Toronto • Singapore

Copyright © 1992 by John Wiley & Sons, Inc.

All rights reserved. Published simultaneously in Canada.

Reproduction or translation of any part of this work
beyond that permitted by Section 107 or 108 of the
1976 United States Copyright Act without the permission
of the copyright owner is unlawful. Requests for
permission or further information should be addressed to
the Permissions Department, John Wiley & Sons, Inc.

Library of Congress Cataloging in Publication Data:

Computational methods of signal recovery and recognition / edited by
  Richard J. Mammone.
      p.   cm. — (Wiley series in telecommunications)
    "A Wiley-Interscience publication."
    Includes bibliographical references and index.
    ISBN 0-471-85384-4
    1. Signal processing—Digital techniques.   I. Mammone, Richard J.
  II. Series.
  TK5102.5.C636   1992
  621.382'2—dc20
                                                      91-31951
                                                        CIP

Printed and bound in the United States of America by Braun-Brumfield, Inc.

10 9 8 7 6 5 4 3 2 1

CONTRIBUTORS

Richard J. Mammone   Department of Electrical and Computer Engineering, Rutgers University. Piscataway, New Jersey

John F. Doherty   Department of Electrical and Computer Engineering, Iowa State University, Ames, Iowa

Kevin Farrell   Center for Computer Aids for Industrial Productivity, Rutgers University, Piscataway, New Jersey

Adam B. Fineberg   Center for Computer Aids for Industrial Productivity, Rutgers University, Piscataway, New Jersey

Shyh-shiaw Kuo   AT&T Bell Laboratories, Murray Hill, New Jersey

Christine I. Podilchuk   AT&T Bell Laboratories, Murray Hill, New Jersey

K. Venkatesh Prasad   California Institute of Technology, Pasadena, California

Ananth Sankar   AT&T Bell Laboratories, Murray Hill, New Jersey

CONTENTS

LIST OF FIGURES

LIST OF TABLES

PREFACE

Computational methods of signal processing are quickly evolving to meet the ever-increasing demands of newly emerging intelligent systems. Applications such as the enhancement of image and speech signals, speech recognition, machine vision, and machine recognition of fingerprints and genetic codes require the most efficient utilization of processing resources. The computational power of digital signal-processing (DSP) chips has increased dramatically over the past several years. Digital signal processors are no longer limited to the implementation of conventional DSP operations such as digital filtering and fast Fourier transforms. The genesis of the new computational methods can be found in the fields of linear algebra, numerical analysis, optimization, and functional analysis, as well as conventional DSP.

This book is intended to introduce the student to the field of computational methods of signal processing. The level of presentation is appropriate for an advanced undergraduate or first-year graduate student in electrical engineering or computer science. Engineers and scientists working in industry and government will also find the book a good introduction to and reference for up-to-date methods of solving practical signal-processing problems.

The material presented here forms the basis of a graduate course in electrical engineering that I have taught at Rutgers University for the past three years. Although the book is multiauthored, I believe that it is well balanced and integrated. My coauthors have all studied their respective areas of expertise while working in the Digital Signal Processing Laboratory at the Center for Computer Aids for Industrial Productivity (CAIP) at Rutgers University. The resulting close teamwork led naturally to the present text. The presentation is enriched by the variation of styles of the different coauthors, while the strong technical interaction of the group has helped to weave a highly consistent text.

The selection of the computational methods presented here is not meant to be exhaustive. We apologize for the absence of many useful contributions which have been excluded due to space limitations. However, we have tried to refer to many of the basic books and seminal journal articles where these topics can be explored. The computational methods discussed here have been found, in our experience, to offer the best solution to the problems we have encountered in our research work.

The book is divided into four parts. Part 1 provides an overview of vector-space methods and their application to signal processing. This part consists of two introductory chapters. Chapter 1 introduces the concept of the vector representation of a signal. The fact that a signal can be accurately represented by a finite number of samples is demonstrated. This is fundamental to the new approach and yet is frequently not explicitly discussed in many texts. It is shown that the concept of signals as vectors in N-dimensional space permits the use of many elementary, but powerful, mathematical tools. The basic problems to be addressed are defined. Image recovery is seen to be an inverse problem, where the measured signal is missing signal components. Adaptive signal recovery is the same process performed in a sample-by-sample manner. Signal recognition is similar to signal recovery, with the added constraint that the signal be known *a priori* to be one of N possible signals. These types of problems arise in digital communications and pattern recognition. Chapter 2 describes how linear systems can be modeled as matrices. This system matrix is the discrete linear operator which maps the input vector into the output vector. Various effects of this linear mapping are discussed. Of particular relevance for the subject at hand are the effects of inverse systems. The difficulties in restoring a input signal vector from a measured output signal vector are enumerated.

The remaining three parts focus on the three types of applications introduced in Chapter 1: signal recovery, adaptive recovery and signal recognition.

Part 2 is centered on signal recovery. In Chapter 3, the problem of image recovery is discussed. The difficulty of signal-recovery problems is reviewed. Methods of regularization, the pseudo-inverse, and row-action projection techniques are presented. In Chapter 4, the use of row-action projection methods for image recovery is dealt with in more detail. Numerical examples using realistic images, which clearly demonstrate the benefits of this approach, are presented. In Chapter 5, the problem of the recovery of 3-D object information from a 2-D blurred image using simulated annealing is discussed. This is called 3-D image restoration.

Part 3 is on adaptive signal processing. Chapter 6 gives a vector-space interpretation of adaptive filtering. The row-action projection method is shown to be more general than the conventional least-mean-square methods in several ways. In Chapter 7, the problem of image recovery is revisited, this time via an adaptive approach. The ability to adapt is shown to yield a significant improvement in performance over conventional recovery methods. In Chapter 8, a computationally efficient method of calculating the discrete Fourier transform spectrum on a sample-by-sample basis is presented. The new method is demonstrated to have clear advantages over conventional methods of time-frequency decomposition. In Chapter 9, methods of adaptive recovery are applied to the problem of channel equalization for digital data. The results are particularly encouraging for time-varying communication channels, such as those encountered in mobile communications. In Chapter 10, adaptive recovery methods are applied to the beamforming problem. Beamforming is frequently used in a communications receiver to spatially filter out signals emanating from directions other than that known to be from the transmitter.

Part 4 deals with signal recognition. In Chapter 11, the use of neural networks for signal recognition is discussed. The popular multilayer perceptron (MLP) architecture and backpropagation training algorithm are presented. The drawbacks of this approach are shown to be the need for the correct MLP architecture to be known *a priori*, slow learning, and a penchant for getting trapped in local minima. In Chapter 12, neural tree networks are introduced to meliorate the drawbacks of MLPs for signal recognition. Chapter 13 presents a new method of recovering multicomponent signals, such as speech, when they have undergone nonlinear contractions and dilations. Traditional approaches to this problem generally attempt to find signal features which are invariant to the distortions. The AM-FM approach introduced here appears to offer significant advantages.

Linear algebra is reviewed in Appendix A. Computer programs for the various computational methods presented in the text are described in Appendix B. A collection of selected problems, exercises, and computer projects is given in Appendix C.

This book was written to be as self-contained as possible. I recommend that the two introductory chapters be read first, then any of the remaining chapters dealing with specific applications could be studied next. The amount of the material covered is intentionally generous for a one-semester graduate course, in order to give the individual instructor as much freedom in presentation as possible.

RICHARD MAMMONE

Piscataway, New Jersey

ACKNOWLEDGMENTS

I would like to thank all my graduate students and colleagues who have helped form the ideas expressed in this text, particularly my former students John Doherty, Kevin Farrell, Adam Fineberg, Steven Gay, S. Kuo, Christine Podilchuk, V. Prasad, Rainer Rothacker, Ananth Sankar, and J. Yogeshwar. I would like to give special thanks to those mentioned above who have generously agreed to contribute their work to this text. It has been my privilege to work with this fine group at the research center for Computer Aids for Industrial Productivity (CAIP) at Rutgers University over the past several years.

I also appreciate the industrial and governmental support for our research efforts through the CAIP center over the years. I would like to thank the past director of the center, Prof. Herbert Freeman, and the present director, Prof. James Flanagan, for their support.

A special note of thanks goes to the series editor, Prof. Donald L. Schilling, whose patience and support have been an inspiration to me not only in the preparation of this text but also in my approach to life. I would like to express my deep gratitude to A. C. Surendran for his gracious efforts in typing from my incomprehensible notes. In addition, the Wiley Interscience editor, George Telecki, has been extremely patient and kind in waiting for the manuscript.

PART 1

THE VECTOR-SPACE APPROACH

1

INTRODUCTION TO THE VECTOR-SPACE APPROACH TO SIGNAL PROCESSING

Richard J. Mammone
Department of Electrical and Computer Engineering
Rutgers University
Piscataway, New Jersey 08855
email: mammone@caip.rutgers.edu

1.1 DIGITAL SIGNAL PROCESSING

The field of digital signal processing (DSP) [1, 2] has its roots in the simulation studies of simple frequency-selective analog filters and the simulation of fairly sophisticated signal-estimation techniques [3]. The conventional tools used in DSP—such as the fast Fourier transform (FFT), z-transform, frequency response, and difference equations—are motivated by the frequency-selective filtering effects of linear shift-invariant systems. Over the last three decades, electrical engineers working in the field have transferred many of the mathematical concepts used in analog filter and system design into the digital domain. The first textbooks promoting this approach (Freeman [4] and Kaiser [5]) appeared in the 1960s. In the 1970s, the seminal books by Oppenheim and Schafer [1] and Rabiner and Gold [2] acted as catalysts to encourage many universities to adopt DSP as a core course in electrical engineering.

Today, it is generally accepted that digital circuits can be designed to perform as the equivalent of any given linear analog circuit. Electrical engineering students are routinely taught how to map a desired analog frequency response into a digital filter-design problem.

The motivations to replace analog circuits with digital circuits include:

3

1. *Modularity*—DSP chips are manufactured so that they are interchangeable in terms of hardware but can implement different signal-processing functions via software.

2. *Precision*—It is difficult and expensive to fabricate resistors, capacitors, inductors, etc. with very high tolerances. Digital circuits have extremely high precision since they typically use floating point arithmetic.

3. *Physical limitations*—Analog circuits requiring negative resistances are not realizable. Circuits requiring large values of inductances can become large and consume much power. Digital methods can implement transfer functions regardless of the size or sign of the equivalent electronic elements.

4. *Computational flexibility*—Some nonlinear operations such as clipping and modulation are routinely used in analog circuits. However, analog circuits are primarily used for linear shift-invariant (LSI) operations. Digital signal processors allow far more computational flexibility than analog processors.

Computational methods that were previously appropriate only for large-scale computers are now feasible for real-time microsystems, i.e., DSP chips. The goal of this book is to help promote the use of more general numerical methods for signal processing applications. The methods advanced here are those that we feel are appropriate at the present time to make the transition from large-scale computers to the realm of real-time DSPs.

The advent of very large–scale integration (VLSI) technology offers a low-cost implementation of DSP chips. The generic DSP chip has evolved to look more like a general-purpose processor with fast multipliers and provision for analog-to-digital converters (ADC) as well as digital-to-analog converters (DAC). The basic DSP system is shown in Figure 1.1.

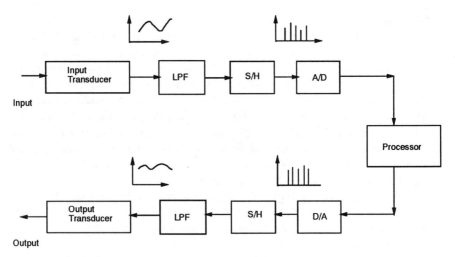

Figure 1.1 DSP system.

The input transducer is a device such as a microphone, TV camera, or thermistor that converts some form of energy into electricity. The output of the transducer is an analog signal. This signal is usually enhanced by passing it through a low-pass filter (with cutoff $f_c = f_{max}$, the ideal maximum frequency of the signal). This is sometimes called an antialiasing filter. The low-pass filter guarantees that there is no energy above the assumed maximum frequency f_{max}. The filtered signal is then sampled at the Nyquist rate ($2f_{max}$) or higher. Each sampled value is then quantized (i.e., converted into a finite number of bits of accuracy) by the ADC. The number of bits used depends on the signal-to-noise ratio (SNR) of the signal. The rule of thumb is that one bit is needed for every 6 dB of SNR. More exactly, if the quantization noise is assumed to be uniformly distributed, then it can be shown [1, 2] that the signal-to-noise ratio due to quantization effects is

$$SNR_q = 6.02B - 1.25. \tag{1.1}$$

The argument is usually given that the SNR_q should be set equal to the SNR of the analog signal. Otherwise, the additional bits used are just quantizing the noise.

The processor then receives a B bit number every sample time. The processor generally performs some operation on several of the input samples and produces an output sample. If the system is to perform automatically—that is, if the processed signal is not the desired output—then the output is used to form a decision. When the decision is determined by the presence of one of a set of signals, we refer to this process as signal recognition.

If the desired output is an analog signal, then the input stages are reversed by converting the B bit binary output values into analog voltage levels using a DAC. Many DACs hold the value for the duration of the sampling. This hold time can be reduced with a sample and hold circuit that holds the value for about 10% of the sampling time. These samples are then passed through a low-pass filter (i.e., a signal reconstruction filter). The analog signal can then be converted back to the original form by the output transducer. A loudspeaker might be used to reproduce a modified version of the input speech or a video monitor might reproduce the modified input image. The process is called real-time if an output sample is produced at each time instant that an input sample arrives; that is, if a one-to-one correspondence between the input and output is assured (although there is generally a time delay between the input and output samples). It is frequently desired to modify the measured input signal so as to undo some degradation such as reverberation or blurring. The output signal will then be a restored version of the input. The method used by the processor in this case is called a signal recovery method. Signal recognition and signal recovery methods will be discussed more fully in the following chapters. The methods described in this text will not be limited to linear shift-invariant (LSI) systems. LSI systems are of great interest due to their widespread use and the powerful mathematical tools available to analyze LSI systems. In addition, they can be

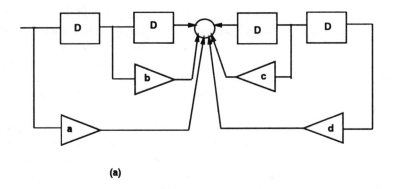

(a)

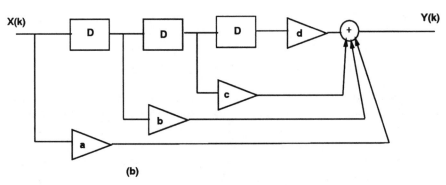

(b)

Figure 1.2 Implementation of LSI systems. (a) Finite impulse response (FIR); (b) infinite impulse response (IIR).

implemented easily by the block diagram shown in Figure 1.2. Digital filters are usually categorized as finite impulse response (FIR) or infinite impulse response (IIR) filters.

The primary objective of this book is to introduce processing methods that are not constrained to be LSI. The linear shift-varying (LSV) signal-processing methods draw heavily on vector-space concepts. LSI methods will be seen to be a subset of the more general LSV methods. The vector-space approach will also lead to the use of convex sets in order to specify nonlinear constraints on the signal.

In the following section, we shall show that a finite signal which is effectively bandlimited can be represented by an N-dimensional vector of N samples of the signal. This result is more realistic than the Nyquist sampling criterion, which states that we only need sample at twice the highest frequency of the signal. The Nyquist sampling theorem requires an infinite number of samples because a bandlimited signal must be of infinite duration. This is not a very practical result. It is shown that, for a finite-duration signal of length T effectively bandlimited to a frequency W, the signal can be accurately reconstructed from $N = 2WT + 1$ samples. This result is of practical importance since it states that signals can be represented to an arbitrary degree of accuracy by a finite number of samples. This vector-space representation allows the use of many powerful vector-analysis tools.

1.2 VECTOR REPRESENTATION OF FINITE-DURATION SIGNALS

Digital processing of signals usually involves sampling of a continuous-time bandlimited signal at a rate greater than twice the maximum frequency (the Nyquist rate). But, in practice, signals are time limited and hence can not be bandlimited due to the uncertainty principle of signal processing [6]. In this section, we shall show that the signal can be effectively represented by a finite number of samples if we assume that

1. the signal is of finite duration:

$$f(t) = 0 \qquad \text{for} \quad |t| > T/2; \tag{1.2}$$

2. the energy of the signal is bounded:

$$\int_{-\infty}^{\infty} |F(f)|^2 \, df = 1; \tag{1.3}$$

3. and $f(t)$ is effectively bandlimited to W:

$$\int_{-W}^{W} |F(f)|^2 \, df = 1 - \eta_W^2, \tag{1.4}$$

where η_W^2 is assumed to be small.

It has been shown [7, 8] that we can approximate $f(t)$ by the series expansion

$$f(t) \approx \tilde{f}(t) = \sum_{i=0}^{L-1} F_i \psi_i(t), \tag{1.5}$$

where $F_i \equiv \int_{-\infty}^{\infty} f(t)\psi_i(t) \, dt$ and $\psi_i(t)$ are the prolate spheroidal wavefunctions (pswfs). The rms error is given by

$$\epsilon_{\text{rms}} \equiv \sqrt{\int_{-T/2}^{T/2} \left[f(t) - \sum_{i=0}^{L-1} F_i \psi_i(t) \right]^2 \, dt}. \tag{1.6}$$

If the number of terms L in the expansion is taken to be the largest integer which is less than or equal to $2WT + 1$, then the rms error can be shown [7, 8] to be bounded by $\epsilon_{rms} < \sqrt{12}\,\eta_W$. Thus ϵ_{rms} can be made arbitrarily small by increasing the effective bandwidth W and subsequently increasing the number of terms in the expansion, L.

Therefore, $\tilde{f}(t)$ is completely determined by the L independent samples $\{f(t_0), f(t_1), \ldots, f(t_{L-1})\}$ of $f(t)$. If the samples are uniformly spaced, the number will be $T/T_s = L \approx 2WT$ for $2WT \gg 1$, which yields $T_s = 1/2W$, or $f_s = 2W$ (i.e., L samples at the Nyquist rate for the duration T).

1.3 VECTOR-SPACE METHODS OF SIGNAL ANALYSIS

A vector **x** is a set of N elements or coordinates such that

$$\mathbf{x} = \begin{pmatrix} x(1) \\ x(2) \\ \vdots \\ x(N) \end{pmatrix}. \tag{1.7}$$

A three-element vector can easily be visualized in three-dimensional space. An N-dimensional vector is more difficult to illustrate. However, most of the concepts familiar to us in two- and three-dimensional vectors can be generalized to N dimensions. The resulting generalization is called an N-dimensional vector space [9].

A vector space is a set of vectors for which vector addition and scalar multiplication are defined. The vector addition of two vectors **x** and **y** is simply

$$\mathbf{x} + \mathbf{y} = \begin{pmatrix} x_1 + y_1 \\ x_2 + y_2 \\ \vdots \\ x_n + y_n \end{pmatrix}, \tag{1.8}$$

and scalar multiplication is defined by

$$\alpha\mathbf{x} = \begin{pmatrix} \alpha x(1) \\ \alpha x(2) \\ \vdots \\ \alpha x(N) \end{pmatrix}. \tag{1.9}$$

The types of vector space of interest here are

$$\mathbf{R}^n = \left\{ \begin{pmatrix} x(1) \\ x(2) \\ \vdots \\ x(N) \end{pmatrix} \,\middle|\, x_1,\ldots,x_n \text{ real numbers} \right\}, \tag{1.10}$$

where $\mathbf{R}^n$ is the set of all n-tuples $(x_1, x_2, \ldots, x_n)$ with real entries and R is the set of real scalars, and

$$\mathbf{C}^n = \left\{ \begin{pmatrix} x(1) \\ x(2) \\ \vdots \\ x(N) \end{pmatrix} \,\middle|\, x_1,\ldots,x_n \text{ complex numbers} \right\}, \tag{1.11}$$

where $\mathbf{C}^n$ is the set of all n-tuples $(x_1, x_2, \ldots, x_n)$ with complex entries and C is the set of complex scalars.

Another vector space of interest is the space of real-valued functions $f(x)$, $a \leq x \leq b$, that are continuous in range. This linear vector space is denoted by $C[a, b]$. The reconstruction of an analog signal from a finite set of samples makes use of the functions $f_i(t) \in C[a, b]$, where the continuous functions are the prolate spheroidal wavefunctions between $t = a = -T/2$ and $t = b = T/2$ and zero elsewhere. We have also chosen a finite-dimensional space (i.e., the first L pswfs); thus, we have $f_i(t) \in C[-T/2, T/2]$, $i = 1, \ldots, L$. The original analog function $f(t)$ can be reconstructed from the L-dimensional vector. Recall that $f(t)$ can approximated by the series expansion

$$f(t) \approx \bar{f}(t) = F_0\psi_0(t) + F_1\psi_1(t) + \cdots + F_{L-1}\psi_{L-1}(t). \tag{1.12}$$

If we take L uniform samples, then the following L equations can be written

$$f(t_0) = F_0\psi_0(t_0) + F_1\psi_1(t_0) + \cdots + F_{L-1}\psi_{L-1}(t_0),$$

$$f(t_1) = F_0\psi_0(t_1) + F_1\psi_1(t_1) + \cdots + F_{L-1}\psi_{L-1}(t_1),$$

$$\vdots$$

$$f(t_{L-1}) = F_0\psi_0(t_L) + F_1\psi_1(t_L) + \cdots + F_{L-1}\psi_{L-1}(t_{L-1}). \tag{1.13}$$

The L coefficients F_i $(i = 0, 1, 2, \ldots, L-1)$ can then be calculated by solving the L equations for the L unknowns. The fact that the above equations have a unique solution has been shown using discrete prolate spheroidal wavefunctions [10].

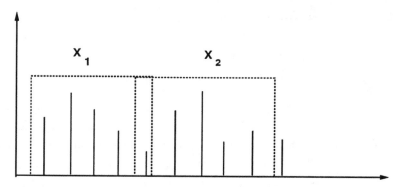

Figure 1.3 Signal vectors.

The concept of signal vectors is important for modern signal processing methods. A signal of long duration is usually segmented into vectors before processing. This is indicated in Figure 1.3. Note that the segmented vectors can be acquired by overlapping or nonoverlapping windows. This topic is discussed in greater detail in the DSP literature [1, 2]. In practice, the data is usually grouped into vectors.

It will be helpful to review some vector-analysis methods that will be useful when discussing modern computational methods. Vector norms are used to measure the size of a vector. The Euclidean L_2 norm is defined for R^n as

$$\|\mathbf{x}\|_2 = \sqrt{x_1^2 + \cdots + x_n^2}, \qquad \mathbf{x} \in R^n. \tag{1.14}$$

One measure of the closeness of two vectors can be given by $\|\mathbf{x} - \mathbf{y}\|_2$. The inner product of two vectors, $\mathbf{x}, \mathbf{y} \in R^n$, is defined by

$$(\mathbf{x}, \mathbf{y}) = \sum_{i=1}^{n} x_i y_i = \mathbf{x}^T \mathbf{y} = \mathbf{y}^T \mathbf{x} \tag{1.15}$$

and, for vectors $\mathbf{x}, \mathbf{y} \in C^n$, the inner product is defined as

$$(\mathbf{x}, \mathbf{y}) = \mathbf{y}^* \mathbf{x}, \tag{1.16}$$

where the superscript * indicates the conjugate transpose.

Two vectors are orthogonal iff $(\mathbf{x}, \mathbf{y}) = 0$. The angle between $\mathbf{x}$ and $\mathbf{y}$ is defined by

$$\Theta(x, y) = \cos^{-1}\left[\frac{(\mathbf{x}, \mathbf{y})}{\|\mathbf{x}\|_2 \|\mathbf{y}\|_2}\right]. \tag{1.17}$$

Note that this can be rewritten as

$$(\mathbf{x}, \mathbf{y}) = \|\mathbf{x}\|_2 \|\mathbf{y}\|_2 \cos \theta. \qquad (1.18)$$

The projection of the vector $\mathbf{x}$ onto the vector $\mathbf{y}$ is given by

$$P_y(\mathbf{x}) = \mathbf{x} \cdot \frac{\mathbf{y}}{\|\mathbf{y}\|_2} = \|\mathbf{x}\|_2 \cos \theta. \qquad (1.19)$$

The projection of the vector $\mathbf{y}$ onto the vector $\mathbf{x}$ is

$$P_x(\mathbf{y}) = \mathbf{y} \cdot \frac{\mathbf{x}}{\|\mathbf{x}\|_2} = \|\mathbf{y}\|_2 \cos \theta. \qquad (1.20)$$

Any vector in R^n can be written as a linear combination of N linearly independent vectors; that is, we can write

$$\mathbf{x} = a_1\mathbf{x}_1 + a_2\mathbf{x}_2 + \cdots + a_N\mathbf{x}_N, \qquad (1.21)$$

where the a_i are scalars and the $\{\mathbf{x}_i\}$ are linearly independent vectors; that is, the equation

$$b_1\mathbf{x}_1 + b_2\mathbf{x}_2 + \cdots + b_N\mathbf{x}_N = 0 \qquad (1.22)$$

holds only if the scalar coefficients are all zero ($b_1 = b_2 = \cdots = b_N = 0$).

The set of linearly independent vectors $\{\mathbf{x}_i\}$ is called a basis and is said to span the vector space R^n. A set that can always be selected such that it spans R^n is the set

$$\mathbf{x}_1 = \begin{pmatrix} 1 \\ 0 \\ \vdots \\ 0 \end{pmatrix}, \qquad \mathbf{x}_2 = \begin{pmatrix} 0 \\ 1 \\ \vdots \\ 0 \end{pmatrix}, \qquad \ldots, \qquad \mathbf{x}_N = \begin{pmatrix} 0 \\ 0 \\ \vdots \\ 1 \end{pmatrix}.$$

Thus, any vector $\mathbf{x}$ can be written in terms of the simple basis set as

$$\mathbf{x} = x_1\mathbf{x}_1 + x_2\mathbf{x}_2 + \cdots + x_N\mathbf{x}_N. \qquad (1.23)$$

There are many possible basis vector sets that can be selected. Sometimes it is desirable to transform a vector into a different set of basis vectors. We may have selected some set of linearly independent vectors $\{\mathbf{v}_i\}$ which span R^n and want to find the scalars $\{a_i\}$ such that

$$\mathbf{x} = a_1\mathbf{v}_1 + a_2\mathbf{v}_2 + \cdots + a_N\mathbf{v}_N. \qquad (1.24)$$

Taking scalar products on both sides for each vector v_i, we have

$$(v_1, x) = a_1(v_1, v_1) + a_2(v_1, v_2) + \cdots + a_N(v_1, v_N), \qquad (1.25)$$

$$(v_2, x) = a_1(v_2, v_1) + a_2(v_2, v_2) + \cdots + a_N(v_2, v_N), \qquad (1.26)$$

$$\vdots$$

$$(v_N, x) = a_1(v_N, v_1) + a_2(v_N, v_2) + \cdots + a_N(v_N, v_N). \qquad (1.27)$$

Note that we have N equations and N unknowns and, if the v_i are linearly independent, the system has an inverse and can be solved for the coefficients $\{a_i\}$. The vector a is the vector x transformed onto the v_i coordinates. If the vectors were chosen to be orthogonal, i.e.,

$$(v_i, v_j) = \begin{cases} \|v\|_2 & \text{if } i = j \\ 0 & \text{if } i \neq j, \end{cases} \qquad (1.28)$$

then the solution for the transformed coordinates of the vectors is simply

$$a_i = \frac{(v_i, x)}{(v_i, v_i)}. \qquad (1.29)$$

The $\{a_i\}$ are the projections of the vector x onto the basis vectors.

There are times when it is desirable to find a basis set of L vectors from a set of M known vectors where $M \geq L$. The Gram–Schmidt orthogonalization method allows for the selection of L linearly independent vectors $\{v_i\}$ from a set of M vectors $\{x_i\}$. The Gram–Schmidt algorithm is as follows:

1. Set $v_1 = x_1$.

2. Then set $v_i = x_i - \sum_{j=1}^{i-1} (x_i, v_j) \dfrac{v_j}{\|v_j\|_2^2}$ for $i = 2, \ldots, L$.

3. If $|v_i| = 0$ for some i, replace x_i with the next vector x_{i+1} and continue until there are L orthogonal vectors v_i.

The resulting set of orthogonal vectors $\{v_i\}$ can be normalized by replacing each vector v_i with the unit vector

$$u_i = \frac{v_i}{\|v_i\|_2}.$$

The set of vectors u_i is called an orthonormal basis. Every finite-dimensional vector space V where the inner product is defined has an orthonormal basis [9]. An arbitrary vector x can be represented using different basis vectors. The change of basis corresponds to a rotation of coordinates. We have seen that, in changing to an orthonormal basis, the elements of the new vector **a** are projections of the original vector **x** onto the basis vectors. The discrete Fourier transform (DFT) will be shown in Chapter 2 to be a transformation of basis vectors onto the orthogonal set of vectors $\{\mathbf{w}_i\}$, where $w_i = e^{j2\pi i/N}$, $i = 0, 1, \ldots, N-1$.

1.4 SIGNAL SPACE

A vector in R^n can be thought of as a vector starting from the origin and ending at some point given by the n-tuple $\{x_1, x_2, \ldots, x_N\}$. Each signal can also be represented by the point in space given by the terminating point of the vector $\{x_1, x_2, \ldots, x_N\}$. A group of signals that are "close" to each other will form a set of points which are "close." This compact set of points in N-dimensional space is called a subspace and represents a set of similar signals. The set is convex if the set of points is such that a line drawn through any two points $\mathbf{x}_1$ and $\mathbf{x}_2$ will never go outside the set (i.e., if $\mathbf{x}_1, \mathbf{x}_2 \in S$, then $\mathbf{y} = \lambda\mathbf{x}_1 + (1-\lambda)\mathbf{x}_2 \in S$, where $0 < \lambda < 1$).

The fact that a signal is known to belong to a convex set in signal space will be seen to play an important role in signal-recovery problems. Ideally, the unknown point **x** may belong to many convex sets, $S_1, S_2, \ldots, S_N$. The intersection of these sets would then contain the desired point **x** (see Figure 1.4).

There are many problems in which the unknown signal **x** is known to be one of a finite number of N discrete signals. This type of problem arises in pattern recognition [11] and in communications [12, 13]. The problem is illustrated in Figure 1.5 [5]. Ideally, the measured signal will always be one of the N possible points or signals. However, any real measurement will have some uncertainty associated with it. This uncertainty is generally modeled as zero-mean noise. A more practical model is illustrated in Figure 1.6. The uncertainties widen the subspace of possible measurements. The problem of signal recognition is to find the best estimate of the underlying signal given a noisy measurement.

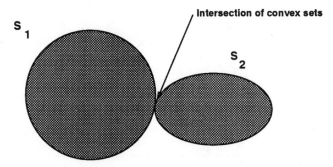

Figure 1.4 Intersection of convex sets.

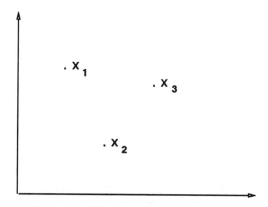

Figure 1.5 Three possible signals as three points in vector space.

1.5 SIGNAL RECOVERY, ADAPTIVE SIGNAL RECOVERY, AND SIGNAL RECOGNITION

The estimation of a signal from a degraded noisy measurement is a frequently encountered problem in practice. The signal might be a speech waveform that is to be estimated from the received signal over a noisy telephone channel. Another scenario might be the restoration of an image that has been degraded by poor-quality optics.

The problem is particularly difficult when some components of the original signal have been removed by the degradation process. Methods which attempt to recover these missing signal components are called signal recovery methods. Signal recovery is only possible if there exists supplementary information that is known *a priori* about the signal; for example, restricting image estimates to consist of only nonnegative light intensity values. This information, along

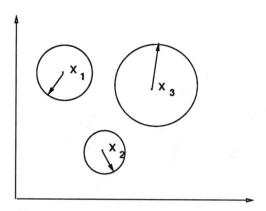

Figure 1.6 Three possible signals with uncertainty.

with the measured signal components, can be used to restore the lost signal components.

Signal recovery problems arise in many fields. In communications, a binary signal might be transmitted through a channel that removes selected frequencies from the signal. There are an infinite number of possible signals if the transmitted signal is not known to be binary. The receiver can take advantage of the fact that the signal at any given time consists of only one of the two values for each bit period. This fact will limit the number of feasible signals to 2^B, where B is the number of bits. Signal estimation methods that constrain the signal to be one of a finite number of possibilities are called signal recognition methods. The generic problem of interest is to find the vector x such that

$$\mathbf{y} = \mathbf{Hx} + \mathbf{n}, \tag{1.30}$$

where $\mathbf{x} \in \mathbf{S}$, the vectors $\mathbf{y}$ and $\mathbf{H}$ are known, and $\mathbf{n}$ is a vector of noise. The set $\mathbf{S}$ can be continuous or discrete.

1.6 SUMMARY AND CONCLUSION

The field of digital signal processing has primarily been concerned with linear shift-invariant systems. Computational power has increased to the point where more sophisticated methods can be considered for signal-processing applications. Vector space techniques are appropriate for practical signal processing applications. The basic properties of vector- and signal-space operators have been reviewed. These properties will be used throughout the text to describe modern signal-processing techniques.

REFERENCES

1. A. V. Oppenheim and R. W. Schafer. *Discrete-Time Signal Processing*. Englewood Cliffs, N.J.: Prentice-Hall, 1989.

2. L. R. Rabiner and B. Gold. *Theory and Application of Digital Signal Processing*. Englewood Cliffs, N.J.: Prentice-Hall, 1975.

3. R. E. Kalman. A new approach to linear filtering and prediction problems. *Trans. ASME Basic Eng.*, **82**:35–45, 1960.

4. H. Freeman. *Discrete Time Systems*. New York: Wiley, 1965.

5. J. F. Kaiser. Digital filters. In F. F. Kuo and J. F. Kaiser, eds., *System Analysis by Digital Computer*. New York: Wiley, 1966.

6. A. Papoulis. *Signal Analysis*. New York: McGraw-Hill, 1977.

7. D. Slepian and H. O. Pollak. Prolate spheroidal wave functions, Fourier analysis and uncertainty—I. *Bell Sys. Tech. J.*, **40**:43–63, 1961.

8. H. J. Landau and H. O. Pollak. Prolate spheroidal wave functions, Fourier analysis and uncertainty—II. *Bell Sys. Tech. J.*, **40**:65–84, 1961.

9. B. Noble and J. W. Daniel. *Applied Linear Algebra*. Englewood Cliffs, N.J.: Prentice-Hall, 1988.

10. D. Slepian. Prolate spheroidal wave functions, Fourier analysis, and uncertainty—V: The discrete case. *Bell Sys. Tech. J.*, 40(57):1371–1430, 1978.

11. R. O. Duda and P. E. Hart. *Pattern Classification and Scene Analysis*. New York: Wiley, 1973.

12. J. M. Wozencraft and I. M. Jacobs. *Principles of Communication Engineering*. New York: Wiley, 1965.

13. H. Taub and D. Schilling. *Principles of Communication Systems*. New York: McGraw-Hill, 1986.

DISCRETE LINEAR SYSTEMS AND SIGNALS—A VECTOR-SPACE APPROACH

Richard J. Mammone

Department of Electrical and Computer Engineering
Rutgers University
Piscataway, New Jersey 08855
email: mammone@caip.rutgers.edu

2.1 DISCRETE APPROXIMATION OF CONTINUOUS-TIME SYSTEMS

We have seen in Chapter 1 that a discrete signal of N samples can be visualized as a vector or a point in N-dimensional space. In this chapter, we shall present a representation of linear systems that is consistent with the vector-space approach. Continuous-time systems can be represented in various ways. An input-output relationship might be specified by a set of differential equations or integral equations or by a combination of the two. These systems can be modeled by discrete systems by using various numerical techniques [1]. The resulting discrete linear system can then be represented by the use of matrices. For example, a convolution relationship, i.e.,

$$y(t) = \int_{-\infty}^{\infty} h(t - \epsilon)x(\epsilon)\, d\epsilon \tag{2.1}$$

$$= h * x \tag{2.2}$$

might be approximated by

$$\mathbf{y} = \mathbf{Hx}. \tag{2.3}$$

The vectors **y** and **x** represent the sampled signals, and the matrix **H** represents the piecewise approximation of the integral convolution operator.

2.2 MATRIX REPRESENTATION OF DISCRETE LINEAR SYSTEMS

A discrete linear system can be represented in matrix form as

$$y = Hx, \tag{2.4}$$

where the ith column vector of the matrix **H** is the response of the system to an impulse applied at the ith instant of time:

$$
\begin{pmatrix} y_1 \\ y_2 \\ \vdots \\ y_N \end{pmatrix}
=
\begin{pmatrix}
h_{11} & h_{12} & \cdots & h_{1m} \\
h_{21} & h_{22} & \cdots & h_{2m} \\
\vdots & \vdots & \ddots & \vdots \\
h_{n1} & h_{n2} & \cdots & h_{nm}
\end{pmatrix}
\begin{pmatrix} x_1 \\ x_2 \\ \vdots \\ x_n \end{pmatrix}
\tag{2.5}
$$

$$
=
\begin{pmatrix} h_{11} \\ h_{21} \\ \vdots \\ h_{n1} \end{pmatrix} x_1
+
\begin{pmatrix} h_{12} \\ h_{22} \\ \vdots \\ h_{n2} \end{pmatrix} x_2
+ \cdots +
\begin{pmatrix} h_{1m} \\ h_{2m} \\ \vdots \\ h_{nm} \end{pmatrix} x_3
\tag{2.6}
$$

$$
= \begin{bmatrix} \mathbf{h}_1 & \mathbf{h}_2 & \cdots & \mathbf{h}_m \end{bmatrix} \mathbf{x}.
\tag{2.7}
$$

Clearly, the output vector **y** is zero when all the elements of the input vector **x** are zero. If one element of **x**—for example, x_2—becomes nonzero, the output is $x_2 \mathbf{h}_2$. If, in addition, x_3 was nonzero, then the output **y** would be the superposition of weighted values of the column vectors of $\mathbf{h}_2$ and $\mathbf{h}_3$. From this discussion we see that the column vectors of **H** are similar to the impulse-response sequence used in conventional linear system analysis. If each successive column $\mathbf{h}_i$ is a shifted version of the preceding vector $\mathbf{h}_{i-1}$, as is the case when **H** is a Toeplitz matrix, (2.4) represents a *linear shift-invariant system* and, in particular, a linear convolution operation. The matrix representation is more general, since it represents not only LSI systems but any discrete linear system.

2.3 DISCRETE SPECTRAL DECOMPOSITION

Spectral analysis is one of the most important tools used in signal processing. The decomposition of signals into sinusoids has several motivations. First, it is helpful to recall that any signal with finite energy can be represented by a Fourier decomposition. Second, sinusoids are the eigenvalues of linear shift-invariant (LSI) systems [2]. That is, the output of an LSI system due to a sinusoidal input at some frequency is a sinusoid at the same frequency; only the amplitude and phase will be different. Third, spectral decomposition occurs naturally in many diverse phenomenon—such as color separation by a prism, x-ray diffraction patterns, and Fourier optics [3]—and has proved useful in studying the processing of speech [4] and image signals [5].

There has been much work on developing methods of extracting the Fourier spectrum of a signal. A popular numerical approximation of the Fourier transform is the discrete Fourier transform (DFT). The DFT can be written in matrix notation as

$$\mathbf{X} = \mathbf{W}\mathbf{x}, \tag{2.8}$$

where $\mathbf{X}$ is the complex-valued vector representing the approximation to the Fourier transform; $\mathbf{x}$ is a vector, usually of real-valued time samples; and $\mathbf{W}$ is a matrix of sampled complex sinusoids given by

$$\mathbf{W} = \begin{pmatrix} W_N^0 & \cdots & W_N^0 & \cdots & W_N^0 \\ W_N^0 & \cdots & W_N^{j-1} & \cdots & W_N^{N-1} \\ \vdots & \ddots & \vdots & \ddots & \vdots \\ W_N^0 & \cdots & W_N^{(i-1)*(j-1)} & \cdots & W_N^{(i-1)*(N-1)} \\ \vdots & \ddots & \vdots & \ddots & \vdots \\ W_N^0 & \cdots & W_N^{(N-1)*(j-1)} & \cdots & W_N^{(N-1)*(N-1)} \end{pmatrix}, \tag{2.9}$$

where $W_N = e^{j2\pi/N}$. Due to the symmetry of the $\mathbf{W}$ matrix, the matrix vector multiplication in (2.8) can be implemented using fewer numerical operations, i.e., multiplications and additions, than the N^2 required by direct computation. There are a number of fast algorithms called fast Fourier transforms (FFT) that require on the order of $N \log N$ multiplications [6]. The numerical approximation to the inverse Fourier transform is given by

$$\mathbf{W}^{-1} = \frac{1}{N}\mathbf{W}^T. \tag{2.10}$$

Note that $\mathbf{W}^T\mathbf{W} = \mathbf{W}\mathbf{W}^T = N\mathbf{I}_N$. Thus, $\mathbf{W}$ is a orthogonal matrix—i.e., the columns and rows are orthogonal—and preserves the L_2 norm of a vector within a scalar multiplier when premultiplied by $\mathbf{W}$—i.e.,

$$\frac{\|\mathbf{X}\|_2^2}{N} = \frac{\|\mathbf{W}\mathbf{x}\|_2^2}{N} = \|\mathbf{x}\|_2^2,$$

which is Parseval's theorem for the DFT in matrix form.

The DFT matrix can be used to diagonalize a cyclic matrix $\mathbf{H}_c$. A cyclic matrix can always be decomposed into the form [7]

$$\mathbf{H}_c = \mathbf{W}^T\mathbf{D}\mathbf{W}, \tag{2.11}$$

where $\mathbf{W}$ is the DFT matrix whose order is the same as that of the cyclic matrix and $\mathbf{D}$ is a diagonal matrix whose diagonal values are the eigenvalues of the system. Hence, (2.4) can be rewritten as

$$y = \mathbf{W}^{-1}\mathbf{D}\mathbf{W}x. \qquad (2.12)$$

Premultiplying both sides by $\mathbf{W}$ yields

$$\mathbf{W}y = \mathbf{D}\mathbf{W}x. \qquad (2.13)$$

If $\mathbf{Y} = \mathbf{W}y$ and $\mathbf{X} = \mathbf{W}x$, then (2.13) becomes

$$\mathbf{Y} = \mathbf{D}\mathbf{X}, \qquad (2.14)$$

which demonstrates that the spectrum of the output at a given frequency depends only on the component of the input at the same frequency. The operation indicated by (2.14) represents a cyclic convolution in the DFT transform domain. The DFT diagonalizes only cyclic matrices. Thus, the filter interpretation of (2.14) is relevant only in the case of a cyclic matrix.

The idea of decoupling the system equations with a transform can be generalized to include a symmetric matrix, i.e., a matrix where

$$\mathbf{H}^T = \mathbf{H}.$$

In this case we can use a similarity transformation. That is, we have

$$\mathbf{H} = \mathbf{U}\mathbf{D}\mathbf{U}^T, \qquad (2.15)$$

where $\mathbf{U}\mathbf{U}^T = \mathbf{U}^T\mathbf{U} = \mathbf{I}$. Thus, we can define a system-dependent transform matrix $\mathbf{U}$ such that

$$y = \mathbf{U}\mathbf{D}\mathbf{U}^T x. \qquad (2.16)$$

Let $\mathbf{Y} = \mathbf{U}^T y$ and $\mathbf{X} = \mathbf{U}^T x$. Then

$$\mathbf{Y} = \mathbf{D}\mathbf{X}, \qquad (2.17)$$

and, since $\mathbf{U}$ is unitary, we have

$$\|\mathbf{X}\|_2 = \|\mathbf{U}^T x\|_2 = \|x\|_2, \qquad (2.18)$$

and Parseval's theorem holds.

A further generalization can be made for any arbitrary matrix $\mathbf{H}$. We can use the singular-value decomposition (SVD) of the matrix $\mathbf{H}$, that is,

$$\mathbf{H} = \mathbf{U}\mathbf{D}\mathbf{V}^T, \qquad (2.19)$$

where $\mathbf{UU}^T = \mathbf{U}^T\mathbf{U} = \mathbf{I}$, $\mathbf{VV}^T = \mathbf{V}^T\mathbf{V}$, $\mathbf{H}^T\mathbf{H} = \mathbf{UDU}^T$, and $\mathbf{HH}^T = \mathbf{VDV}^T$.

Substituting (2.19) into (2.4), we obtain

$$\mathbf{y} = \mathbf{UDV}^T\mathbf{x}. \tag{2.20}$$

Let $\mathbf{Y} = \mathbf{U}^T\mathbf{y}$ and $\mathbf{X} = \mathbf{V}^T\mathbf{x}$. Then

$$\mathbf{Y} = \mathbf{DX}, \tag{2.21}$$

and, since $\mathbf{U}$ and $\mathbf{V}$ are unitary, we have

$$\|\mathbf{Y}\|_2 = \|\mathbf{U}^T\mathbf{y}\|_2 = \|\mathbf{y}\|_2, \tag{2.22}$$

$$\|\mathbf{X}\|_2 = \|\mathbf{V}^T\mathbf{x}\|_2 = \|\mathbf{x}\|_2, \tag{2.23}$$

which is, again, analogous to Parseval's theorem.

Thus, we see that the discrete transform matrices $\mathbf{U}$ and $\mathbf{V}$ can be used to decompose arbitrary discrete linear operators. While the column and row vectors of the DFT and IDFT can be viewed as sampled sinusoids, there is no such simple interpretation for the more general unitary transforms. Although the general transforms are also made up of eigenvectors of linear operators, the linear systems are not restricted to be shift invariant and the transforms are specific to the particular system matrix $\mathbf{H}$.

2.3.1   The Geometry of Linear Operators

It will be shown that the orthogonal transformations discussed in the previous section can be viewed as an orthogonal rotation of coordinates. Consider (2.4), where the system matrix $\mathbf{H}$ is replaced with the spectral decomposition

$$\mathbf{y} = \mathbf{UDU}^T\mathbf{x}. \tag{2.24}$$

The term $\mathbf{U}^T\mathbf{x}$ is a rotation of the coordinates of the point $\mathbf{x}$. This is followed by a stretching or shortening of coordinates by the diagonal matrix $\mathbf{D}$, which is followed by a reverse rotation $\mathbf{U}$. If some elements of $\mathbf{D}$ are zero, then the corresponding eigenvectors are removed from the input vector $\mathbf{x}$. This elimination of some components of $\mathbf{x}$ is called a projection operation. This process is similar to that used in computer graphics, where a 3-D representation of an object is rotated and projected onto a 2-D screen.

The transformation can be written in the form

$$\mathbf{X} = \mathbf{U}^T\mathbf{x} = \begin{pmatrix} \mathbf{u}_{c1} \\ \mathbf{u}_{c2} \\ \vdots \\ \mathbf{u}_{cn} \end{pmatrix} \mathbf{x}, \tag{2.25}$$

where $\mathbf{u}_{ci}$ is the ith column vector of $\mathbf{U}$. Given a vector $\mathbf{x}$ and recalling that $|\mathbf{u}_{ci}| = 1$, we find that its projection onto the rotated coordinate vector u_{ci} is

$$X_i = \frac{\mathbf{u}_{ci} \cdot \mathbf{x}}{\|\mathbf{u}_{ci}\|_2} = \mathbf{u}_{ci} \cdot \mathbf{x}. \tag{2.26}$$

Thus, the vector $\mathbf{x}$ is projected onto the unit vectors $\{\mathbf{u}_{ci}\}$ to give the projection onto the rotated coordinates $u_{c1}, u_{c2}, \ldots, u_{cN}$. Note that the rotated coordinate system is orthogonal; that is, the vectors $u_{c1}, u_{c2}, \ldots, u_{cN}$ are orthogonal to each other. Since $\mathbf{U}$ is unitary, we can return to the original coordinate system by projecting $\mathbf{x}_i$ onto the row vector of $\mathbf{U}$:

$$\mathbf{x}_i = \mathbf{u}_{ri} \cdot \mathbf{X}. \tag{2.27}$$

2.4 INVERSE SYSTEMS

A common problem encountered in many disciplines is the inversion of a linear system of equations. That is, given the values of $\mathbf{y}$ and $\mathbf{H}$ and the relationship $\mathbf{y} = \mathbf{H}\mathbf{x}$, we wish to find $\mathbf{x}$. The most obvious approach would be to use the inverse matrix, i.e.,

$$\mathbf{x}_{\text{inv}} = \mathbf{H}^{-1}\mathbf{y}.$$

There are many well-known ways to invert a matrix [1]. If the matrix is a symmetric nonnegative definite one, a square root method (such as Cholesky decomposition) can be used to reduce the complexity of the computation. If the matrix is Toeplitz, then Levinson's recursion can be used. If $\mathbf{H}$ is cyclic, then the DFT can be used.

There are several reasons why the inverse may not be useful in practice. The first reason applies to the case where the equations are inconsistent. The additional problem of rank deficiency will be described in the next section.

Consider the following set of inconsistent equations:

$$x_1 + 2x_2 + x_3 = 2, \tag{2.28}$$

$$x_1 - x_2 + x_3 = 1, \tag{2.29}$$

$$2x_1 + x_2 + 2x_3 = 2. \tag{2.30}$$

The first two equations can be added to obtain an equation with the same coefficients as found in the third equation. However, the values found on the right-hand side are not the same, since the addition of (2.28) and (2.29) yields

$$2x_1 + x_2 + 2x_3 = 3, \tag{2.31}$$

which is inconsistent with (2.30). The problem of inconsistencies within a set of equations can be circumvented by introducing extra variables to compensate for the error. Thus, we can write the set of linear equations as

$$x_1 + 2x_2 + x_3 + e_1 = 2, \qquad (2.32)$$

$$x_1 - x_2 + x_3 + e_2 = 1, \qquad (2.33)$$

$$2x_1 + x_2 + 2x_3 + e_3 = 2. \qquad (2.34)$$

The resulting set of equations is underdetermined, i.e., there are more unknowns than equations. Thus, there are many solutions. We have transformed a problem that had no solutions into one that has an infinite number of solutions. The situation has improved by going to a higher dimension (six unknowns instead of three). In order to select one of the many solutions we are free to choose some additional constraints. A good condition might be to constrain the error vector **e** to have the smallest energy. Ideally, we would like this quantity to be zero. In this case, we would like to find some **x** such that the following conditions are satisfied:

$$\text{minimize} \quad \sum_{i=1}^{3} \mathbf{e}_i^2 \qquad (2.35)$$

$$\text{subject to} \quad \mathbf{y} = \mathbf{H}\mathbf{x} + \mathbf{e}. \qquad (2.36)$$

One may consider the minimization of some other cost function, such as $\sum \mathbf{e}_i$. However, this cost function is not a good measure of the error, since there may be some large positive and negative errors that cancel each other, thus giving an incorrect indication of the size of the error.

The solution to the least squares problem with linear constraints can be solved exactly in closed form. The solution can be obtained by solving for **e**; that is,

$$\mathbf{e} = \mathbf{y} - \mathbf{H}\mathbf{x}. \qquad (2.37)$$

Substituting this expression into the objective function, we obtain

$$E = \sum (\mathbf{y} - \mathbf{H}\mathbf{x})^2. \qquad (2.38)$$

We can then take the partial derivatives with respect to the unknowns x_i and find the stationary point

$$\frac{\partial E}{\partial x_i} = 0. \qquad (2.39)$$

For example, consider the problem given by

$$\text{Minimize} \qquad \epsilon_1^2 + \epsilon_2^2 + \epsilon_3^2 \tag{2.40}$$

$$\text{subject to} \qquad y_1 = h_{11}x_1 + h_{12}x_2 + \epsilon_1 \tag{2.41}$$

$$y_2 = h_{21}x_1 + h_{22}x_2 + \epsilon_2 \tag{2.42}$$

$$y_3 = h_{31}x_1 + h_{32}x_2 + \epsilon_3. \tag{2.43}$$

We can now substitute into the objective function to get

Minimize

$$E = (y_1 - h_{11}x_1 - h_{12}x_2)^2 + (y_2 - h_{21}x_1 - h_{22}x_2)^2 + (y_3 - h_{31}x_1 - h_{32}x_2)^2. \tag{2.44}$$

The resulting objective function is not constrained; it is a function of only the unknowns x_1 and x_2. Recall that $\mathbf{y}$ and $\mathbf{H}$ are given. Thus, we can find the optimum value for $\mathbf{x}$ by elementary calculus, i.e.,

$$\frac{\partial E}{\partial x_i} = 0,$$

$$2h_{11}e_1 + 2h_{21}e_2 + 2h_{31}e_3 = 0,$$

$$2h_{12}e_1 + 2h_{22}e_2 + 2h_{33}e_3 = 0,$$

$$\begin{pmatrix} h_{11} & h_{21} & h_{31} \\ h_{12} & h_{22} & h_{33} \end{pmatrix} \begin{pmatrix} e_1 \\ e_2 \\ e_3 \end{pmatrix} = \begin{pmatrix} 0 \\ 0 \end{pmatrix}, \tag{2.45}$$

$$\mathbf{H}^T \mathbf{e} = 0.$$

This set of equations is called the normal equations. We can interpret the first equation as the dot product of the first column vector of $\mathbf{H}$ with the error vector, i.e.,

$$\mathbf{h}_{c1} \cdot \mathbf{e} = 0,$$

and the second equation can be written as

$$\mathbf{h}_{c2} \cdot \mathbf{e} = 0.$$

Thus, the error vector is normal or perpendicular to the column vectors of the matrix $\mathbf{H}$. The normal equations can be rewritten in matrix form as

$$\mathbf{H}^T [\mathbf{y} - \mathbf{H}\mathbf{x}] = 0, \tag{2.46}$$

$$\mathbf{H}^T \mathbf{y} - \mathbf{H}^T \mathbf{H}\mathbf{x} = 0, \tag{2.47}$$

$$\mathbf{H}^T\mathbf{y} = \mathbf{H}^T\mathbf{H}\mathbf{x}. \tag{2.48}$$

Recall that we originally started with a set of equations given by

$$\mathbf{y} = \mathbf{H}\mathbf{x}. \tag{2.49}$$

Then artificial variables were introduced, which were eliminated to obtain (2.44). The least squares solution is given by

$$\mathbf{x}_{LS} = \left[\mathbf{H}^T\mathbf{H}\right]^{-1}\mathbf{H}^T\mathbf{y} = \mathbf{H}_{LS}\mathbf{y}, \tag{2.50}$$

where

$$\mathbf{H}_{LS} = \left[\mathbf{H}^T\mathbf{H}\right]^{-1}\mathbf{H}^T \tag{2.51}$$

is the least squares inverse matrix. This matrix depends only on $\mathbf{H}$ and can be precomputed for a specific system $\mathbf{H}$. The minimum error vector (in the L_2-norm sense) can now be computed by

$$\epsilon = [\mathbf{y} - \mathbf{H}\mathbf{x}_{LS}] = [\mathbf{I} - \mathbf{H}\mathbf{H}_{LS}]\mathbf{y} = \mathbf{G}\mathbf{y}, \tag{2.52}$$

where the matrix $\mathbf{G} = [\mathbf{I} - \mathbf{H}\mathbf{H}_{LS}]$.

The L_2 norm of the error vector, or the rms error value, is

$$\|\epsilon\|_2 = \|\mathbf{G}\mathbf{y}\|_2 \leq |\mathbf{G}|\|\mathbf{y}\|_2, \tag{2.53}$$

where $|\mathbf{G}|$ indicates some norm of the matrix $\mathbf{G}$.

Thus, the size of the gain matrix dictates the amount of noise amplification in the least squares estimate. If the gain matrix is diagonalized, it is easy to see that

$$\|\epsilon\|_2 \leq \frac{\lambda_{max}}{\lambda_{min}}\|\mathbf{y}\|_2, \tag{2.54}$$

where λ_{max} and λ_{min} are the maximum and minimum eigenvalues of the gain matrix. The ratio of maximum to minimum eigenvalues is called the condition number of the matrix, $\mathbf{G}$. If the condition number is large ($> 10^2$), the matrix is said to be ill-conditioned. An ill-conditioned system matrix $\mathbf{H}$ usually indicates that the noise will be amplified in the inversion process to a degree that is unacceptable in practice. The limiting case occurs when $\lambda_{min} = 0$. This case is called rank deficient since, when one or more of the eigenvalues of $\mathbf{H}$ are zero, neither the inverse matrix $\mathbf{H}^{-1}$ nor $[\mathbf{H}^T\mathbf{H}]^{-1}$ exists. In this case, we can use the pseudo-inverse solution.

2.4.1   Pseudo-Inverse Solution

The inverse of a matrix $\mathbf{H}$ can be expressed in diagonalized form by writing the singular-value decomposition [8] of $\mathbf{H}$,

$$\mathbf{H} = \mathbf{UDV}^T, \tag{2.55}$$

where

$$\mathbf{D} = \begin{pmatrix} \lambda_1 & 0 & \cdots & 0 \\ 0 & \lambda_2 & \cdots & 0 \\ \vdots & \vdots & \ddots & \vdots \\ 0 & 0 & \cdots & \lambda_N \end{pmatrix}.$$

Then

$$\mathbf{H}^{-1} = \mathbf{VD}^{-1}\mathbf{U}^T, \tag{2.56}$$

where

$$\mathbf{D}^{-1} = \begin{pmatrix} \frac{1}{\lambda_1} & 0 & \cdots & 0 \\ 0 & \frac{1}{\lambda_2} & \cdots & 0 \\ \vdots & \vdots & \ddots & \vdots \\ 0 & 0 & \cdots & \frac{1}{\lambda_N} \end{pmatrix}.$$

This definition of the inverse matrix has been generalized to include rank-deficient matrices by defining the pseudo-inverse matrix

$$\mathbf{H}^{\dagger} = \mathbf{VD}^{\dagger}\mathbf{U}^T, \tag{2.57}$$

where

$$\mathbf{D}^{\dagger} = \begin{pmatrix} \frac{1}{\lambda_1} & 0 & \cdots & & 0 \\ 0 & \frac{1}{\lambda_2} & \cdots & & 0 \\ \vdots & \vdots & \ddots & & \vdots \\ & & \cdots & \frac{1}{\lambda_r} & \cdots \\ & & & 0 & \\ & & & & 0 \end{pmatrix}.$$

The diagonal matrix $\mathbf{D}^{\dagger}$ consists of reciprocals of the nonzero eigenvalues and zeros where the eigenvalues are zero. $\mathbf{H}^{\dagger}$ is called the pseudo-inverse matrix.

The pseudo-inverse solution is given by

$$\mathbf{x}^{\dagger} = \mathbf{H}^{\dagger}\mathbf{y}. \tag{2.58}$$

The pseudo-inverse solution can be shown to be a least squares solution of the original set of equations. In the rank-deficient case, there can be an infinite set of least squares solutions (see the Fredholm alternative in Appendix A). The pseudo-inverse solution is the one with minimum energy. We can gain some insight into the pseudo-inverse solution by substituting (2.49) into (2.58), i.e.,

$$\mathbf{x}^\dagger = \mathbf{H}^\dagger \mathbf{H} \mathbf{x} = \mathbf{V} \mathbf{D}^\dagger \mathbf{U}^T \mathbf{U} \mathbf{D} \mathbf{V}^T \mathbf{x}, \tag{2.59}$$

$$\mathbf{x}^\dagger = \mathbf{V} \mathbf{D}^\dagger \mathbf{D} \mathbf{V}^T \mathbf{x}, \tag{2.60}$$

$$\mathbf{x}^\dagger = \mathbf{P} \mathbf{x}. \tag{2.61}$$

The matrix $\mathbf{P}$ is called a projection matrix. The diagonal elements consist of r ones and $n - r$ zeros. The pseudo-inverse solution is the projection of x onto the eigenvectors of $\mathbf{H}$ whose corresponding eigenvalues are not zero. This subspace of R^N is called the range space of $\mathbf{H}$. The complementary space is called the null space. The projection operator onto the null space is given by

$$\mathbf{Q} = \mathbf{V} \begin{pmatrix} 0 & & & & & & \\ & \ddots & & & & & \\ & & 0 & & & & \\ & & & \ddots & & & \\ & & & & 1 & & \\ & & & & & \ddots & \\ & & & & & & 1 \end{pmatrix} \mathbf{V}^T = \mathbf{I} - \mathbf{P}. \tag{2.62}$$

Thus, the family of solutions to (2.4) is given by

$$\mathbf{x}_{LS} = \mathbf{x}^\dagger + \mathbf{Q} \mathbf{z}, \tag{2.63}$$

where z is any arbitrary vector. This form shows that the pseudo-inverse solution vector $\mathbf{x}^\dagger$ is similar to the particular solution and that the vector $\mathbf{Q}\mathbf{z}$ is analogous to a homogeneous solution since

$$\mathbf{y}_{LS} = \mathbf{H} \mathbf{x}_{LS} = \mathbf{H} \mathbf{x}^\dagger + \mathbf{H} \mathbf{Q} \mathbf{z} = \mathbf{H} \mathbf{x}^\dagger. \tag{2.64}$$

The vectors $\mathbf{x}^\dagger$ and $\mathbf{Q}\mathbf{z}$ lie in perpendicular subspaces:

$$\|\mathbf{x}_{LS}\|_2 = \|\mathbf{x}^\dagger\|_2 + \|\mathbf{Q}\mathbf{z}\|_2. \tag{2.65}$$

Thus, we see that the minimum-norm least squares solution is given by

$$\|\mathbf{x}_{LS}\|_{2\min} = \|\mathbf{x}^\dagger\|_2, \tag{2.66}$$

when $z = 0$.

Note that the pseudo-inverse solution is consistent with the least squares solution given by

$$\mathbf{H}_{LS} = [\mathbf{H}^T\mathbf{H}]^{-1}\mathbf{H}^T\mathbf{y} = [\mathbf{U}^T\mathbf{D}^{-2}\mathbf{V}]\mathbf{V}^T\mathbf{D}\mathbf{U}, \qquad (2.67)$$

$$\mathbf{H}_{LS} = \mathbf{V}^T\mathbf{D}^{-1}\mathbf{U}. \qquad (2.68)$$

If $\mathbf{H}^T\mathbf{H}$ is rank deficient, then we can use the pseudo-inverse solution of $\mathbf{H}^T\mathbf{H}$ and obtain

$$\mathbf{H}_{LS}^\dagger = \mathbf{V}^T\mathbf{D}^\dagger\mathbf{U} = \mathbf{H}^\dagger.$$

Thus, we see that the pseudo-inverse solution is a least squares solution to (2.4) and, from (2.67), we get the least squares solution with the least norm.

2.5 LEAST SQUARES ESTIMATION OF THE PARAMETERS OF AN AUTOREGRESSIVE MODEL

Autoregressive (AR) models are frequently used in signal processing. Applications can be found in speech processing, array processing, spectral estimation, and seismology. An autoregressive model is of the form

$$s(k) = a_1 s(k-1) + a_2 s(k-2) + \cdots + a_N s(k-N). \qquad (2.69)$$

That is, the kth sample is taken to be a linear combination of the N previous samples. In practice, the model will not fit a measured signal exactly. In this case, a error term is added to the model, i.e.,

$$s(k) = a_1 s(k-1) + a_2 s(k-2) + \cdots + a_N s(k-N) + e(k), \qquad (2.70)$$

where $e(k)$ represents the unpredictable or random part of the signal. This model is recognized as a difference equation with an output $s(k)$ and an input given by $e(k)$. The corresponding transfer function is

$$H(z) = \frac{s(z)}{e(z)} = \frac{1}{a_1 z^{-1} + a_2 z^{-2} + \cdots + a_N z^{-N}}. \qquad (2.71)$$

For this reason, the AR model is sometimes called an all-pole model. We shall consider a speech-processing application in order to see how such a model can be used. Analysis of a signal implies decomposition into simpler components. In the case of the AR model, the analysis is called linear prediction. The analysis consists of finding N coefficients $\{a_i\}$ and a sequence $e(k)$ such that (2.70) holds for a given speech sequence $s(k)$. The speech signal can be reconstructed from the set of coefficients $\{a_i\}$ and the sequence $e(k)$ using the difference equation given by (2.70). This process is called synthesis.

A good estimate of the $\{a_i\}$ will yield an error vector $e(k)$ that is uncorrelated noise. That is, $R_{ee} = \sigma^2 \delta(n)$, or, equivalently, the power spectrum is white, i.e., $S_{ee}(\omega) = \sigma^2$. Thus, the Fourier power spectrum of the output speech signal $s(k)$ is given by the frequency response of the model

$$H(\omega) = H(z)\,|_{z=e^{j\omega T_s}}\,. \tag{2.72}$$

Speech is approximated well by a 10th-order AR model corresponding to five resonant frequencies (two complex conjugate poles for each resonant frequency) [9]. The resonant frequencies and bandwidth around each are called formants in speech. It turns out that any white noise–like signal can be used as input to the synthesis difference equation (2.70) and produce an acceptable-sounding reconstruction of the original speech signal. Thus, the linear-prediction model effectively offers a very compact representation of the speech signal with 10 values of the AR coefficients for each speech sound. In practice, 100–200 samples of a speech waveform can be modeled by the ten coefficients. This requires 10–20 times fewer data bits to represent the same speech sound. The use of the AR model to reduce the number of bits is called linear predictive coding (LPC). A full understanding of LPC would require discussions of voiced and unvoiced sounds, pitch extractions, and practical problems which we must relegate to the references [4, 9]. In this chapter, we are interested in the mathematical methods used in LPC analysis. This can be understood by considering a small numerical example.

We wish to find a second-order AR model that best fits some vector of measured speech samples. There are two approaches commonly used to solve this signal analysis problem: the covariance and autocorrelation methods. Both methods begin by writing the linear equations for the AR model. Thus, using the covariance method, we write

$$s(3) = a_1 s(2) + a_2 s(1), \tag{2.73}$$

$$s(4) = a_1 s(3) + a_2 s(2), \tag{2.74}$$

$$s(5) = a_1 s(4) + a_2 s(3), \tag{2.75}$$

or, in the matrix form,

$$\begin{pmatrix} s(3) \\ s(4) \\ s(5) \end{pmatrix} = \begin{pmatrix} s(2) & s(1) \\ s(3) & s(2) \\ s(4) & s(3) \end{pmatrix} \begin{pmatrix} a_1 \\ a_2 \end{pmatrix}, \tag{2.76}$$

or

$$\mathbf{s} = \boldsymbol{\mathcal{S}}\mathbf{a}. \tag{2.77}$$

For example, if the measured samples were

$$s(1) = 1, \tag{2.78}$$

$$s(2) = 2, \tag{2.79}$$

$$s(3) = 3, \tag{2.80}$$

$$s(4) = 5, \tag{2.81}$$

$$s(5) = 8, \tag{2.82}$$

we would have

$$\begin{pmatrix} 3 \\ 5 \\ 8 \end{pmatrix} = \begin{pmatrix} 2 & 1 \\ 3 & 2 \\ 5 & 3 \end{pmatrix} \begin{pmatrix} a_1 \\ a_2 \end{pmatrix}. \tag{2.83}$$

The least squares solution to (2.77) is

$$\mathbf{a} = \left[\mathbf{S}^T \mathbf{S} \right]^{-1} \left[\mathbf{S}^T \mathbf{s} \right]. \tag{2.84}$$

In this case,

$$\mathbf{S}^T \mathbf{s} = \begin{pmatrix} 2 & 3 & 5 \\ 1 & 2 & 3 \end{pmatrix} \begin{pmatrix} 3 \\ 5 \\ 8 \end{pmatrix} = \begin{pmatrix} 61 \\ 37 \end{pmatrix}, \tag{2.85}$$

$$\mathbf{\Phi} = \mathbf{S}^T \mathbf{S} = \begin{pmatrix} 38 & 23 \\ 23 & 14 \end{pmatrix}. \tag{2.86}$$

The matrix $\mathbf{\Phi}$ is called the covariance matrix. It is always symmetric positive semidefinite. Thus, it can be factored using the Cholesky decomposition

$$\mathbf{\Phi} = \mathbf{U}^T \mathbf{U}, \tag{2.87}$$

where $\mathbf{U}$ is an upper triangular matrix. The coefficients of $\mathbf{U}$ can be obtained by the recursion

$$u_{ii} = \left(a_{ii} - \sum_{k=1}^{i-1} u_{ii}^2 \right)^{1/2}, \qquad i = 1, 2, \ldots, n, \tag{2.88}$$

$$u_{ij} = \frac{1}{u_{ii}} \left(a_{ij} - \sum_{k=1}^{i-1} u_{ki} u_{kj} \right), \qquad j > i; \tag{2.89}$$

for $i = 1$, we get

$$u_{11} = \sqrt{a_{11}} = 6.1644, \tag{2.90}$$

$$u_{12} = a_{12}/u_{11} = 3.731; \tag{2.91}$$

for $i = 2$,

$$u_{22} = \sqrt{a_{22} - u_{12}^2} = 0.28. \tag{2.92}$$

Thus,

$$U = \begin{pmatrix} u_{11} & u_{12} \\ 0 & u_{22} \end{pmatrix} = \begin{pmatrix} 6.1644 & 3.731 \\ 0 & 0.28 \end{pmatrix}. \tag{2.93}$$

The Cholesky decomposition is used to solve the least squares problem, i.e.,

$$\begin{pmatrix} 61 \\ 37 \end{pmatrix} = \begin{pmatrix} 38 & 23 \\ 23 & 14 \end{pmatrix} \begin{pmatrix} a_1 \\ a_2 \end{pmatrix} \tag{2.94}$$

$$= \begin{pmatrix} 6.1644 & 0 \\ 3.731 & 0.28 \end{pmatrix} \begin{pmatrix} 6.1644 & 3.731 \\ 0 & 0.28 \end{pmatrix} \begin{pmatrix} a_1 \\ a_2 \end{pmatrix}$$

$$= U^T U a. \tag{2.95}$$

By defining a dummy vector **b** such that

$$b = Ua, \tag{2.96}$$

we have

$$\begin{pmatrix} 61 \\ 37 \end{pmatrix} = \begin{pmatrix} 6.1644 & 0 \\ 3.731 & 0.28 \end{pmatrix} \begin{pmatrix} b_1 \\ b_2 \end{pmatrix} = U^T b. \tag{2.97}$$

This can be solved directly for the b_i from the first equation, i.e.,

$$b_1 = \frac{61}{6.1644} = 9.8955, \tag{2.98}$$

$$b_2 = \frac{37 - (3.731) \cdot (9.8955)}{0.28} = 0.2853. \tag{2.99}$$

Thus, we can solve

$$\begin{pmatrix} 9.8999 \\ 0.2853 \end{pmatrix} = \begin{pmatrix} 6.1644 & 3.731 \\ 0 & 0.28 \end{pmatrix} \begin{pmatrix} a_1 \\ a_2 \end{pmatrix}$$

for the a_i to get

$$a_1 = 1.0189 \approx 1, \tag{2.100}$$

$$a_2 = 0.9886 \approx 1. \tag{2.101}$$

The use of Cholesky decomposition can be shown to reduce the complexity of the matrix inverse from $O(n^3)$ to $O(n)$ operations. This is due to the symmetry of Φ. The solution can be checked by noting that the initial choices of samples were the elements of a Fibonacci series.

The other method used to find the AR parameters [4, 9] is the autocorrelation method. In this method, the data is extrapolated outside the window of observations with zeros. The AR equations are thus

$$\begin{pmatrix} s(2) \\ s(3) \\ s(4) \\ s(5) \\ 0 \\ 0 \end{pmatrix} = \begin{pmatrix} s(1) & 0 \\ s(2) & s(1) \\ s(3) & s(2) \\ s(4) & s(3) \\ s(5) & s(4) \\ 0 & s(5) \end{pmatrix} \begin{pmatrix} a_1 \\ a_2 \end{pmatrix}. \tag{2.102}$$

This can be rewritten more compactly in matrix form as

$$\hat{s} = \hat{S}\hat{a}. \tag{2.103}$$

The least squares solution is then

$$\hat{a} = [\hat{S}^T\hat{S}]^{-1}\hat{S}^T\hat{s}. \tag{2.104}$$

The numerical data in this case are

$$\hat{s} = \begin{pmatrix} 2 \\ 3 \\ 5 \\ 8 \\ 0 \\ 0 \end{pmatrix}, \tag{2.105}$$

$$\hat{S} = \begin{pmatrix} 1 & 0 \\ 2 & 1 \\ 3 & 2 \\ 5 & 3 \\ 8 & 5 \\ 0 & 8 \end{pmatrix}, \tag{2.106}$$

$$R_{ss} = [\hat{S}^T\hat{S}] = \begin{pmatrix} 103 & 63 \\ 63 & 103 \end{pmatrix}. \tag{2.107}$$

Note that $\mathbf{R}_{ss}$ is the autocorrelation matrix of the data sequence $s(k)$. It is a symmetric Toeplitz nonnegative definite matrix. The Toeplitz structure (i.e., the values along any diagonal are all equal) allows for the use of fast inverse algorithms. The Durbin algorithm which is based on the Levinson algorithm is typically used:

$$\mathbf{r}_{ss} = \hat{\mathbf{S}}^T \hat{\mathbf{s}} = \begin{pmatrix} 63 \\ 37 \end{pmatrix}. \tag{2.108}$$

The least squares solution is the solution to the system of two equations with two unknowns given by

$$\begin{pmatrix} 63 \\ 37 \end{pmatrix} = \begin{pmatrix} 103 & 63 \\ 63 & 103 \end{pmatrix} \begin{pmatrix} \hat{a}_2 \\ \hat{a}_1 \end{pmatrix}. \tag{2.109}$$

The solution is then

$$\hat{a}_1 = 0.6262, \tag{2.110}$$

$$\hat{a}_2 = -0.02378. \tag{2.111}$$

The autocorrelation method is guaranteed to give coefficients $\{a_i\}$ that yield a stable difference equation.

The AR model using the autocorrelation method does not fit the data as well as the covariance method due to the large number of zeros relative to the actual data used in the initial equations. The autocorrelation and covariance solutions approach one another as the number of data samples goes to infinity. For a small number of samples, the difference can be significant, as illustrated in the previous numerical example.

2.6 STATISTICAL LEAST SQUARES

The previous discussions were based on deterministic concepts of linear algebra. The unknown variables $\mathbf{x}$ and $\mathbf{e}$ were assumed to be deterministic. An alternate approach is to assume that either $\mathbf{x}$ or $\mathbf{e}$ or both are random vectors. In the case where both are random vectors, we can minimize the expected value of the sum of the squared error over the ensemble of the random variables (r.v.) $\mathbf{x}$ and $\mathbf{e}$. That is, we can

$$\text{minimize} \quad E\left[\sum \mathbf{e}^2\right]$$

$$\text{subject to} \quad \mathbf{y} = \mathbf{H}\mathbf{x} + \mathbf{e}, \tag{2.112}$$

where $E(\cdot)$ is the expectation operator and $\mathbf{y}$ and $\mathbf{e}$ are vectors of random variables [10]. The equivalent problem is

$$\text{minimize} \qquad E[(\mathbf{y} - \mathbf{Hx})^T(\mathbf{y} - \mathbf{Hx})],$$

or, equivalently,

$$\frac{\partial E(\sum \mathbf{e}^2)}{\partial \mathbf{x}} = E \frac{\partial \sum \mathbf{e}^2}{\partial \mathbf{x}} = \mathbf{0}, \qquad (2.113)$$

$$E[-2\mathbf{H}^T\mathbf{y} + 2\mathbf{H}^T\mathbf{Hx}] = 0. \qquad (2.114)$$

Thus

$$E[\mathbf{H}^T\mathbf{y}] = E[\mathbf{H}^T\mathbf{H}]\mathbf{x}, \qquad (2.115)$$

$$\mathbf{x}_{ALS} = E[\mathbf{H}^T\mathbf{H}]^{-1} E[\mathbf{H}^T\mathbf{y}]. \qquad (2.116)$$

We shall refer to the statistical least squares estimate $\mathbf{x}_{ALS}$ as the average least squares solution. An important application of statistical least squares is the autoregressive modeling of a process. Here the ith value is predicted by a linear combination of the previous samples, i.e.,

$$y(4) = a_3y(3) + a_2y(2) + a_1y(1) + e(1), \qquad (2.117)$$

$$y(5) = a_3y(4) + a_2y(3) + a_1y(2) + e(2), \qquad (2.118)$$

$$\vdots$$

$$y(N) = a_3y(N-1) + a_2y(N-2) + a_1y(N-3) + e(N). \qquad (2.119)$$

If $\mathbf{H}$ is the Toeplitz matrix given by

$$\mathbf{H} = \begin{pmatrix} y(3) & y(2) & y(1) \\ y(4) & y(3) & y(2) \\ & \vdots & \\ y(N-1) & y(N-2) & y(N-3) \end{pmatrix}, \qquad (2.120)$$

then we see that $E[\mathbf{H}^T\mathbf{H}] = \mathbf{R}_{yy}$, where $\mathbf{R}_{yy}$ is the autocorrelation matrix of $\mathbf{y}$ and

$$E[\mathbf{H}^T\mathbf{y}] = \mathbf{r}_y, \qquad (2.121)$$

where $\mathbf{r}_y$ is the autocorrelation vector of $\mathbf{y}$. The solution to (2.112) is

$$\mathbf{x}_{ALS} = \mathbf{R}_{yy}^{-1}\mathbf{r}_y. \qquad (2.122)$$

This is the Wiener-filter solution. In this case, $\mathbf{x}_{ALS}$ is the vector of filter coefficients which will give the best prediction of $\mathbf{y}$ based on the previous values. It should be understood that we mean "best" in the average least squares sense. Thus, the best on average may be a poor estimate for a specific case.

The required matrix $\mathbf{R}_{yy}$ can be approximated [11] in practice by making L measurements of the vector $\mathbf{y}$, where L is a large number, and then computing

$$\hat{\mathbf{R}}_{yy} = \frac{1}{L} \sum_{i=1}^{L} \mathbf{y}_i \mathbf{y}_i^T. \tag{2.123}$$

The vector $\mathbf{r}_y$ is estimated by

$$\mathbf{r}_y^{(k)} = \frac{1}{L} \sum_{i=1}^{L} y_i(n) y_i(n+k). \tag{2.124}$$

In the case where we are restricted to one measurement, i.e., $L = 1$, the approximation yields the same result as the deterministic least squares:

$$\mathbf{x}_{ALS} = [\mathbf{H}^T \mathbf{H}]^{-1} \mathbf{H}^T \mathbf{y}. \tag{2.125}$$

The approximations

$$E[\mathbf{H}^T \mathbf{H}] \approx \mathbf{H}^T \mathbf{H}, \tag{2.126}$$

$$E[\mathbf{H}^T \mathbf{y}] \approx \mathbf{H}^T \mathbf{y}, \tag{2.127}$$

are called stochastic approximations [11] and are frequently used in practice.

Note that the computation given by (2.123) forms an $N \times N$ symmetric matrix by adding rank-1 matrices. The resulting matrix may or may not be full rank. If $\hat{\mathbf{R}}_{yy}$ is rank deficient, then we can use the pseudo-inverse in place of the inverse. However, this result is not the actual average least squares solution.

Although we shall show that it is frequently appropriate if L is sufficiently large and if the noise is white, $\hat{\mathbf{R}}_{yy}$ will not be rank deficient. This can be seen in the following.

Let

$$\mathbf{y} = \mathbf{H}\mathbf{x} + \mathbf{n} = \mathbf{y}' + \mathbf{n}. \tag{2.128}$$

Assuming that $\mathbf{y}$ and $\mathbf{n}$ are independent and uncorrelated yields

$$\mathbf{R}_{yy} = \mathbf{R}_{y'y'} + \mathbf{R}_{nn}. \tag{2.129}$$

If the noise is independent and identically distributed (i.i.d.), then

$$\mathbf{R}_{nn} = \sigma_n^2 \mathbf{I} \tag{2.130}$$

and we get

$$\mathbf{R}_{yy} = \mathbf{R}_{y'y'} + \sigma_n^2 \mathbf{I} \tag{2.131}$$

$$= \mathbf{U}\mathbf{D}\mathbf{U}^T + \mathbf{U}[\sigma_n^2 \mathbf{I}]\mathbf{U}^T \tag{2.132}$$

$$= \mathbf{U}\begin{pmatrix} \lambda_1 + \sigma^2 & & & \\ & \lambda_2 + \sigma^2 & & \\ & & \ddots & \\ & & & \lambda_N + \sigma^2 \end{pmatrix}\mathbf{U}^T. \tag{2.133}$$

Thus, the inverse of $\hat{\mathbf{R}}_{yy}$ is stabilized by the measurement noise. That is, if there is noise ($\sigma^2 \neq 0$), the diagonal elements may always have a positive value even when the underlying process may have zero eigenvalues for $L > r$.

However, the average least squares solution has the undesirable effect of amplifying eigenvectors present in the measurement by a random number since

$$\mathbf{x}_{ALS} = \hat{\mathbf{R}}_{yy}\mathbf{r}_{yy} = \mathbf{U}\begin{pmatrix} \frac{1}{\lambda_1+\sigma^2} & & & \\ & \frac{1}{\lambda_2+\sigma^2} & & \\ & & \ddots & \\ & & & \frac{1}{\lambda_N+\sigma^2} \end{pmatrix}\mathbf{U}^T\mathbf{r}_y. \tag{2.134}$$

If some λ_i are zero, then the corresponding eigenvectors of $\mathbf{U}$ are eliminated from the output vector $\mathbf{y}$ by the operation

$$\mathbf{y} = \mathbf{H}\mathbf{x} = \mathbf{U}\begin{pmatrix} \lambda_1 & & & \\ & \lambda_2 & & \\ & & \ddots & \\ & & & 0 \end{pmatrix}\mathbf{U}^T\mathbf{x}. \tag{2.135}$$

Yet the average least squares solution will retain and actually amplify the eigenvectors by a factor of $1/\sigma^2$.

The process by which we deemphasize the effects of small eigenvalues in the inverse process is called regularization. The pseudo-inverse solution is a form of regularization that employs only two possible weights for the $1/\lambda_i$ terms of the inverse. The two weights are zero and one. Thus, the pseudo-inverse solution is a regularized solution. We recommend that, in practice, the pseudo-inverse be used in place of $\hat{\mathbf{R}}_{yy}^{-1}$ whenever possible to avoid this amplification of noise.

2.6.1   Maximum-Likelihood (ML) Approach

Another statistical least squares approach is to assume that $\mathbf{e}$ is a multivariate Gaussian vector whose distribution is given by

$$P_{e_1,e_2,\ldots,e_n}(e_1, e_2, \ldots, e_n) = \frac{1}{\sqrt{(2\pi)^n \det(\mathbf{R}_{ee})}} e^{-(\mathbf{e}^T\mathbf{R}_{ee}^{-1}\mathbf{e})/2}. \tag{2.136}$$

where $\mathbf{R}_{ee}$ is the $N \times N$ autocorrelation matrix

$$\mathbf{R}_{ee} = E[\mathbf{e}\mathbf{e}^T]. \tag{2.137}$$

The autocorrelation matrix is symmetric and positive semi-definite. Thus $\det(\mathbf{R}_{ee}) \neq 0$ or $\mathbf{R}_{ee}$ is full rank. The maximum-likelihood (ML) estimate of $\mathbf{e}$, given the measurement $\mathbf{y}$, can be obtained by writing

$$\mathbf{y} = \mathbf{H}\mathbf{x} + \mathbf{e} = \mathbf{y}' + \mathbf{e}, \tag{2.138}$$

where

$$\mathbf{e} = \mathbf{y} - \mathbf{y}' \tag{2.139}$$

and $\mathbf{y}'$ indicates the true value. Equation (2.139) indicates that, given an r.v. with mean value y_i', we can rewrite (2.136):

$$P_{y'/y}(\mathbf{y}'/\mathbf{y}) = \frac{1}{\sqrt{(2\pi)^n \det(\mathbf{R}_{ee})}} e^{-((\mathbf{y}-\mathbf{y}')^T \mathbf{R}_{ee}^{-1}(\mathbf{y}-\mathbf{y}'))/2}. \tag{2.140}$$

The most likely value for $\mathbf{y}'$, given $\mathbf{y}$, is the value which maximizes the above probability. The probability is maximized when the argument of the exponent is minimized. Thus, we wish to find a $\mathbf{y}'$ such that the cost function

$$M_2 = (\mathbf{y} - \mathbf{y}')^T \mathbf{R}_{ee}^{-1}(\mathbf{y} - \mathbf{y}') \tag{2.141}$$

is minimum. The form of objective function to be minimized in numerical optimization is called a quadratic cost function. M_2 is a weighted least squares cost function. This measure of the error is also sometimes called the Mahalanobis metric of the error. The M_2 metric of the error is more general than the normal least squares criterion in that, if $\mathbf{R}_{ee}^{-1}$ is not a diagonal matrix, the M_2 norm contains cross terms of the error such as $e_1 e_2$ and $e_2 e_3$. The $\mathbf{R}_{ee}^{-1}$ factor is useful if the noise is correlated. The matrix $\mathbf{R}_{ee}^{-1}$ in the M_2 metric is said to whiten, or decorrelate, the estimated noise. If the noise is white, i.e.,

$$\mathbf{R}_{ee} = \sigma^2 \mathbf{I}, \tag{2.142}$$

then the M_2 norm becomes equivalent to the least squares, or L_2 norm, of the error. In the common case of Gaussian white noise, we shall now show how the above weighted least squares problem can be solved using the statistical least squares method.

The above weighted least squares problem can be formally posed in the following way:

$$\text{minimize} \qquad \mathbf{e}^T \mathbf{R}_{ee}^{-1} \mathbf{e} \tag{2.143}$$

$$\text{subject to} \quad \mathbf{y} = \mathbf{Hx} + \mathbf{e}. \tag{2.144}$$

Note that

$$\mathbf{R}_{ee}^{-1} = \mathbf{UD}^2\mathbf{U}^T.$$

Since the eigenvalues are positive, $\mathbf{D}$ is composed of the squares of positive numbers. Thus, we can find the square root of $\mathbf{D}$ to obtain

$$\mathbf{R}_{ee}^{-1} = \mathbf{UDDU}^T = \mathbf{UDD}^T\mathbf{U}^T. \tag{2.145}$$

If we define a new variable

$$\boldsymbol{\eta} = \mathbf{D}^T\mathbf{U}^T\mathbf{e}, \tag{2.146}$$

we can rewrite (2.123) as

$$\text{minimize} \quad \boldsymbol{\eta}^T\boldsymbol{\eta} \tag{2.147}$$

$$\text{subject to} \quad \mathbf{y} = \mathbf{Hx} + \mathbf{D}^{-1}\mathbf{U}\boldsymbol{\eta}. \tag{2.148}$$

Multiplying the constraints by $\mathbf{DU}^T$, we obtain

$$\text{minimize} \quad \boldsymbol{\eta}^T\boldsymbol{\eta} \tag{2.149}$$

$$\text{subject to} \quad \mathbf{DU}^T\mathbf{y} = \mathbf{DU}^T\mathbf{Hx} + \boldsymbol{\eta}. \tag{2.150}$$

We can thus transform a weighted least squares problem, where $\mathbf{R}_{ee}^{-1}$ is positive definite, by substituting

$$\mathbf{y}' = \mathbf{DU}^T\mathbf{y} \tag{2.151}$$

and

$$\mathbf{H}' = \mathbf{DU}^T\mathbf{H} \tag{2.152}$$

and finding the least squares solutions to the transformed problem

$$\mathbf{y}' = \mathbf{H}'\mathbf{x} + \mathbf{e}. \tag{2.153}$$

Thus, the method of deterministic least squares can be justified on various statistical grounds. The least squares method is probably the most widely used method in practice; this is, in part, due to the many ways of deriving the method.

Note that the statistical approaches try implicitly to enforce both statistical constraints,

$$E[\mathbf{e}] = 0 \tag{2.154}$$

and

$$E[\mathbf{e}\mathbf{e}^T] = \mathbf{R}_{ee}. \tag{2.155}$$

The second constraint is usually followed by the assumption that

$$\mathbf{R}_{ee} = \sigma^2 \mathbf{I}. \tag{2.156}$$

In general, statistical methods such as the Wiener filter and the ML estimate will give estimates for both $\mathbf{x}$ and $\mathbf{e}$, where $\mathbf{e}$ will closely approximate the conditions given by (2.154)–(2.156).

2.7 THE ROW-ACTION PROJECTION METHOD

The previous methods of least squares were all block methods. That is, a block of data (the $\mathbf{y}$ vector) was transformed by a matrix to form the estimate:

$$\mathbf{x}_{\text{estimate}} = \mathbf{A}\mathbf{y}, \tag{2.157}$$

where $\mathbf{A}$ might be $\mathbf{H}^{-1}$, $\mathbf{H}^{\dagger}$, $[\mathbf{H}^T\mathbf{H}]^{-1}$, or $\mathbf{R}_{yy}^{-1}\mathbf{r}_y$. Thus, a block of measured data is processed to obtain a block of estimated values, $\mathbf{x}_{\text{estimate}}$. In some applications, this approach can be prohibitive in terms of computation and memory requirements. Thus, we might look for methods that operate on one subset of $\mathbf{H}$ at a time.

In this section, we shall introduce a method of finding an estimate by operating only on the rows of $\mathbf{H}$ at any given time. Note that there are also matrix-inversion methods based on column operators and matrix-perturbation methods, such as the Greville method [12], Woodbury [13], and the Sherman–Morrison formula [13]. However, a row-operator approach is more appropriate for adaptive signal-processing applications since each row or equation corresponds to one data sample. Row-action methods—such as ART (algebraic reconstruction technique) [14]—can perform an operation for each sample as it is acquired, while column-based methods still require a block of data to perform each operation. In addition, the row-action projection (RAP) method has the characteristic of working for both inconsistent and rank-deficient matrices. These properties make the RAP method a very attractive method for sample-by-sample estimation of signals.

The RAP method can be illustrated by writing out the set of input-output equations of a discrete linear system:

$$y_1 = h_{11}x_1 + h_{12}x_2 + h_{13}x_3, \tag{2.158}$$

$$y_2 = h_{21}x_1 + h_{22}x_2 + h_{23}x_3, \tag{2.159}$$

$$y_M = h_{M1}x_1 + h_{M2}x_2 + h_{M3}x_3. \tag{2.160}$$

Each of the M equations can be viewed as a hyperplane in N-dimensional space ($N = 3$ in this case). We can rewrite the ith equation as

$$y_i = \langle \bar{\mathbf{h}}_i \cdot \mathbf{x} \rangle, \tag{2.161}$$

where $\langle \bar{\mathbf{h}}_i \cdot \mathbf{x} \rangle$ denotes the dot product and $\bar{\mathbf{h}}_i$ is the vector corresponding to the ith equation.

In N-dimensional space, the locus of points that satisfy (2.161) is a hyperplane whose surface normal is the vector $\hat{\mathbf{h}}_i$ and whose perpendicular displacement from the origin is y_i, as shown in Figure 2.1.

A set of M algebraic equations will give a set of M hyperplanes in N-dimensional space, as shown in Figure 2.2. Note that, usually, $M \geq N$ in practice. That is, we are generally interested in overdetermined systems.

The basic idea of RAP is to start with some initial guess $\mathbf{x}_0$ for $\mathbf{x}$. The point $\mathbf{x}_0$ is projected onto the hyperplane given by one of the equations. Consider the case of projecting onto the first hyperplane. We can find the projections $\mathbf{x}'$ from $\mathbf{x}_0$ by moving along the unit normal to the hyperplane (i.e., $\mathbf{h}_1/|\mathbf{h}_1|$) by the perpendicular distance d. Thus, the update is given by

$$\mathbf{x}_1 = \mathbf{x}_0 + d\frac{\mathbf{h}_1}{\|\mathbf{h}_1\|_2}. \tag{2.162}$$

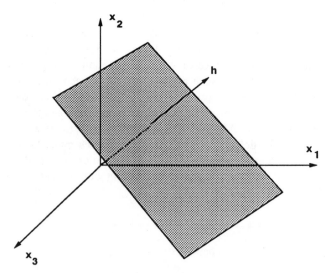

Figure 2.1 A hyperplane in 3-D space with surface normal h.

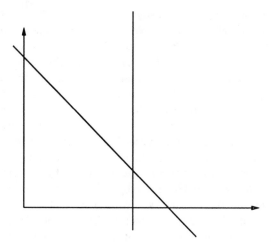

Figure 2.2 Two equations in two-dimensional space.

The perpendicular distance d can be rewritten as the projection of the vector $\mathbf{x}_1 - \mathbf{x}_0$ onto the unit normal, i.e.,

$$d = \frac{\mathbf{h}_1}{\|\mathbf{h}_1\|_2} \cdot (\mathbf{x}_1 - \mathbf{x}_0) \tag{2.163}$$

$$= \frac{\mathbf{h}_1 \cdot \mathbf{x}_1 - \mathbf{h}_1 \cdot \mathbf{x}_0}{\|\mathbf{h}_1\|_2} \tag{2.164}$$

$$= \frac{\mathbf{y}_1 - \mathbf{h}\mathbf{x}_0}{\|\mathbf{h}_1\|_2} \tag{2.165}$$

$$= \frac{\mathbf{e}_1}{\|\mathbf{h}_1\|_2}, \tag{2.166}$$

where $\mathbf{e}_1 = \mathbf{y}_1 - \mathbf{h}\mathbf{x}_0$.

Thus (2.162) becomes

$$\mathbf{x}_1 = \mathbf{x}_0 + \frac{\mathbf{e}_1}{\|\mathbf{h}_1\|_2^2}\mathbf{h}_1. \tag{2.167}$$

The next step is to project the updated point $\mathbf{x}_1$ onto the hyperplane given by the second equation, then the third, and so on. The termination point is the point common to all the hyperplanes, i.e., the solution to the linear equations given by (2.160). The general RAP method is given by

$$\mathbf{x}_{i+1} = \mathbf{x}_i + \mu\frac{\mathbf{e}_j}{\|\mathbf{h}_j\|_2^2}\mathbf{h}_j, \tag{2.168}$$

where μ is a step-size parameter bounded by $0 < \mu < 2$. For $0 < \mu < 1$, the operator moves the point toward the hyperplane. Only when $\mu = 1$ is the point projected onto the hyperplane. For $1 < \mu < 2$, the point is moved through the hyperplane. The first case, $0 < \mu < 1$, is called underrelaxation. Underrelaxation is useful when the data vector is noisy and confidence in the measured value is low. Overrelaxation, $1 < \mu < 2$, is useful when there is little noise and we wish to speed up convergence. Thus, the selection of μ determines the tradeoff between convergence speed and accuracy. We may adopt a strategy in which we start with a large value for μ and decrease it as we get closer to the solution. We shall indicate some strategies for changing the step size in subsequent chapters.

The reason for the use of the two indices i and j is that orderings of the equations, or cycling strategies, other than projecting sequentially on each equation can be used to update the estimate x_i. We might want to reuse old equations or skip new equations that seem to overshoot the estimate. Numerical examples where these cycling strategies are useful will be seen in the following chapters.

It has been shown that the RAP method converges asymptotically to the pseudo-inverse solution for rank-deficient matrices H [15]. That is,

$$\lim_{i \to \infty} x_i = x^\dagger, \tag{2.169}$$

which is a desirable characteristic. RAP is also quite tolerant of inconsistencies. RAP can be modified to include constraints on the unknown vector x such as positivity, upper and lower bounds, and finite energy. These constraints will be shown to form convex sets that guarantee convergence.

More details on the RAP method will be given in several chapters of this book. The method of projection onto convex sets (POCS) is a generalization of RAP in which the hyperplanes are replaced with any convex set. This will also be described in later chapters for various applications.

2.8   A MEASURE OF THE DISTANCE BETWEEN TWO VECTORS

It is frequently desirable to calculate a measure of the distance between two vectors. For example, we may have N prestored vectors x_i, $i = 1, 2, \ldots, N$ and, given a vector y, would like to find the x_i closest to it. This problem is typically found in pattern-recognition applications. For example, the x_i might correspond to phonemes, letters, or images of objects such as cars or tanks. The problem can be stated as follows: given some measurement Y, find the most likely object x. This is similar to the problem

$$\text{minimize} \quad \sum \epsilon^2 \tag{2.170}$$

given

$$y = Hx + \epsilon, \tag{2.171}$$

where $\mathbf{H} = \mathbf{I}$ and $\mathbf{x}$ is constrained to be one of the N vectors $\mathbf{x}_i$ $(i = 1, 2, \ldots, N)$ that are known *a priori*. Then the optimum $\mathbf{x}_i$ is such that

$$\|\mathbf{y} - \mathbf{x}_{opt}\|_2^2 = \min_{i=1,2,\ldots,N} \|\mathbf{y} - \mathbf{x}_i\|_2^2. \tag{2.172}$$

One approach is to compute the N vectors given by $\mathbf{y} - \mathbf{x}_i$ and select the x_i which yields the minimum value of $|\mathbf{y} - \mathbf{x}_i|^2$. Another approach is to first normalize the template vectors $\mathbf{x}_i$ and the input vector such that

$$\|x_i\|_2 = 1 \qquad \text{for} \quad i = 1, 2, \ldots, N$$

and

$$\|\mathbf{y}\|_2 = \mathbf{1}.$$

The distance between the vectors becomes

$$\|\mathbf{y} - \mathbf{x}_i\|_2^2 = \|\mathbf{y}^2 - 2\mathbf{x}_i \cdot \mathbf{y} - \mathbf{x}_i^2\|_2 \tag{2.173}$$

$$= \|\mathbf{y}\|_2^2 - 2(\mathbf{x}_i \cdot \mathbf{y}) + \|\mathbf{x}_i^2\|_2 \tag{2.174}$$

$$= 1 - 2(\mathbf{x}_i \cdot \mathbf{y}) + 1 \tag{2.175}$$

$$= 2 - 2(\mathbf{x}_i \cdot \mathbf{y}). \tag{2.176}$$

Thus, the minimum distance between the two vectors, $|\mathbf{y} - \mathbf{x}_i|$, corresponds to the maximum value of the scalar product

$$(\mathbf{x}_i \cdot \mathbf{y}) = \|\mathbf{x}_i\|_2 \|\mathbf{y}\|_2 \cos \theta = \cos \theta, \tag{2.177}$$

where $\cos \theta$ is the cosine of the angle between the vectors. If $\theta = 0$, then $\cos \theta = 1$ and

$$\|\mathbf{x}_i \cdot \mathbf{y}\|_2 = \mathbf{1}$$

is the highest value of the scalar product, in which case the distance is

$$\|\mathbf{y} - \mathbf{x}_i\|_2 = \mathbf{2} - \mathbf{2}(\mathbf{x}_i \cdot \mathbf{y}) = \mathbf{0}. \tag{2.178}$$

The $\mathbf{x}_i$ which maximizes the scalar product $(\mathbf{x}_i \cdot \mathbf{y})$ is $\mathbf{x}_{opt}$ and corresponds to the minimum distance $\|\mathbf{x}_i - \mathbf{y}\|_2$. The dot-product approach can be interpreted as finding the projection of the input vector $\mathbf{y}$ onto the vectors $\{\mathbf{x}_i\}$ for $i = 1, 2, \ldots, N$. The $\mathbf{x}_i$ with the largest projection is the one closest to $\mathbf{y}$.

This type of implementation of the minimum-distance rule is known as a matched-filter, cross-correlator, template-matching implementation [16]. If we are free to select the coordinate vectors x_i (wave-shape design), as in radar and communications, then it is desirable to take the x_i to be orthonormal, i.e.,

$$\left(x_i \cdot x_j \right) = \begin{cases} 0, & \text{for } i \neq j, \\ 1, & \text{for } i = j. \end{cases} \tag{2.179}$$

Thus, if $y = x_i$, the output for the correct vector is a "1"—i.e., the maximum possible (since it is the cosine of an angle)—and the output of all the other x_i is "0." Recall that $|y - x|^2$ and $|y \cdot x|$ must be nonnegative. Therefore, the lowest possible value for $(y \cdot x_i)$ is zero. This waveform design allows for the least amount of corruption of the output due to noise. This property is due to the fact that the outputs of the correct and incorrect inputs are as far apart as possible.

Thus, orthonormal vector sets $\{x_i\}$, where each x_i represents a symbol (i.e., those satisfying (2.179)) are desirable for communication systems [17]. For example, if each vector x_i pertains to 3 bits of information, then we would need $N = 8$ different orthonormal vectors.

The DFT matrix can also be interpreted as a matched filter where the basis functions are sampled sinusoids.

In radar systems [18], a signal is transmitted through a medium and reflected back to a receiver that is colocated with the transmitter. The parameter of interest here is the time of return of the signal. Thus, we can think of each vector x_i as a shifted version of the transmitted vector x_{i-1}.

The system could be constructed with a number of discrete tests of the signal to see if it has arrived at the ith time instant. The test with the largest output value determines the time slot selected. This sequence of tests can be carried out as a convolution,

$$z_i = x_i * y, \tag{2.180}$$

with a solution corresponding to that time instant $i = j$ such that

$$z_j = \max_{1 \leq i \leq N} \{z_i\}. \tag{2.181}$$

The optimum pulse shape should be selected to be orthonormal with shifted versions of itself, i.e.,

$$\left(x_i \cdot x_j \right) = \begin{cases} 1, & j = 0, \\ 0, & j \neq 0. \end{cases} \tag{2.182}$$

The previous discussions can be reformulated in a statistical framework by taking expectations of the various vector quantities. Thus, the average square error $E[(y - x_i)^2]$ can be minimized by maximizing the cross correlation

$$\mathbf{R}_{x,y} = E(\mathbf{y} \cdot \mathbf{x}_i), \tag{2.183}$$

and the condition for optimal symbol waveshapes is placed on the correlation between the signal vectors. That is,

$$E(\mathbf{x}_i \cdot \mathbf{x}_j) = \delta[i - j], \tag{2.184}$$

where $\delta(\cdot)$ is the Kronecker delta function:

$$\delta(i - j) = \begin{cases} 1, & \text{for } i = j, \\ 0, & \text{for } i \neq j. \end{cases}$$

2.9 RANDOM SEARCH STRATEGIES

The standard least squares problem with linear constraints is to find the vector $\mathbf{x}$ which will

$$\text{minimize} \quad \sum e^2 \tag{2.185}$$

$$\text{subject to} \quad \mathbf{y} = \mathbf{Hx} + \mathbf{e}. \tag{2.186}$$

The case where the values of $\mathbf{x}$ are integers over a finite range is more general than it might first appear to be. There is always the implicit assumption in computational methods that $\mathbf{x}$ is of finite precision. Since $\mathbf{x}$ consists only of integer elements over a finite range, there are a finite number of possible $\mathbf{x}_i$. We could enumerate each value of $\mathbf{x}$ and calculate the cost,

$$e_i = \sum (\mathbf{y} - \mathbf{hx}_i)^2, \tag{2.187}$$

for each. The $\mathbf{x}_i$ with the least error e_i would be selected as optimal. However, the total number of possible $\mathbf{x}_i$ denoted by N could be enormous in practice. One way to reduce the number of computations is to sample a small subset of $\mathbf{x}_i$ randomly and select the $\mathbf{x}_i$ with the minimum e_i. Even for fairly small samples, we can get a reasonable estimate of e_{min} from this random search procedure. We shall see in Chapter 5 that random searches such as simulated annealing can be useful for image restoration. In addition, random searches are frequently used to find a good initial estimate for gradient-descent methods.

2.10 ALTERNATE ERROR NORMS

So far we have considered the minimization of only the least squares or the weighted least squares error criterion. We have seen [19] that if the error vector consists of a set of Gaussian i.i.d. random variables, then the least squares solution is also the ML estimate. If the PDF of the error variable is Laplacian— that is, of the form

$$p_L(n) = K_L e^{-\sum |n|},\tag{2.188}$$

where K_L is the appropriate normalization constant—then, in this case, the ML estimate is given by the minimization of the L_1 norm, where the L_1 norm is the sum of the absolute values. The resulting estimate is called the least absolute value (LAV) estimate. As is the case for a uniformly distributed error, the ML estimate is given by minimization of the L_∞ norm or by minimization of the largest error value in the error vector. The LAV estimate is frequently preferable to the least squares estimate when the data contains statistical outliers, that is, if a relatively small number of errors have large values.

In order to see why it is more desirable to use the L_1 norm on data with statistical outliers, note that the tails of the Gaussian PDF, i.e.,

$$p_y(n) = k_y e^{-\sum (n)^2},\tag{2.189}$$

fall off much more rapidly than the Laplacian $p_L(n)$. Thus, the Gaussian PDF gives a smaller probability that a large error will occur than the Laplacian PDF. Another way of seeing this is that, in the least squares method, the squared error terms emphasize the larger errors and deemphasize the smaller errors. Thus, the least squares method tends to find an estimate with a set of estimated error values such that no value is much greater than the other estimated values.

The L_1 norm of the error does not overly emphasize large errors by squaring them; thus, LAV estimates of the error may have a few relatively large values (statistical outliers). The property of allowing a few relatively large errors in the estimate is called robustness. The LAV estimate is robust in the sense that if a few large-noise samples are superimposed on the measurement, the estimated signal will correctly identify those samples as due to the noise rather than the signal.

A classic example showing that the LAV is more robust than the least squares estimate can be seen in linear curve fitting with statistical outliers. The difference between least squares and LAV can be seen by examining a numerical example. Given the underlying relationship that $y = mx$ and the measured value of y and x, find the m that best fits the observed data.

Thus, given the underlying relationship

$$y = mx$$

and the data

$$\begin{pmatrix} 0.01 \\ 0.99 \\ 2.01 \\ 4 \end{pmatrix} = m \begin{pmatrix} 0 \\ 1 \\ 2 \\ 3 \end{pmatrix}, \tag{2.190}$$

the least squares for m is

$$m = \frac{\mathbf{x}^T\mathbf{y}}{\mathbf{x}^T\mathbf{x}} = \frac{\mathbf{x} \cdot \mathbf{y}}{\mathbf{x} \cdot \mathbf{x}} = \frac{0.99 + 4.02 + 12}{1 + 4 + 9} = \frac{17.01}{14} = 1.215. \tag{2.191}$$

The slope is off by 21.5% from the true value $m = 1$. We shall discuss a method to find the minimum L_1-norm estimate in Section 2.11. This minimum L_1-norm method can be applied here to obtain $m = 1.01$, which is off by only 1%.

Thus, we see that the robustness of the LAV estimate results in better estimates when large errors are likely. In the following two sections, we shall introduce gradient and block methods of finding the LAV estimate.

2.10.1 The LAV Estimate Using Gradient Descent

The least absolute value estimate can be obtained by a gradient descent method of the form

$$\mathbf{x}^{i+1} = \mathbf{x}^i = \mu \nabla E_1, \tag{2.192}$$

where $\nabla E_1 = \sum_{i=1}^{N} |e|$ and μ is the step size,

$$\nabla E_1 = \left(\frac{\partial \sum |e_1|}{\partial \mathbf{x}_j} \right) = \left(\frac{\partial \sum |\mathbf{y} - \mathbf{h}\mathbf{x}_i|}{\partial \mathbf{x}_j} \right) = \text{sign}(e_j)\mathbf{h}_j, \tag{2.193}$$

where $\mathbf{h}_j$ is the jth column of $\mathbf{H}$ and

$$\text{sign}(e_j) = \begin{cases} +1, & e_j > 0, \\ 0, & e_j = 0, \\ -1, & e_j < 0. \end{cases}$$

Thus, the gradient-descent equation for LAV becomes

$$\mathbf{x}^{i+1} = \mathbf{x}^i + \mu \, \text{sign}(e_j)\mathbf{h}_j. \tag{2.194}$$

The gradient of the L_1 norm is not defined at $e_j = 0$. However, this is not a problem since the gradient-descent method is not updated when the error is zero, i.e., $e_j = 0$.

2.11   BLOCK LEAST ABSOLUTE VALUE

The block least absolute value (LAV) estimate can be formulated as

$$\text{find some } \mathbf{x} \in R^N \text{ so as to}$$

$$\text{minimize} \quad \sum_{i=1}^{N} |\eta| \tag{2.195}$$

$$\text{subject to} \quad \mathbf{y} = \mathbf{Hx} + \boldsymbol{\eta}. \tag{2.196}$$

It can be shown that if the rank of $\mathbf{H}$ is r, then the LAV estimate will satisfy r equations exactly [19]; that is, r values of the vectors $\boldsymbol{\eta}$ will be zero for $\mathbf{x} = \mathbf{x}_{LAV}$.

To illustrate this point, consider the following set of equations:

$$1m = 1, \tag{2.197}$$

$$2m = 2.1, \tag{2.198}$$

$$3m = 5. \tag{2.199}$$

In this case, the rank is 1 (i.e., $r = 1$) and $N = 3$. The last equation is a statistical outlier, i.e., a large error has been added. The least squares solution is

$$m = \frac{1 \cdot 1 + 2 \cdot 2.1 + 3 \cdot 5}{1 \cdot 1 + 2 \cdot 2 + 3 \cdot 3} = 1.44286. \tag{2.200}$$

The LAV estimate can be obtained by solving each equation for m and selecting the value which yields the minimum $\sum |e_i|$. Thus, we find

$$\text{from (2.197)}, \quad m_1 = 1, \tag{2.201}$$

$$\text{from (2.198)}, \quad m_2 = 1.05, \tag{2.202}$$

$$\text{from (2.199)}, \quad m_3 = 1.67. \tag{2.203}$$

Now, calculating $\sum e_i$ for each m_i, we see that

$$\text{for } m_1, \quad E_1 = |1 - 1| + |2.1 - 2| + |5 - 3| = 2.1, \tag{2.204}$$

$$\text{for } m_2, \quad E_2 = |1 - 1.05| + |2.1 - 2 \cdot 1.05| + |5 - 3 \cdot 1.05| = 1.9, \tag{2.205}$$

$$\text{for } m_3, \quad E_3 = |1 - 1.67| + |2.1 - 2 \cdot 1.67| + |5 - 3 \cdot 1.67| = 1.91. \tag{2.206}$$

The second value, m_2, corresponds to the minimum E, so it is the LAV estimate, i.e.,

$$m_{LAV} = 1.05.$$

Note that the LAV estimate is not greatly influenced by the outlier. This approach—testing all the possible subsets of r equations from the total number (N)—is, in general, prohibitively expensive. If we wish to set K equations to be exactly satisfied out of N equations, there are

$$\binom{N}{K} = \frac{N!}{(N-K)!K!}$$

ways to test for this. An exhaustive search is not practical for even a small number of equations.

There are search methods that go from one estimate to the next in such a way as to ensure a decrease in the L_1 norm of the error. Linear programming has been found to be useful in practice for image restoration [19] and spectral estimation [20] where the restored signal is sparse, i.e., where it consists of a large number of zeros or constant values.

Linear programming is useful for solving problems of the form

$$\text{minimize} \quad \mathbf{Cz} \tag{2.207}$$

$$\text{subject to} \quad \mathbf{y} = \mathbf{Az} \tag{2.208}$$

$$\text{where } \mathbf{z} \geq \mathbf{1}. \tag{2.209}$$

The LAV problem can be transformed into a LP problem by making the substitution

$$\eta = \eta^+ - \eta^-,$$

where η^+ consists of all the positive values of η, with zeros in those positions where η is negative, and η^- consists of positive values where η is negative and zero elsewhere.

Optionally, we can do the same for $\mathbf{x}$; however, we shall assume $\mathbf{x}$ to be nonnegative here for ease of presentation.

The problem then becomes

$$\text{minimize} \quad \eta^+ + \eta^- + \cdots + \eta_N^+ + \eta_N^- \tag{2.210}$$

$$\text{subject to} \quad \mathbf{y} = \mathbf{Hx} + \eta^+ - \eta^-, \tag{2.211}$$

where $\mathbf{x} \geq \mathbf{0}$.

We define the following:

$$\mathbf{C} = \left[\underbrace{1 \cdots 1}_{2N} \; \underbrace{0 \cdots 0}_{N} \right], \tag{2.212}$$

$$\mathbf{z} = \begin{pmatrix} \eta^+ \\ \eta^- \\ x \end{pmatrix},$$
(2.213)

and

$$\mathbf{A} = [\mathbf{H} \,|\, \mathbf{I} \,|\, -\mathbf{I}].$$
(2.214)

Then we can solve the LAV problem by using an LP algorithm like the simplex method. See [19] for an example.

The minimum L_∞-norm solution can also be formulated as an LP problem. The minimax, or L_∞-norm, estimate is such that the largest value of the error vector is the smallest possible value. Let us denote this value by the scalar E. Then the LP formulation is given by

$$\text{minimize} \quad E$$
(2.215)

$$\text{subject to} \quad \mathbf{y} - \mathbf{Hx} \geq E[\mathbf{I}]$$
(2.216)

$$\text{and} \quad -\mathbf{y} + \mathbf{Hx} \leq E[\mathbf{I}],$$
(2.217)

where $\mathbf{I}$ is the identity vector, i.e., a vector of all ones.

2.12 SUMMARY AND CONCLUSIONS

We have introduced the idea of modeling an arbitrary linear system as a matrix operating on an input signal vector. The matrix operator can be diagonalized using orthogonal matrices similar to the transforms used in conventional system theory. The geometrical interpretations of a matrix operation are shown to consist of rotations and scaled projections of the input vector.

Inverse systems were discussed. The problems of inconsistent equations and rank deficiency were addressed. The pseudo-inverse solution was shown to offer a reasonable estimate in most circumstances. The statistical least squares, MAP, and RAP approaches were briefly reviewed. These methods and their applications will be discussed more fully in the following chapters.

A measure of the distance between two vectors was introduced. This concept is vital to signal-recognition methods. Gradient-descent and random-search strategies were discussed.

ACKNOWLEDGMENTS

The research reported here was made possible, in part, through the support of the New Jersey Commission on Science and Technology and the Center for Computer Aids for Industrial Productivity (CAIP) at Rutgers University.

REFERENCES

1. K. E. Atkinson. *An Introduction to Numerical Analysis*. New York: Wiley, 1978.

2. T. Kailath. *Linear Systems*. Englewood Cliffs, N.J.: Prentice-Hall, 1980.

3. J. W. Goodman. *Introduction to Fourier Optics*. New York: McGraw-Hill, 1968.

4. J. L. Flanagan. *Speech Analysis, Synthesis, and Perception*, 2nd ed. New York: Springer-Verlag, 1972.

5. W. K. Pratt. *Digital Image Processing*. New York: Wiley, 1978.

6. A. V. Oppenheim and R. W. Schafer. *Discrete-Time Signal Processing*. Englewood Cliffs, N.J.: Prentice-Hall, 1989.

7. R. Bellman. *Introduction to Matrix Analysis*, 2nd ed. New York: McGraw-Hill, 1970.

8. G. H. Golub and C. F. Van Loan. *Matrix Computations*, 2nd ed. Baltimore: Johns Hopkins, 1989.

9. L. R. Rabiner and R. W. Schafer. *Digital Processing of Speech Signals*. Englewood Cliffs, N.J.: Prentice-Hall, 1978.

10. A. Papoulis. *Probability, Random Variables, and Stochastic Processes*. New York: McGraw-Hill, 1984.

11. H. Robbins and S. Monro. A stochastic approximation method. *Ann. Math. Statist.*, **22**:400–407, 1951.

12. A. Ben-Israel and T. Greville. *Generalized Inverses: Theory and Applications*. New York: Wiley, 1974.

13. W. H. Press, B. P. Flannery, S. A. Teukolsky, and W. T. Vetterling. *Numerical Recipes*. New York: Cambridge, 1986.

14. G. T. Herman. *Image Reconstruction from Projections: The Fundamentals of Computerized Tomography*. New York: Academic Press, 1980.

15. K. Tanabe. Projection method for solving a singular system of linear equations and its applications. *Numer. Math.*, **17**:203–214, 1971.

16. R. O. Duda and P. E. Hart. *Pattern Classification and Scene Analysis*. New York: Wiley, 1973.

17. H. Taub and D. Schilling. *Principles of Communication Systems*. New York: McGraw-Hill, 1986.

18. M. K. Skolnik. *Introduction to Radar Systems*, 2nd ed. New York: McGraw-Hill, 1980.

19. R. J. Mammone. Image restoration using linear programming. In Henry Stark, ed., *Image Recovery: Theory and Recovery*, pp. 127–156. London: Academic Press, 1987.

20. R. J. Mammone and G. Eichmann. Restoration of discrete Fourier spectra using linear programming. *J. Opt. Soc. Am.*, **72**(8), 1982.

PART 2

SIGNAL RECOVERY

3

IMAGE RECOVERY: A VECTOR-SPACE APPROACH

Christine I. Podilchuk

AT&T Bell Laboratories
600 Mountain Avenue
Murray Hill, New Jersey 07947
email: chrisp@research.att.com

3.1 APPLICATIONS

Image recovery—or, more generally, signal recovery—is a fundamental problem encountered in many scientific disciplines. Image recovery encompasses many different problems—reconstructing medical images using x-ray computerized tomography, recovering the structure of molecules, and uncovering distant stars. Applications for image restoration include interferometric imaging. Astronomers, together with mathematicians, have devised interferometric imaging techniques in order to obtain very high resolution images. However, such techniques require knowledge of both the magnitude and the phase of the Fourier transform of the image field. Many times, one can only measure the magnitude of the image-field distribution and image-recovery techniques are needed to recover the phase information. In the fields of chemistry and biology, the finest details of molecular structure, which are not visible with the most powerful microscopes, appear as the result of image-recovery techniques. Another need for image recovery arises when x-ray or electron-diffraction patterns are used to study the microstructure of matter. The intensity is easily measured, but restoration techniques are needed to recover the phase information so that the complex three-dimensional structure can be restored. Image-recovery algorithms can also be used as an aid in reconstructing a two-dimensional image from its line integrals, as in x-ray computerized tomography, electron microscopy, and radio astronomy.

One reason for developing image-recovery algorithms is the limitations imposed by all physical imaging instruments. An imaging system, no matter how well designed, is limited in the amount of data it can collect. Many times, the high-frequency information—or, equivalently, the fine details in the object—are lost. A simple example is the finite aperture of a lens, which limits the number of high spatial frequencies of an object that can enter the lens. Because the lens attenuates high spatial frequencies, the degradation due to the finite size of the aperture can be modeled by a low-pass filter. Image-restoration techniques have been developed that attempt to increase the effective aperture size in telescopes, cameras, and other optical imaging systems, without requiring that one design and build more expensive—and, perhaps, physically unrealizable—systems.

3.2 PROBLEM FORMULATION

Image recovery may be viewed as an estimation process in which operations are performed on an observed, or measured, field to estimate the ideal image field that would be observed if no image degradation were present. In order to design a digital image-recovery system effectively, it is necessary to characterize the image-degradation effects of the physical imaging system, the image digitizer, and the image display. The basic idea is to model the image-degradation effects as accurately as possible and then perform operations to undo the degradation and obtain a restored image. Note that when the degradation model is not sufficiently accurate, even the best recovery algorithms will not yield satisfactory results.

There are two basic approaches to modeling the image degradation operator, *a priori* modeling and *a posteriori* modeling. In *a priori* modeling, measurements are made on the physical imaging system to determine the response for an arbitrary image field. For certain applications, the system response is deterministic; for other applications, the system response may only be determined in a stochastic sense. The *a posteriori* modeling approach for image degradation is based on measurements of the specific image to be recovered.

For the continuous image formation model, a continuous-image light distribution denoted as $C(x, y, t, \lambda)$ dependent on spatial coordinates (x, y), time (t), and spectral wavelength (λ), is the driving force of a physical imaging system both subject to point and spatial degradation effects and corrupted with deterministic and stochastic disturbances. The degradations may include diffraction in the optical system, sensor nonlinearities, optical system aberrations, film nonlinearities, atmospheric turbulence effects, image motion blur, and geometric distortion. The physical imaging system produces a set of output image fields $I_O^{(i)}(x, y, t) = \mathcal{O}_P \{C, x, y, t, \lambda; C(x, y, t, \lambda)\}$, where $\mathcal{O} \{\cdot\}$ represents a general operator that is dependent on x, y, t, λ, and the amplitude of the light distribution, C. The physical imaging system may either have a time memory or be memoryless. Time memory results in the interaction of observed pixel values at the same coordinates (x, y) at different time frames. Assuming that the imaging system is memoryless, the observed image can be expressed as

$$I_O^{(i)}(x,y,t) = \mathcal{O}_P \{C,x,y,\lambda; C(x,y,t,\lambda)\}. \tag{3.1}$$

For many imaging systems, the point-wavelength response assumes the form of a wavelength-weighted integration of the input light distribution denoted as

$$I_O^{(i)}(x,y,t) = \mathcal{O}_P \left\{ C,x,y,t; \int_0^\infty C(x,y,t,\lambda)S_i(\lambda) \, d\lambda \right\}, \tag{3.2}$$

where $S_i(\lambda)$ is the sensor wavelength response. Image-degradation models can often be represented as a cascade of spatial $\mathcal{O}_S\{\cdot\}$ and point intensity $\mathcal{O}_C\{\cdot\}$ effects.

When the spatial image degradation is linear, it can be expressed as a superposition operator. When the imaging system is memoryless with separable point and spatial degradation, the observed image can be modeled by

$$I_O^{(i)}(x,y,t) = \mathcal{O}_C \left\{ C,\lambda; \iint\limits_{-\infty}^{\infty} C(\alpha,\beta,t,\lambda)J(x,y,\alpha,\beta,t,\lambda) \, d\alpha \, d\beta \right\}, \tag{3.3}$$

where $J(\cdot)$ is the impulse response of the spatial degradation. The impulse response is known as the point-spread function in optical systems. When the point-spread function is linear and shift invariant, the observed image is related to the image light distribution by a convolution integral,

$$I_O^{(i)}(x,y,t) = \mathcal{O}_C \left\{ C,t; \iint\limits_{-\infty}^{\infty} C(\alpha,\beta,t,\lambda)J(x-\alpha;y-\beta;t,\lambda) \, d\alpha \, d\beta \right\}. \tag{3.4}$$

Each observed image field, $I_O^{(i)}(x,y,t)$, is digitized to produce an array of image samples, $I_S(m_1,m_2,t)$, at each time instant t. It is assumed that continuous image fields are sampled at a high enough rate to satisfy the Nyquist criterion. The array of image samples can be processed using a restoration algorithm, after which the restored image samples can be interpolated by the image-display system to produce a continuous image estimate.

For the case of an optical image-formation system, the imaging device provides a deterministic transformation of the input spatial light distribution to an output spatial light distribution. The basic concepts of geometric optics can be applied to optical imaging systems. It is assumed that light rays always travel in straight-line paths in a homogeneous medium. Therefore, a bundle of rays passing through a clear aperture onto a screen should produce a geometric light projection of the aperture. However, the light distribution at the boundary between the light and dark areas on the screen is not sharp and becomes fuzzier as the aperture size decreases. For a pinhole aperture, the entire screen becomes diffusely illuminated. This is due to the finite aperture size, which causes a bending of rays known as diffraction. Diffraction of light can be quantitatively characterized by considering light as electromagnetic radiation that satisfies Maxwell's equations.

In most optical imaging systems, the optical radiation emitted by an object is transmitted or reflected light from an incoherent light source. The image radiation is assumed to be quasimonochromatic; that is, the spectral bandwidth of the image radiation detected at the image plane is smaller than the center wavelength of the radiation. Under these assumptions, the optical system will behave linearly and can be represented by the superposition integral equation,

$$I_O(x_i, y_i) = \iint\limits_{-\infty}^{\infty} D(x_i, y_i; x_0, y_0) I_I(x_0, y_0) \, dx_0 \, dy_0, \tag{3.5}$$

where $D(x_i, y_i; x_0, y_0)$ represents the image-intensity response to a point source of light or point-spread function and $I_I(x, y)$ represents the ideal image of the object. When the point-spread function is space invariant, (3.5) can be replaced with

$$I_O(x_i, y_i) = \iint\limits_{-\infty}^{\infty} D(x_i - x_0, y_i - y_0) I_I(x_0, y_0) \, dx_0 \, dy_0, \tag{3.6}$$

which, in the Fourier domain, is represented by the scalar multiplication

$$\mathcal{I}_O(w_x, w_y) = \mathcal{D}(w_x, w_y) \mathcal{I}_I(w_x, w_y). \tag{3.7}$$

The Fourier transform of the point-spread function $D(x_i, y_i)$, denoted as $\mathcal{D}(w_x, w_y)$, is called the optical transfer function (OTF) and is defined by

$$\mathcal{D}(w_x, w_y) = \frac{\iint_{-\infty}^{\infty} D(x, y) \exp\left\{-i(w_x x + w_y y)\right\} \, dx \, dy}{\iint_{-\infty}^{\infty} D(x, y) \, dx \, dy}. \tag{3.8}$$

The absolute value of the OTF is known as the modulation transfer function (MTF). A commonly used optical image-formation system is a circular thin lens.

The physical digitizer used in many imaging systems may introduce errors because the continuous image reconstructed from the physical samples may not be identical to the image field sampled. This discussion assumes that the sampling rate satisfies the Nyquist criterion so that aliasing error does not become a problem. Sampling the image with a finite width sampling pulse blurs the image to produce

$$I_S(m_1, m_2, t) = \iint\limits_{-\infty}^{\infty} I_O(\alpha, \beta, t) P(\alpha - m_1 \, \Delta x, \beta - m_2 \, \Delta y) \, d\alpha \, d\beta, \tag{3.9}$$

where $P(x, y)$ is the pulse shape of the sampling pulse and Δx and Δy are the two-dimensional sampling periods. Image-quantization errors can be an important source of image degradation.

The image-display system produces a continuous image-field estimate $\hat{I}_I(x, y, t)$ by interpolating the output array of the digital restoration processor $I_R(k_1, k_2, t)$ with a two-dimensional interpolation function $B(x, y)$, which is expressed as

$$\hat{I}_I(x, y, t) = \sum_{k_1=-K_1}^{K_1} \sum_{k_2=-K_2}^{K_2} I_R(k_1, k_2, t)B(x - k_1 \Delta x, y - k_2 \Delta y). \tag{3.10}$$

Ideally, the spatial interpolation function $B(x, y)$ should consist of two-dimensional sinc or Bessel functions. Unfortunately, such interpolation functions usually have negative lobes and cannot be implemented exactly by the summation of positive light quantities.

A discrete formulation of the degradation operator can be modeled as a cascade of the degradation operators due to the imaging system, digitizer, and display system, along with noise terms due to external and internal sources.

General restoration algorithms that encompass every type of degradation and noise source have not been developed. It is necessary at this point to introduce the specific image-degradation model that is addressed in developing the image recovery algorithms introduced in this chapter and Chapter 4. The model consists of an image field subject to a linear blur with an additive Gaussian noise term that is independent of the image field. Assume that the spatially truncated image data array $\mathbf{F}(n_1, n_2)$ of dimension $N_1 \times N_2$ is blurred with the spatially truncated discrete point-spread function $h(l_1, l_2; m_1, m_2)$ of dimension $L_1 \times L_2$ to produce an output image $\mathbf{G}(m_1, m_2)$ of dimension $M_1 \times M_2$. In general, the point-spread function is space variant so that

$$\mathbf{G}(m_1, m_2) = \sum_{n_1=1}^{m_1} \sum_{n_2=1}^{m_2} \mathbf{F}(n_1, n_2)h(m_1 - n_1 + 1, m_2 - n_2 + 1; m_1, m_2) \tag{3.11}$$

for $m_1 = 1, 2, \ldots, M_1$ and $m_2 = 1, 2, \ldots, M_2$. The dimensions of the ideal image, the point-spread function, and the observed image are related by $M_1 = N_1 + L_1 - 1$ and $M_2 = N_2 + L_2 - 1$. By column or row scanning the original two-dimensional data, a vector-space relationship can be developed for the image-restoration problem. Let $\mathbf{g}$ represent the $M = M_1 \times M_2$ lexicographically ordered two-dimensional observed image $\mathbf{G}(m_1, m_2)$ and $\mathbf{f}$ represent the $N = N_1 \times N_2$ lexicographically ordered ideal image $\mathbf{F}(n_1, n_2)$. The $M \times N$ matrix $\mathbf{H}$ represents the finite area-superposition operator and is composed of the elements of the point-spread function $h(l_1, l_2; m_1, m_2)$. Therefore, the expression given by (3.11) is a discrete linear system and can be rewritten in vector-space notation as

$$\mathbf{g} = \mathbf{Hf}. \tag{3.12}$$

The degradation matrix $\mathbf{H}$ can be decomposed into submatrices of dimension $M_1 \times N_1$, where

$$\mathbf{H} = \begin{pmatrix} H_{1,1} & 0 & \cdots \\ H_{2,1} & H_{2,2} & 0 & \cdots \\ \vdots & \vdots \\ H_{L_2,1} & H_{L_2,2} \\ 0 & H_{L_2+1,2} \\ \vdots & 0 \\ & \vdots \\ & & & & H_{M_2,N_2} \end{pmatrix} \qquad (3.13)$$

and

$$H_{m_2,n_2}(m_1, n_1) = h(m_1 - n_1 + 1, m_2 - n_2 + 1; m_1, m_2), \qquad (3.14)$$

$$1 \leq n_1 \leq N_1,$$

$$1 \leq n_2 \leq N_2,$$

$$n_1 \leq m_1 \leq n_1 + L_1 - 1,$$

$$n_2 \leq m_2 \leq n_2 + L_2 - 1.$$

Note that $\mathbf{H}$ is quite structured and very sparse. If the point-spread function is space invariant, then

$$\mathbf{H}_{m_2,n_2} = \mathbf{H}_{m_2+1,n_2+1}. \qquad (3.15)$$

The additive noise term due to measurement errors and external and internal noise sources is represented by a vector $\mathbf{n}$ of length M; thus the vector-space representation of the image-recovery problem in the presence of noise is

$$\mathbf{g} = \mathbf{Hf} + \mathbf{n}. \qquad (3.16)$$

Finding a solution $\mathbf{f}$ for the system of linear equations described by (3.16) seems to be straightforward. However, in practice, (3.16) usually consists of a very large set of equations which are ill conditioned or rank deficient and $\mathbf{n}$ introduces inconsistencies. All these problems must be adequately addressed in order to obtain a useful solution.

3.3 INVERSE FILTERING TECHNIQUES

The earliest attempts to recover images were based on the concept of inverse filtering, where the degradation operator is inverted to restore the desired image

[1]. In the following analysis, the image-recovery problem will be described by the continuous two-dimensional convolution given by

$$I_O(x, y) = \int_{-\infty}^{\infty} \int_{-\infty}^{\infty} I_I(\alpha, \beta) D(x - \alpha, y - \beta) \, d\alpha \, d\beta + N(x, y), \qquad (3.17)$$

where $I_O(x, y)$ corresponds to the continuous blurred output image represented by **g** in the discrete formulation, $I_I(x, y)$ represents the continuous ideal image represented by **f** in the discrete formulation, and $D(x, y)$ represents the filter coefficients for the shift-invariant blurring function represented by the Toeplitz matrix **H** in the discrete formulation. The inverse filtering approach is described by looking at (3.17) in the Fourier domain,

$$\mathcal{I}_O(w_x, w_y) = \mathcal{I}_I(w_x, w_y) \mathcal{D}(w_x, w_y) + \mathcal{N}(w_x, w_y), \qquad (3.18)$$

so that an estimate of the Fourier transform of the ideal image $\hat{\mathcal{I}}_I(w_x, w_y)$ is calculated as

$$\hat{\mathcal{I}}_I(w_x, w_y) = [\mathcal{I}_I(w_x, w_y) \mathcal{D}(w_x, w_y) + \mathcal{N}(w_x, w_y)] \mathcal{D}^{-1}(w_x, w_y). \qquad (3.19)$$

From (3.19), one can see that the inverse-filtering approach is successful in the absence of noise and when the degradation operator is accurately modeled. However, in most cases of practical interest, noise and measurement error are present in the image-recovery problem. The additive noise term in (3.19) can become quite large at spatial frequencies where the degradation operator $\mathcal{D}(w_x, w_y)$ is small. Typically, both $\mathcal{D}(w_x, w_y)$ and $\mathcal{I}_\mathcal{I}(w_x, w_y)$ are small at high spatial frequencies; thus, direct inversion of the degradation operator leads to severely degraded image quality at high spatial frequencies which correspond to the detailed regions of the image. Another problem with the direct inversion technique is that, for most applications, the image-recovery problem is extremely ill conditioned. This means that a small change in the noise term $N(x, y)$ can lead to a large change in the estimate of $I_I(x, y)$. The ill-conditioned problem is explained by letting $Z(x, y)$ represent a dither function added to the ideal image $I_I(x, y)$,

$$I_I^*(x, y) = I_I(x, y) + Z(x, y). \qquad (3.20)$$

There may be many dither functions for which

$$\left| \int\!\!\!\int_{-\infty}^{\infty} Z(\alpha, \beta) D(x - \alpha, y - \beta) \, d\alpha \, d\beta \right| < N(x, y). \qquad (3.21)$$

For such dither functions, the perturbed image given by (3.20) may satisfy the integral convolution equation given by (3.17) to within the given measurement accuracy determined by $N(x, y)$. Ill-conditioned systems can produce two kinds of problems for image recovery. High-frequency image information may be masked by observation noise, or a small amount of observation noise may lead to a restoration that contains very large false high-frequency components. The other potential problem with the inverse-filtering approach is that, when the system of equations is rank deficient, the problem of inverting zero-valued terms in the transfer function arises.

Improved restoration quality is possible with Wiener-filtering techniques that incorporate *a priori* statistical knowledge of the noise [2] in the inverse-filtering estimate. For a continuous Wiener-filtering system, the impulse response of the restoration filter is chosen to minimize the mean square error, as defined by

$$\mathcal{E} = E\left\{(I_I - \hat{I}_I)^2\right\},$$ (3.22)

where $\hat{I}_I$ denotes the estimate of the ideal image I_I and $E\{\cdot\}$ denotes the expected value. It is assumed that the noise is zero mean and independent of the image and that its power spectral density $\mathcal{N}_W(w_x, w_y)$ is known. The transfer function for the Wiener restoration filter is

$$\mathcal{D}_W^{-1}(w_x, w_y) = \frac{\mathcal{D}(w_x, w_y)}{|\mathcal{D}(w_x, w_y)|^2 + \mathcal{N}_W(w_x, w_y)}.$$ (3.23)

For the no-noise case, the Wiener filter reduces to the inverse filter. Including the noise power spectral density provides a smooth rolloff in the inversion process so that high-frequency noise terms cannot severely distort the image estimate. If additional information about the image, such as the spatial correlation, is available, the Wiener filter can be further improved. Assuming that the ideal image $I_I(x, y)$ has a known two-dimensional power spectral density $\mathcal{I}_W(w_x, w_y)$ and that the image process is zero mean, the restoration filter becomes

$$\mathcal{D}_W^{-1}(w_x, w_y) = \frac{\mathcal{D}(w_x, w_y)\mathcal{I}_W(w_x, w_y)}{|\mathcal{D}(w_x, w_y)|^2\mathcal{I}_W(w_x, w_y) + \mathcal{N}_W(w_x, w_y)}$$

$$= \frac{\mathcal{D}(w_x, w_y)}{|\mathcal{D}(w_x, w_y)|^2 + \mathcal{N}_W(w_x, w_y)/\mathcal{I}_W(w_x, w_y)}.$$ (3.24)

The signal-to-noise ratio is defined as

$$\text{SNR}(w_x, w_y) = \frac{\mathcal{I}_W(w_x, w_y)}{\mathcal{N}_W(w_x, w_y)},$$ (3.25)

so that the Wiener restoration filter for stochastic image processes becomes

$$\mathcal{D}_W^{-1}(w_x, w_y) = \frac{\mathcal{D}(w_x, w_y)}{|\mathcal{D}(w_x, w_y)|^2 + \frac{1}{\text{SNR}(w_x, w_y)}}.$$ (3.26)

If the image is uncorrelated, (3.26) reduces to (3.23).

When the degradation operator changes its characteristics in a random fashion, e.g., the blur caused by imaging through a turbulent atmosphere, the Wienerfilter can be modified [3] by considering the impulse response of the blurring filter to be a sample of a two-dimensional stochastic process with known mean and known power spectral density. Other researchers have developed parametric estimation filters based on the Wiener-filter concept.

The filtering methods discussed so far have been developed for continuous image fields but can easily be adapted and applied to the restoration of discrete images. Assume that all continuous functions are sampled at a rate high enough to satisfy the Nyquist sampling criterion for the original image. The restoration filter's point-spread function is truncated to produce an $L \times L$ array, $h_W^{-1}(j,k)$. The observed image field $I_O(x,y)$ is also truncated to produce an $M_1 \times M_2$ array $G(j,k)$. Performing two-dimensional DFTs on the arrays $h_W^{-1}(j,k)$ and $G(j,k)$ yields

$$\mathcal{H}_W^{-1}(u,v) = \frac{1}{N}\sum_{j=0}^{L-1}\sum_{k=0}^{L-1} h_W^{-1}(j,k)\exp\left\{\frac{-2\pi i}{J}(uj+vk)\right\} \qquad (3.27)$$

and

$$\mathcal{G}(u,v) = \frac{1}{N}\sum_{j=0}^{M_1}\sum_{k=0}^{M_2} G(j,k)\exp\left\{\frac{-2\pi i}{J}(uj+vk)\right\}, \qquad (3.28)$$

$$u,v = 0,1,2,\ldots,J-1.$$

The inverse DFT of the result is

$$\hat{F}(j,k) = \frac{1}{N}\sum_{u=0}^{J-1}\sum_{v=0}^{J-1}\mathcal{G}(u,v)\mathcal{H}_W^{-1}(u,v)\exp\left\{\frac{2\pi i}{J}(uj+vk)\right\}, \qquad (3.29)$$

where $\hat{F}(j,k)$ denotes the estimate of the ideal image. The only approximation that has been made so far is the spatial truncation of the restoration filter. The truncation error can be made as small as desired, at the expense of dimensionality and processing complexity. The major problem of the conversion from the continuous to the discrete form of the Wiener filtering technique is the accurate determination of the values of the discrete samples of the restoration filter's point-spread function. A simple approach is to find the inverse Fourier transform of $\mathcal{H}_W^{-1}(w_x,w_y)$ analytically and sample $H_W^{-1}(x,y)$ at the Nyquist rate. However, it is often difficult to obtain the inverse Fourier transform analytically. Another approach would be to convert $\mathcal{H}_W^{-1}(w_x,w_y)$ directly to the discrete Fourier transform domain to obtain $\mathcal{H}_W^{-1}(u,v)$. This may also be difficult. The last alternative is to form the inverse DFT of $\mathcal{H}_W^{-1}(w_x,w_y)$ with $w_x = 2\pi u/J\Delta$ and $w_y = 2\pi v/J\Delta$, where Δ is the sample spacing. The resulting spatial domain point-spread function is windowed to produce an $L \times L$ array. The DFT of the truncated array produces the desired $\mathcal{H}_W^{-1}(u, v)$. Failure to truncate the point-spread function will result in wraparound error.

3.4 THE PSEUDO-INVERSE TECHNIQUE OF IMAGE RECOVERY

Wiener-type filters attempt to minimize the noise amplification obtained in a direct inverse by providing a taper determined by the statistics of the image and noise process under consideration. The pseudo-inverse image restoration technique also attempts to minimize the noise effects by inverting only part of the degradation operator. The pseudo-inverse method is applied directly to the discrete vector-space representation of the image-recovery problem, as stated in (3.16). A solution is sought that minimizes the least squares modeling error given by

$$\mathcal{E}_m = (\mathbf{n})^T(\mathbf{n}) = (\mathbf{g} - \mathbf{Hf})^T(\mathbf{g} - \mathbf{Hf}). \tag{3.30}$$

For a consistent set of equations, a solution is sought that minimizes the least squares estimation error; that is,

$$\begin{aligned} \mathcal{E}_e &= (\mathbf{f} - \hat{\mathbf{f}})^T(\mathbf{f} - \hat{\mathbf{f}}) \\ &= \text{tr}\{(\mathbf{f} - \hat{\mathbf{f}})(\mathbf{f} - \hat{\mathbf{f}})^T\}, \end{aligned} \tag{3.31}$$

where $\mathbf{f}$ is the desired image vector and $\hat{\mathbf{f}}$ is the estimate. The least squares solution is not unique when the rank of the $M \times N$ matrix $\mathbf{H}$ is $r < N \le M$. However, the Moore–Penrose generalized inverse or pseudo-inverse [4] provides a unique least squares solution that has the minimum norm of all the least squares solutions. The generalized inverse provides an optimum solution that minimizes the estimation error for a consistent set of equations and the modeling error for an inconsistent set of equations. Therefore, the generalized inverse provides an optimum solution for both consistent and inconsistent sets of equations, as defined by the performance functions $\mathcal{E}_e$ and $\mathcal{E}_m$, respectively. The least squares inverse solution satisfies the normal equations

$$\mathbf{H}^T\mathbf{g} = \mathbf{H}^T\mathbf{Hf}. \tag{3.32}$$

The Moore–Penrose generalized inverse, pseudo-inverse, or least squares solution with minimum norm is defined as

$$\mathbf{f}^\dagger = (\mathbf{H}^T\mathbf{H})^\dagger\mathbf{H}^T\mathbf{g} = \mathbf{H}^\dagger\mathbf{g}, \tag{3.33}$$

where the dagger (†) denotes the pseudo-inverse[1] and the rank of $\mathbf{H}$ is $r \le N \le M$.

---

[1]See Chapter 2 and Secction 3.9.

For the case of an inconsistent problem set, the pseudo-inverse solution can be expressed as

$$\mathbf{f}^\dagger = \mathbf{H}^\dagger \mathbf{g} = \mathbf{H}^\dagger \mathbf{H} \mathbf{f} + \mathbf{H}^\dagger \mathbf{n}, \tag{3.34}$$

where $\mathbf{f}^\dagger$ is the minimum-norm least squares solution. If the set of equations are overdetermined, with rank $r = N < M$, the term $\mathbf{H}^\dagger \mathbf{H} = \mathbf{I}_N$; thus the pseudo-inverse solution becomes

$$\mathbf{f}^\dagger = \mathbf{f} + \mathbf{H}^\dagger \mathbf{n}$$

$$= \mathbf{f} + \Delta\mathbf{f}. \tag{3.35}$$

A straightforward result from numerical analysis is the bound on the relative error,

$$\frac{\|\Delta\mathbf{f}\|}{\|\mathbf{f}\|} < \|\mathbf{H}^\dagger\| \cdot \|\mathbf{H}\| \frac{\|\mathbf{n}\|}{\|\mathbf{g}\|}, \tag{3.36}$$

where the product $\|\mathbf{H}^\dagger\| \cdot \|\mathbf{H}\|$ is the condition number of $\mathbf{H}$. This quantity determines the relative error in the estimate in terms of the ratio of the vector norm of the noise to the vector norm of the observed image. The condition number of $\mathbf{H}$ can be calculated from

$$\mathcal{C}_H = \|\mathbf{H}^\dagger\| \cdot \|\mathbf{H}\| = \frac{\sigma_{\max}}{\sigma_{\min}}, \tag{3.37}$$

where $\sigma_{\max}$ and $\sigma_{\min}$ denote the largest and smallest singular values[2] of the matrix $\mathbf{H}$, respectively. The larger the condition number, the greater the sensitivity to noise perturbations. A matrix with a large condition number ($\mathcal{C} > 100$) results in an ill-conditioned system.

When the degradation operator is separable—that is, when the column and row blurring functions act independently of each other—the pseudo-inverse solution can be obtained with great computational savings. For a separable degradation operator,

$$\mathbf{H} = \mathbf{H}_C \otimes \mathbf{H}_R, \tag{3.38}$$

where $\mathbf{H}_C$ and $\mathbf{H}_R$ represent the column and row degradation operators respectively. Therefore, the generalized inverse is separable so that it can be obtained by

$$\mathbf{H}^\dagger = \mathbf{H}_C^\dagger \otimes \mathbf{H}_R^\dagger. \tag{3.39}$$

---

[2]See Appendix A, Review of Linear Algebra.

The pseudo-inverse solution can be obtained with tremendous computational savings when the problem can be expressed by the circular convolution vector-space relationship

$$\mathbf{g}_e = \mathbf{H}_C \mathbf{f}_e, \tag{3.40}$$

where $\mathbf{H}_C$ is the $J \times J$ matrix with $J \geq M$, and $\mathbf{f}_e$ and $\mathbf{g}_e$ are the $J \times 1$ vectors obtained by zero padding the $N \times 1$ image vector $\mathbf{f}$, and the $M \times 1$ output vector $\mathbf{g}$, respectively. The matrix $\mathbf{H}_C$ has the following structure:

$$\mathbf{H}_C = \begin{pmatrix} \mathbf{C}_{1,1} & 0 & \cdots & \mathbf{C}_{1,J-L+1} & \cdots & \mathbf{C}_{1,J} \\ \mathbf{C}_{2,1} & \mathbf{C}_{2,2} & & & & \mathbf{C}_{2,J} \\ \vdots & & & & & \mathbf{C}_{L-1,J} \\ \mathbf{C}_{L,1} & \mathbf{C}_{L,2} & & & & 0 \\ 0 & \mathbf{C}_{L-1,2} & & & & \\ \vdots & & & & & \\ & & & \mathbf{C}_{J,J-L+1} & & \mathbf{C}_{J,J} \end{pmatrix}, \tag{3.41}$$

where

$$\mathbf{C}_{m_2,n_2}(m_1, n_1) = h_e(k_1, k_2; m_1, m_2).$$

$$1 \leq n_1 \leq M_1,$$

$$1 \leq n_2 \leq M_2,$$

$$1 \leq m_1 \leq M_1,$$

$$1 \leq m_2 \leq M_2,$$

$$k_1 = (m_1 - n_1 + 1) \bmod(M_1),$$

$$k_2 = (m_2 - n_2 + 1) \bmod(M_2).$$

If the point-spread function is spatially invariant and orthogonally separable,

$$\mathbf{H}_C = \mathbf{H}_{C,R} \otimes \mathbf{H}_{C,C}, \tag{3.42}$$

where $\mathbf{H}_{C,R}$ denotes the row-degradation operator and $\mathbf{H}_{C,C}$ denotes the column-degradation operator. The two-dimensional circular convolution can be expressed as

$$\mathbf{G}_e = \mathbf{H}_{C,C} \mathbf{F}_e \mathbf{H}_{C,R}^T. \tag{3.43}$$

Because $\mathbf{H}_C$ is a circular matrix, it can be diagonalized using the DFT matrix.[3] Tremendous savings in computational cost can be achieved by computing the pseudo-inverse for this problem as follows:

- append the point-spread function $\mathbf{H}$ with zeros to make a $J \times J$ matrix, where $J \ge M$, and compute the two-dimensional DFT of the appended $\mathbf{H}$ matrix

$$\mathcal{H}_e = W_J \mathbf{H}_e W_J; \tag{3.44}$$

- append the observed image $\mathbf{G}$ with zeros to form a $J \times J$ matrix and find the two-dimensional DFT of the appended $\mathbf{G}$ matrix

$$\mathcal{G}_e = W_J \mathbf{G}_e W_J; \tag{3.45}$$

- perform the scalar division

$$\hat{\mathcal{F}}_e = \frac{\mathcal{G}_e}{J\mathcal{H}_e}; \tag{3.46}$$

- take the inverse DFT

$$\hat{\mathbf{F}}_e = W_J^{-1} \hat{\mathcal{F}}_e W_J^{-1}; \tag{3.47}$$

- and extract the desired $N_1 \times N_2$ image $\mathbf{F}$ from the $J \times J$ image $\mathbf{F}_e$.

When the degradation model cannot be described by a circulant matrix, singular-value decomposition (SVD) can be used to obtain the pseudo-inverse solution. The SVD of H is

$$\mathbf{H} = \mathbf{U}\Sigma\mathbf{V}^T, \tag{3.48}$$

where $\mathbf{U}$ is a unitary matrix composed of the orthonormal eigenvectors of $\mathbf{H}^T\mathbf{H}$, $\mathbf{V}$ is a unitary matrix composed of the orthonormal eigenvectors of $\mathbf{H}\mathbf{H}^T$, and Σ is a diagonal matrix composed of the singular values of $\mathbf{H}$. The number of nonzero diagonal terms denotes the rank of $\mathbf{H}$. The degradation matrix can be expressed in series form as

$$\mathbf{H} = \sum_{i=1}^{r} \sigma_i \mathbf{u}_i \mathbf{v}_i^T, \tag{3.49}$$

---

[3]See Appendix A, Review of Linear Algebra.

where $\mathbf{u}_i$ and $\mathbf{v}_i$ are the ith columns of $\mathbf{U}$ and $\mathbf{V}$, respectively, and r is the rank of $\mathbf{H}$. The pseudo-inverse of $\mathbf{H}$ can be expressed as

$$\mathbf{H}^\dagger = \mathbf{V}\mathbf{\Sigma}^\dagger\mathbf{U}^T = \sum_{i=1}^{r} \sigma_i^{-1}\mathbf{v}_i\mathbf{u}_i^T; \tag{3.50}$$

thus, the pseudo-inverse solution becomes

$$\mathbf{f}^\dagger = \mathbf{H}^\dagger\mathbf{g} = \mathbf{V}\mathbf{\Sigma}^\dagger\mathbf{U}^T\mathbf{g} \tag{3.51}$$

or

$$\mathbf{f}^\dagger = \sum_{i=1}^{r} \sigma_i^{-1}\mathbf{v}_i\mathbf{u}_i^T\mathbf{g} = \sum_{i=1}^{r} \sigma_i^{-1}(\mathbf{u}_i^T\mathbf{g})\mathbf{v}_i. \tag{3.52}$$

A sequential restoration algorithm based on finding the pseudo-inverse solution using SVD on the degradation operator is

$$\mathbf{f}^{\dagger(k+1)} = \mathbf{f}^{\dagger(k)} + \sigma_k^{-1}(\mathbf{u}_k^T\mathbf{g})\mathbf{v}_k. \tag{3.53}$$

The iterative technique for finding the pseudo-inverse solution is advantageous when dealing with ill-conditioned systems and noise-corrupted data. The iterative form can be terminated before inverting small singular values that may cause the solution to become unstable. Note that when the degradation matrix $\mathbf{H}$ is circulant, it can be diagonalized using the DFT matrix so that the SVD method reduces to the method using only DFTs, as mentioned before.

Regression restoration techniques have been developed [5] that incorporate statistical knowledge of the noise. Assuming that $\mathbf{n}$ is zero mean with the known covariance function $\mathbf{K_n}$, the regression method seeks to minimize the weighted-error performance function,

$$\mathcal{E}_w = (\mathbf{g} - \mathbf{Hf})^T\mathbf{K_n}^{-1}(\mathbf{g} - \mathbf{Hf}). \tag{3.54}$$

The estimate that minimizes the performance function is obtained by setting the gradient of the performance function with respect to $\mathbf{f}$ equal to zero and solving for $\mathbf{f}$:

$$\begin{aligned} \nabla_f\mathcal{E}_w &= \frac{\delta\mathcal{E}_w}{\delta\mathbf{f}} \\ &= -2\mathbf{H}^T\mathbf{K_n}^{-1}(\mathbf{g} - \mathbf{Hf}) \\ &= 0. \end{aligned} \tag{3.55}$$

The restoration matrix $\mathbf{H}_w^\dagger$ becomes

$$\mathbf{H}_w^\dagger = (\mathbf{H}^T\mathbf{K_n}^{-1}\mathbf{H})^\dagger\mathbf{H}^T\mathbf{K_n}^{-1} \tag{3.56}$$

and the weighted least squares solution, or regression solution, is

$$\mathbf{f}_w^\dagger = (\mathbf{H}^T \mathbf{K}_n^{-1} \mathbf{H})^\dagger \mathbf{H}^T \mathbf{K}_n^{-1} \mathbf{g}. \tag{3.57}$$

If the noise is white with variance σ_n^2, then $\mathbf{K}_n = \sigma_n^2 \mathbf{I}$ and the weighted least squares solution reduces to the minimum-norm least squares solution given by (3.33).

Assuming that the ideal image is a random process with known first- and second-order statistics, the Wiener solution minimizes the mean squared estimation error:

$$\mathcal{E}_e = E\{(\mathbf{f} - \hat{\mathbf{f}})^T (\mathbf{f} - \hat{\mathbf{f}})\}, \tag{3.58}$$

where $\mathbf{f}$ denotes the ideal image and $\hat{\mathbf{f}}$ denotes the estimate. The restoration filter can be obtained either by computing the gradient of the cost function with respect to $\mathbf{f}$ and setting the result equal to zero or by employing the orthogonality principle to determine the restoration-filter coefficients and the estimate $\hat{\mathbf{f}}$ that minimizes the cost function. The resulting restoration matrix is

$$\mathbf{H}_W^{-1} = \mathbf{R}_{fg}(\mathbf{R}_{gg})^{-1}, \tag{3.59}$$

where $\mathbf{R}_{gg}$ is the nonsingular covariance matrix of the observation and $\mathbf{R}_{fg}$ is the cross-covariance matrix of the image and observation. When the image is corrupted with an additive noise term with known covariance $\mathbf{K}_n$, the weighted Wiener restoration matrix becomes

$$\mathbf{H}_{wW}^{-1} = \mathbf{R}_{fg}(\mathbf{R}_{gg})^{-1} = (\mathbf{H}^T \mathbf{K}_n^{-1} \mathbf{H} + \mathbf{K}_f^{-1})^\dagger \mathbf{H}^T \mathbf{K}_n^{-1}. \tag{3.60}$$

If the ideal image is assumed to be uncorrelated and the noise process is assumed to be white, then, as the ratio of image energy to noise energy approaches infinity, the Wiener estimate of (3.60) approaches the minimum-norm least squares solution of (3.33).

3.5 REGULARIZATION TECHNIQUES OF IMAGE RECOVERY

Smoothing and regularization techniques [6–8] have been developed in a attempt to overcome the ill-conditioning problems associated with image recovery. These methods attempt to force smoothness on the solution of a least squares error problem. The problem can be formulated in two different ways. One way of stating the problem for regularization is

$$\text{minimize} \quad \mathbf{f}^T \mathbf{S} \tag{3.61}$$

$$\text{subject to} \quad (\mathbf{g} - \mathbf{Hf})^T \mathbf{W}(\mathbf{g} - \mathbf{Hf}) = e, \tag{3.62}$$

where S is a smoothing matrix, W is an error-weighting matrix, and e denotes a residual scalar-estimation error. The error-weighting matrix can be chosen as $W = K_n^{-1}$. The smoothing matrix is typically the first- or second-order difference. The stationary point of the Lagrangian expression

$$L_1(f, \lambda) = f^T S f + \lambda[(g - Hf)W^T(g - Hf)^T - e] \tag{3.63}$$

is sought. By taking derivatives with respect to f and λ and setting them equal to zero, the solution for a nonsingular overdetermined set of equations becomes

$$\hat{f} = \left(H^T W^{-1} H + \frac{1}{\lambda} S\right)^{-1} H^T W g, \tag{3.64}$$

where λ is chosen to satisfy the compromise between residual error and smoothness in the estimate.

Another form of the smoothing problem is

$$\text{minimize} \qquad (g - Hf)^T W(g - Hf) \tag{3.65}$$

$$\text{subject to} \qquad f^T S f = d, \tag{3.66}$$

where d represents a fixed degree of smoothness. The Lagrangian expression for this formulation becomes

$$L_2(f, \gamma) = (g - Hf)W^T(g - Hf)^T + \gamma(f^T S f - d) \tag{3.67}$$

and the solution for a nonsingular overdetermined set of equations is

$$f = (H^T W^{-1} H + \gamma S)^{-1} H^T W g. \tag{3.68}$$

Note that, for the two problem formulations, the results given by (3.64) and (3.68) are identical if $\gamma = 1/\lambda$. The shortcomings of such a regularization technique are that the smoothing function S must be estimated and either the degree of smoothness d or the degree of error e must be known to determine γ or λ.

Constrained restoration techniques have also been developed [5] to overcome the problem of an ill-conditioned system. Linear equality constraints and linear inequality constraints have been enforced to yield one-step solutions similar to those described in this section. All the techniques described so far attempt to overcome the problem, of noise-corrupted data and ill-conditioned systems by forcing some sort of taper on the inverse of the degradation operator. We now examine iterative restoration techniques that are divided into two categories: row-action methods and block-action methods.

3.6 ROW-ACTION TECHNIQUES OF IMAGE RECOVERY

Many researchers in different scientific disciplines have devoted their time to the
development of efficient algorithms to find the solution to a linear set of equations
such as that given by (3.12). We have already shown how the image-restoration
problem can be represented by this vector-space relationship. As early as 1937,
Kaczmarz [9] developed an iterative projection technique to solve (3.12). The
algorithm is expressed as

$$\mathbf{f}^{(k+1)} = \mathbf{f}^{(k)} + \lambda_k \frac{g_{i_k} - (\mathbf{h}_{i_k}, \mathbf{f}^{(k)})}{\|\mathbf{h}_{i_k}\|^2} \mathbf{h}_{i_k}. \tag{3.69}$$

The relaxation parameter λ_k is bound by $0 \le \lambda_k \le 2$; $\mathbf{h}_i$ represents the ith row
of the matrix $\mathbf{H}$, g_i represents the ith element of the vector $\mathbf{g}$; $(\cdot, \cdot)$ denotes the
inner product between two vectors, k denotes the iteration; and $\|\cdot\|$ represents
the Euclidean, or L_2, norm of a vector defined as

$$\|y\| = \left(\sum_{i=1}^{N} y_i^2 \right)^{1/2} \tag{3.70}$$

This algorithm is computationally much simpler than the algorithms described in
the previous section which all depend on some type of inversion of the degradation operator. Kaczmarz proved that (3.69) converges to the unique solution
when the relaxation parameter is unity and $\mathbf{H}$ is a square, nonsingular matrix,
that is, $\mathbf{H}$ has an inverse.

When the control is cyclic, the ith equation becomes $i_k = k \bmod M + 1$, where
M is the total number of equations available and k is the iteration; that is, the
algorithm is repeated in a cyclic fashion through all the equations. If the set of
equations given by (3.12) is consistent, the control is cyclic, and λ_k obeys the
condition

$$\limsup_{k \to \infty} |1 - \lambda_k| < 1, \tag{3.71}$$

then the algorithm given by (3.69) will converge to a solution of $\mathbf{g} = \mathbf{Hf}$. If,
in addition, the initial estimate $\mathbf{f}^{(0)} \in \mathcal{R}(H^T)$, where $\mathcal{R}(\cdot)$ represents the range
space, then $\mathbf{f}^{(k)}$ converges to the minimum-norm, least squares, or pseudo-inverse
solution; that is,

$$\lim_{k \to \infty} \mathbf{f}^{(k)} = \mathbf{H}^\dagger \mathbf{g}. \tag{3.72}$$

The dagger (†) denotes the pseudo-inverse.

In many practical applications, the data is corrupted with noise and measurement inaccuracies, as described in (3.16). This is due to the fact that, in most

applications, the system of equations will be greatly overdetermined; many more equations are available than unknowns ($M \gg N$). The noise term represents measurement inaccuracies, noise corruption of data, and discretization in the model. For an inconsistent set of equations, if the method of Kaczmarz is periodic— that is, $\lambda_k = \lambda_{i_k-1}$, where $i_k = k \bmod M + 1$—then every subsequence $(\mathbf{f}^{(kM+i)})_k$, $0 \le i < M$, converges. The proof for the algorithm's convergence can be found in [10] for the consistent case, in [11] or [12] for the inconsistent case with $\lambda_k = 1$, and in [13] for the general inconsistent case with a periodic relaxation parameter.

In general, the order in which one performs the Kaczmarz algorithm on the M existing equations can differ. *Cyclic control*, where the algorithm iterates through the equations in a periodic fashion, has already been mentioned. *Almost cyclic control* is applied when M sequential iterations of the Kaczmarz algorithm yield exactly one operation per equation in any order. *Remotest set control* is applied when one performs the operations on the most distant equation first; "most distant" in the sense that the projection onto the hyperplane represented by the equation is the one furthest away. The measure of distance is determined by the norm. This type of control is seldom used since it requires comparing measurements that depend on all the equations.

The method of Kaczmarz for $\lambda = 1.0$ can be expressed geometrically as follows: given $\mathbf{f}^{(k)}$ and the hyperplane $\mathcal{H}_{i_k} = \{x \in R^n \mid (\mathbf{h}_{i_k}, x) = g_{i_k}\}$, $\mathbf{f}^{(k+1)}$ is the orthogonal projection of $\mathbf{f}^{(k)}$ onto $\mathcal{H}_{i_k}$. This is illustrated in Figure 3.1. Note that by changing the relaxation parameter, the next iterate can be a point anywhere along the line segment connecting the previous iterate and its orthogonal reflection with respect to the hyperplane.

Since (3.12) is used to model problems in many diverse disciplines, efforts to solve a linear set of equations have been made in seemingly unrelated fields which have resulted in repeated discoveries of the same algorithm. For this reason, the Kaczmarz algorithm appears under different names in different appli-

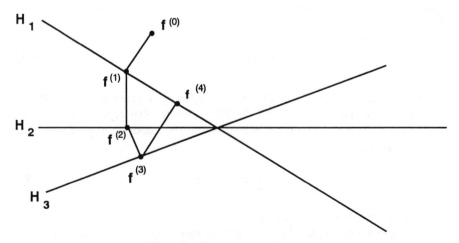

Figure 3.1 The method of Kaczmarz.

cations. This algorithm is used in the field of medical imaging for computerized tomography (CT) [11, 14] and is called the algebraic reconstruction technique (ART). The Kaczmarz algorithm also appears in the literature as the Widrow–Hoff least mean squares (LMS) algorithm [15] and has been successfully applied to channel equalization, echo cancellation, system identification, and adaptive array processing. A generalization of the method of Kaczmarz based on the theory of projection onto convex sets (POCS) is described in this chapter and will be referred to as the row-action projection method, or RAP. This general method can be applied to a variety of signal-recovery problems, some of which are described in other chapters of this book.

To solve linear inequalities where (3.12) is replaced with

$$\mathbf{Hf} \le \mathbf{g}, \tag{3.73}$$

a method very similar to Kaczmarz's algorithm is developed by Agmon, Motzkin, and Schoenberg [16, 17]:

$$\mathbf{f}^{(k+1)} = \mathbf{f}^{(k)} + c^{(k)}\mathbf{h}_{i_k};$$

$$c^{(k)} = \min\left(0, \lambda_k \frac{g_{i_k} - (\mathbf{h}_{i_k}, \mathbf{f}^{(k)})}{\|\mathbf{h}_{i_k}\|^2}\right). \tag{3.74}$$

The relaxation parameter is defined for $0 \le \lambda_k \le 2$. The geometrical interpretation is as follows: let $\mathcal{L}_{i_k} = \left\{x \in R^n \mid (\mathbf{h}_{i_k}, x) \le g_{i_k}\right\}$ so that for unity relaxation, $\mathbf{f}^{(k+1)}$ is the orthogonal projection of $\mathbf{f}^{(k)}$ onto the closed halfspace $\mathcal{L}_{i_k}$, as shown in Figure 3.2. The method of solving linear inequalities by Agmon, Motzkin, and Schoenberg is mathematically identical to the perceptron convergence theorem from the theory of learning machines; see Minsky and Papert [18].

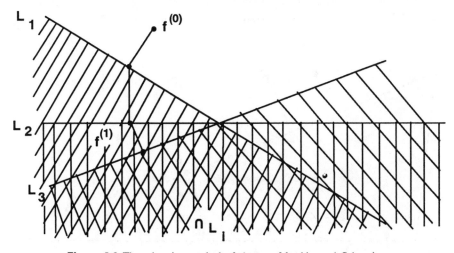

Figure 3.2 The relaxation method of Agmon, Motzkin, and Schoenberg.

Both the Kaczmarz method for solving a set of linear equalities and the Agmon–Motzkin–Schoenberg method for solving inequalities are considered feasible approaches for obtaining a solution. The algorithms seek to find a point $\mathbf{f}$ that lies within a specified vicinity of all the hyperplanes defined by the linear set of either equalities or inequalities. We shall now investigate a method that is considered an optimization approach for finding the solution. In an optimization approach, the algorithm uses a predesignated objective function $O(f)$ in order to choose the best solution from the feasible set of solutions. In both the feasibility approach and the optimization approach, additional information about the particular problem may be included to further restrict the feasible set and narrow the possibilities.

Hildreth's algorithm [19] is a row-action iterative algorithm based on an optimization approach. The optimization problem to be solved can be stated as

$$\text{minimize} \qquad 1/2 \, \|\mathbf{f}\|^2 \tag{3.75}$$

$$\text{subject to} \qquad (\mathbf{h}_i, \mathbf{f}) \le g_i, \qquad \text{for all} \quad i \in M. \tag{3.76}$$

Hildreth's algorithm can be expressed in three steps as

$$\mathbf{f}^{(k+1)} = \mathbf{f}^{(k)} + c^{(k)} \mathbf{h}_{i_k},$$

$$\mathbf{z}^{(k+1)} = \mathbf{z}^{(k)} - c^{(k)} \mathbf{e}_{i_k},$$

$$c^{(k)} = \min \left(z_{i_k}^{(k)}, \lambda_k \frac{g_{i_k} - (\mathbf{h}_{i_k}, \mathbf{f}^{(k)})}{\|\mathbf{h}_{i_k}\|^2} \right). \tag{3.77}$$

The vector $\mathbf{e}_{i_k} \in \mathcal{R}^M$ is the i_kth unit vector. The term $\mathbf{z}^{(0)}$ is initialized as an arbitrary vector in the nonnegative orthant of $\mathcal{R}^M$ and $\mathbf{f}^{(0)}$ is initialized as $-\mathbf{H}^T \mathbf{z}^{(0)}$. The geometrical interpretation of Hildreth's algorithm is illustrated in Figure 3.3. The algorithm behaves similarly to the method of Agmon, Motzkin,

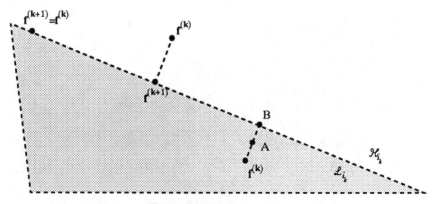

Figure 3.3 Hildreth's method.

and Schoenberg. The difference between the two algorithms is that for Hildreth's algorithm, if $\mathbf{f}^{(k)} \in \text{int} \mathcal{L}_{i_k}$, a move is made perpendicular to $\mathbf{H}_{i_k}$, as illustrated. Either $\mathbf{c}^{(k)} = \mathbf{z}_{i_k}^{(k)}$, which results in $\mathbf{f}^{(k)}$ being projected onto the point marked A in Figure 3.3, or $\mathbf{f}^{(k)}$ is orthogonally projected onto $\mathcal{H}_{i_k}$, marked B in Figure 3.3.

Hildreth's method for interval constraints, introduced by Herman and Lent in [20], solves the following problem:

$$\text{minimize} \qquad 1/2 \, \|\mathbf{f}\|^2 \tag{3.78}$$

$$\text{subject to} \qquad \gamma_i \leq (\mathbf{h}_i, \mathbf{f}) \leq \delta_i, \qquad i \in M. \tag{3.79}$$

The method is summarized in the following steps:

$$\mathbf{f}^{(k+1)} = \mathbf{f}^{(k)} + \mathbf{c}^{(k)} \mathbf{h}_{i_k},$$

$$\mathbf{z}^{(k+1)} = \mathbf{z}^{(k)} - \mathbf{c}^{(k)} \mathbf{e}_{i_k},$$

$$\mathbf{c}^{(k)} = \text{mid} \left(\mathbf{z}_{i_k}^{(k)}, \frac{\delta_{i_k} - (\mathbf{h}_{i_k}, \mathbf{f}^{(k)})}{\|\mathbf{h}_{i_k}\|^2}, \frac{\gamma_{i_k} - (\mathbf{h}_{i_k}, \mathbf{f}^{(k)})}{\|\mathbf{h}_{i_k}\|^2} \right), \tag{3.80}$$

where $\mathbf{e}_{i_k}$ is the i_kth unit vector of length M. The term $\text{mid}(a, b, c)$ stands for the median of the three numbers a, b, and c. The vectors $\mathbf{f}^{(0)}$ and $\mathbf{z}^{(0)}$ are initialized to zero. Convergence of this algorithm has only been proved for the case of unity relaxation.

The geometrical interpretation of Hildreth's method for interval constraints is shown in Figure 3.4 where, given $\mathbf{f}^{(k)}$, $\mathbf{f}^{(k+1)}$ will either be inside the hyperslab or on one of the two boundaries, depending on the value of $\mathbf{c}^{(k)}$.

The use of entropy for image and signal recovery is rigorously founded in some application areas. The argument for using the maximum-entropy approach is based on information theory. Advocates of the maximum-entropy approach

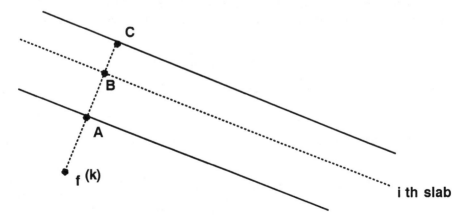

Figure 3.4 Hildreth's method for interval constraints.

argue that, from the standpoint of information theory, the maximum-entropy approach yields a restored image with the lowest information content consistent with the available data without introducing extraneous information. Such an approach seems very attractive, especially when dealing with noise-corrupted data. One such technique which falls under the category of a row-action, iterative restoration algorithm is the multiplicative algebraic reconstruction technique (MART) suggested by Gordon et al. [21]. The problem solved by MART can be stated as

$$\text{minimize} \quad \sum_{j=1}^{N} f_j \ln f_j \tag{3.81}$$

$$\text{subject to} \quad (\mathbf{h}_i, \mathbf{f}) = g_i, \qquad i \in M. \tag{3.82}$$

The method is expressed as

$$f_j^{(k+1)} = \left(\frac{g_{i_k}}{(\mathbf{h}_{i_k}, \mathbf{f}^{(k)})} \right) \lambda_k \mathbf{h}_{i_k j} f_j^{(k)}, \qquad 0 < \lambda_k \le 1, \tag{3.83}$$

with $\mathbf{f}^{(0)} = e^{-1} \mathbf{1}$. Unlike Kaczmarz's algorithm, which has been analyzed for problems dealing with inconsistent equations, MART has only been analyzed for a consistent set of equations. Since most practical applications involve noise-corrupted data, MART is not an attractive choice for many practical problems.

3.7 BLOCK-ACTION TECHNIQUES OF IMAGE RECOVERY

A generalization of the Kaczmarz algorithm introduced in the previous section has been suggested by Eggermont [13]; it can be described as a block iterative algorithm. Recall the set of linear equations describing the image-recovery problem, $\mathbf{g} = \mathbf{Hf}$. In order to describe the generalization of the Kaczmarz algorithm to a block-action method, the matrix $\mathbf{H}$ is partitioned into L blocks of length b:

$$\mathbf{H} = \begin{pmatrix} \mathbf{h}_1^T \\ \mathbf{h}_2^T \\ \vdots \\ \mathbf{h}_{bL}^T \end{pmatrix} = \begin{pmatrix} \mathbf{H}_1 \\ \mathbf{H}_2 \\ \vdots \\ \mathbf{H}_L \end{pmatrix}, \tag{3.84}$$

and $\mathbf{g}$ is partitioned as

$$\mathbf{g} = \begin{pmatrix} g_1 \\ g_2 \\ \vdots \\ g_{bL} \end{pmatrix} = \begin{pmatrix} \mathbf{g}_1 \\ \mathbf{g}_2 \\ \vdots \\ \mathbf{g}_L \end{pmatrix}, \tag{3.85}$$

where $\mathbf{g}_i, i = 1, 2, \ldots, L$, is a vector of length l and the subblocks $\mathbf{H}_i$ are of dimension $b \times N$. The generalized group-iterative variation of the Kaczmarz algorithm is expressed as

$$\mathbf{f}^{(k+1)} = \mathbf{f}^{(k)} + \mathbf{H}_{i_k}^T \Sigma_k (\mathbf{g}_{i_k} - \mathbf{H}_{i_k} \mathbf{f}^{(k)}), \tag{3.86}$$

where $\mathbf{f}^{(0)} \in \mathcal{R}^N$. Eggermont proves that the block-action method of Kaczmarz converges to a solution under certain conditions. If $\{\Sigma_k\}_k$ is a bounded sequence of relaxation matrices, then

$$\limsup_{k \to \infty} \|\mathbf{H}_{i_k}^\dagger (\mathbf{I}_b - \mathbf{H}_{i_k} \mathbf{H}_{i_k}^T \Sigma_k) \mathbf{H}_{i_k}\| < 1, \tag{3.87}$$

with $\mathbf{H}_{i_k}^\dagger$ denoting the Moore–Penrose pseudo-inverse[4] of $\mathbf{H}_{i_k}$ and $\mathbf{I}_b$ denoting a $b \times b$ identity matrix. If the set of equations represented by (3.12) is consistent, the sequence $\{\mathbf{f}^{(k)}\}$ generated by (3.86) converges to a solution. If, in addition, $\mathbf{f}^{(0)} \in \mathcal{R}(H^T)$, then

$$\lim_{k \to \infty} \mathbf{f}^{(k)} = \mathbf{H}^\dagger \mathbf{g}. \tag{3.88}$$

If the control is cyclic—that is, $i_k = k \bmod L + 1$—and the method is periodic with $\Sigma_k = \Sigma_{i_k-1}$, then every subsequence $\{\mathbf{f}^{(kL+i)}\}_k$, $0 \le i < L$, converges even when the set of equations are inconsistent.

A further generalization of Kaczmarz's algorithm led Eggermont [13] to the following form of the general block Kaczmarz algorithm:

$$\mathbf{f}^{(k+1)} = \mathbf{f}^{(k)} + \mathbf{H}_{i_k}^\dagger \Lambda_k (\mathbf{g}_{i_k} - \mathbf{H}_{i_k} \mathbf{f}^{(k)}), \tag{3.89}$$

where, once again, $\mathbf{H}_{i_k}^\dagger$ denotes the Moore–Penrose inverse of $\mathbf{H}_{i_k}$, Λ_k is the $b \times b$ relaxation matrix, and the control is cyclic, that is, $i_k = k \bmod L + 1$. Note that the algorithms described by (3.86) and (3.89) are similar; in fact, they become identical by letting either $\Lambda_k = \mathbf{H}_i \mathbf{H}_i^T \Sigma_k$ or $\Sigma_k = (\mathbf{H}_i \mathbf{H}_i^T)^\dagger \Lambda_k$.

The convergence theorem described by Eggermont for (3.89) is similar to the convergence theorem for (3.86). Let $\{\Lambda_k\}_k$ be a bounded sequence of relaxation matrices satisfying

$$\limsup_{k \to \infty} \|\mathbf{H}_{i_k}^\dagger (\mathbf{I}_b - \Lambda_k) \mathbf{H}_{i_k}\| < 1. \tag{3.90}$$

If the control is cyclic, that is, $i_k = k \bmod L + 1$ and (3.12) is consistent, the sequence generated by the algorithm described in (3.89) converges to a solution.

---

[4]See Appendix A, Review of Linear Algebra.

If, in addition to the above, $\mathbf{f}^{(0)} \in \mathcal{R}(H^T)$, then $\{\mathbf{f}^{(k)}\}_k$ converges to the pseudo-inverse solution; that is,

$$\lim_{k \to \infty} \mathbf{f}^{(k)} = \mathbf{H}^\dagger \mathbf{g}. \tag{3.91}$$

Just as for the row-actiom method of Kaczmarz, if the method is periodic, that is, $\Lambda_k = \Lambda_{i_k-1}$, then every subsequence $\{\mathbf{f}^{(kL+i)}\}_k$, $0 \le i < L$, converges, even when the set of linear equations is inconsistent.

For the case when $\Lambda_k = \lambda_k \mathbf{I}_b$, the condition imposed by (3.90) reduces to

$$\limsup_{k \to \infty} |1 - \lambda_k| \|\mathbf{H}_i^\dagger \mathbf{H}_i\| < 1. \tag{3.92}$$

Since $\|\mathbf{H}_i^\dagger \mathbf{H}_i\| = 1$, (3.92) reduces to the condition given in (3.71) for the row-action Kaczmarz technique. The generalizations given by (3.86) and (3.89) encompass many techniques used in diverse areas of application. Obviously, when the block size b is equal to one, the algorithm given by (3.86) reduces to the row-action Kaczmarz algorithm—also known as ART in the area of image recovery from computerized tomography [14] and as the normalized LMS (Widrow–Hoff) algorithm in the communications area.

When the block size b given in (3.86) is equal to the number of equations M, the algorithm becomes identical to Landweber's iteration [22] for solving Fredholm equations of the first kind. The algorithm becomes

$$\mathbf{f}^{(k+1)} = \mathbf{f}^{(k)} + \mathbf{H}^T \Sigma_k (\mathbf{g} - \mathbf{H}\mathbf{f}^{(k)}), \qquad k = 1, 2, \dots . \tag{3.93}$$

Landweber's iterative algorithm for solving a Fredholm equation of the first kind is motivated by considering the properties of the Fredholm equation of the second kind,

$$\mathbf{g} = \mathbf{H}\mathbf{f} + \alpha \mathbf{f}, \tag{3.94}$$

where $\mathbf{H}$ is a compact square linear operator of full rank and α is a small positive number. If α is neither 0 nor the negative of any eigenvalue of $\mathbf{H}$, then $\{\mathbf{H} + \alpha \mathbf{I}\}^{-1}$ exists and is bounded so that (3.94) has a unique solution,

$$\mathbf{f} = (\mathbf{H} + \alpha \mathbf{I})^{-1} \mathbf{g}. \tag{3.95}$$

Furthermore, (3.94) can be solved using the iterative algorithm of Liouville and Neumann:

$$\mathbf{f}^{(k+1)} = \alpha^{-1} (\mathbf{g} - \mathbf{H}\mathbf{f}^{(k)}), \tag{3.96}$$

with the initial estimate $\mathbf{f}^{(0)}$ equal to zero. The algorithm will converge to the unique solution if the previous conditions are satisfied for α. Landweber's approach is to first premultiply both sides of the Fredholm equation of the first kind by the adjoint of $\mathbf{H}$ to yield the normal equations

$$\mathbf{H}^T\mathbf{H}\mathbf{f} = \mathbf{H}^T\mathbf{g}. \tag{3.97}$$

By adding $\mathbf{f}$ to both sides and converting the equation to an iterative form, the resulting Landweber iteration becomes

$$\mathbf{f}^{(k+1)} = \mathbf{H}^T\mathbf{g} + (\mathbf{I} - \mathbf{H}^T\mathbf{H})\mathbf{f}^{(k)}. \tag{3.98}$$

Note, in this form, that Landweber's iteration is equivalent to the block-action Kaczmarz algorithm in (3.89) with $\Lambda_k = \mathbf{I}_k$. When the operator $\mathbf{H}$ is self-adjoint, (3.98) reduces to

$$\mathbf{f}^{(k+1)} = \mathbf{g} + (\mathbf{I} - \mathbf{H})\mathbf{f}^{(k)}. \tag{3.99}$$

Landweber's iteration as given in (3.98) has properties similar to those obtained by truncating the singular series expansion of the original problem. The singular-value decomposition of the matrix $\mathbf{H}$ is described by $\{\mathbf{u}_k, \mathbf{v}_k; \sigma_k\};$[5] since $\mathbf{H}$ is self-adjoint, $\mathbf{u}_k = \mathbf{v}_k$. Assume that the following two conditions hold:

- $\mathbf{g} \in \{\mathcal{N}(H^T)\}^\perp$, where $\mathcal{N}(\cdot)$ denotes the null space, and
- $\sum_{k=1}^{\infty}(1/\sigma_k)^2|(\mathbf{g}, \mathbf{u}_k)|^2 < \infty.$

Then the solution is given by

$$\mathbf{f} = \sum_{k=1}^{\infty}(1/\sigma_k)(\mathbf{g}, \mathbf{u}_k)\mathbf{v}_k, \tag{3.100}$$

where equality is defined as

$$\lim_{N\to\infty}\left|\mathbf{f} - \sum_{k=1}^{N}(1/\sigma_k)(\mathbf{g}, \mathbf{u}_k)\mathbf{v}_k\right| = 0. \tag{3.101}$$

By taking $\mathbf{f}^{(0)} = 0$ and applying successive substitution to (3.99), the resulting algorithm takes the form

$$\mathbf{f}^{(k+1)} = \sum_{i=0}^{k}(\mathbf{I} - \mathbf{H})^i\mathbf{g}. \tag{3.102}$$

---

[5]appendix A, Review of Linear Algebra

The vector **g** can be written in the form

$$\mathbf{g} = \sum_{j=1}^{\infty} (\mathbf{g}, \mathbf{u}_j)\mathbf{u}_j, \tag{3.103}$$

and, using the eigendecomposition for **H**, we obtain

$$\mathbf{H}\mathbf{u}_j = \sigma_j \mathbf{u}_j. \tag{3.104}$$

From (3.103) and (3.104), Landweber's iteration becomes

$$\mathbf{f}^{(k+1)} = \sum_{j=1}^{\infty} (\mathbf{g}, \mathbf{u}_j)\mathbf{u}_j \sum_{i=0}^{k} (1 - \sigma_j)^i$$

$$= \sum_{j=1}^{\infty} \frac{1}{\sigma_j}[1 - (1 - \sigma_j)^{k+1}](\mathbf{g}, \mathbf{u}_j)\mathbf{u}_j. \tag{3.105}$$

Without loss of generality, the original problem can be scaled so that no singular value σ_k is greater than one. Therefore (3.105) approaches the solution given by the singular series expansion in (3.100). If the linear set of equations are inconsistent, as generally found in most practical applications, inverting all the singular values will lead to an unstable solution. As mentioned before, many researchers have attempted some sort of regularization technique to offset the effects of noise. A simple way of regularizing Landweber's algorithm is to stop the algorithm after an appropriate number of iterations. If the algorithm given by (3.105) is stopped after K iterations, the solution can be expressed as

$$\mathbf{f}^{(k)} = \sum_{j=1}^{\infty} \frac{1}{\sigma_j}[1 - (1 - \sigma_j)^K](\mathbf{g}, \mathbf{u}_j)\mathbf{u}_j. \tag{3.106}$$

Note that the result given by (3.106) is the same as that given by the singular series expansion except for the term in the brackets, $1 - (1 - \sigma_j)^K$, which can be thought of as a regularization taper in the inversion process of the operator **H**. Note that, the smaller the number of iterations K, the more severe the taper. The result provided by Landweber's algorithm, which terminates after K iterations, falls in between the two extremes of taking the inverse and truncating the inverse at a fixed number of singular values. The number of iterations provides a tradeoff between noise amplification and loss of fine detail in the image. As the number of iterations are increased for Landweber's method, the solution approaches the pseudo-inverse solution. In the case of noise-corrupted data, Landweber's iterations improve the estimate at first. After a while, the estimate begins to deteriorate due to the inversion of small singular values that correspond to areas of high noise energy and low signal energy.

The generalized block–Kaczmarz algorithm given in (3.89) with the block size b equal to the number of equations M is similar to the algorithm described by Youla [23] of alternating orthogonal projections. This can be seen by first reformulating the restoration problem as follows: the original image vector $\mathbf{f}$ is known *a priori* to belong to a linear subspace $\mathcal{F}$ of a parent Hilbert space $\mathcal{S}$, but all that is available is the measurement g, which is the projection of $\mathbf{f}$ onto a known linear subspace $\mathcal{H}$ also in $\mathcal{S}$. Assume that the following conditions are met:

1. $\mathcal{F} \bigcap \mathcal{H}^\perp = \mathbf{0}$ and
2. $\angle(\mathcal{F} - \mathcal{H}^\perp) > 0$.

If both conditions 1 and 2 are satisfied for a consistent set of equations, then there exists an effective iterative restoration algorithm that employs only projections onto $\mathcal{F}$ and projections onto the orthogonal complement of $\mathcal{H}$.

Let $\mathcal{P}$ be any closed linear manifold (CLM) in $\mathcal{S}$ and let $\mathcal{P}^\perp = \mathcal{Q}$ be the orthogonal complement to $\mathcal{P}$. According to the projection theorem, every $x \in \mathcal{S}$ possesses a unique decomposition

$$x = x_P + x_Q, \qquad \circ \qquad (3.107)$$

where $x_P \in \mathcal{P}$, $x_Q \in \mathcal{Q}$ and $x_P = Px$, $x_Q = Qx$ are the orthogonal projections onto $\mathcal{P}$ and $\mathcal{Q}$, respectively. The Hilbert space $\mathcal{S}$ can be expressed as $\mathcal{S} = \mathcal{P} + \mathcal{Q}$. The vectors x_P and x_Q are orthogonal to each other; that is, $(x_P, x_Q) = 0$.

It is known that $\mathbf{f} \in \mathcal{S}$ belongs to a CLM $\mathcal{F}$. In the absence of noise, an effective iterative restoration algorithm exists that only requires linear projections onto $\mathcal{F}$ and onto the orthogonal complement of $\mathcal{H}$, $(\mathcal{H}^\perp)$. Therefore the problem given in (3.12) can be expressed as

$$g = P_H \mathbf{f} = P_H P_F \mathbf{f} = (1 - Q_H) P_F \mathbf{f}$$

$$= P_F \mathbf{f} - Q_H P_F \mathbf{f} = \mathbf{f} - Q_H P_F \mathbf{f}, \qquad (3.108)$$

where P_F is the projection operator onto $\mathcal{F}$, P_H is the projection operator onto $\mathcal{H}$, and Q_H is the projection operator onto $\mathcal{H}^\perp$.

The following iterative algorithm was first suggested by von Neumann [24]:

$$\mathbf{f}^{(k+1)} = g + Q_H P_F \mathbf{f}^{(k)}, \qquad (3.109)$$

where alternating orthogonal projections are made onto CLMs. Youla [23] proves that, in the absence of noise, the method of alternating orthogonal projections will converge under any one of the two conditions 1 or 2. The algorithm given by (3.109) is similar to the generalization of Kaczmarz's algorithm stated in (3.89). The pseudo-inverse $\mathbf{f}^\dagger$ in (3.89) is replaced by g in (3.109) and the projection operator P_F is added.

3.8 IMAGE RECOVERY USING POCS

A generalization of restoration techniques has been developed that incorporates many of the existing algorithms and is based on the functional-analysis concepts of projection onto convex sets (POCS) [25]. Many of the algorithms discussed in this chapter are a form of POCS, that includes the Kaczmarz method for solving linear equalities, the Agmon–Motzkin–Schoenberg algorithm for solving linear inequalities, Landweber's method, and the method of alternating orthogonal projections. The basic idea of POCS is to develop a recursive restoration algorithm that incorporates a large number of linear and nonlinear constraints that satisfy certain properties. The more *a priori* image information that one can incorporate into the algorithm in the form of convex constraints, the more effective the algorithm becomes.

A projection operator onto a closed convex set is an example of a nonlinear mapping that is easily analyzed and contains some very useful properties. The projection operators minimize error distance and are nonexpansive; two very important properties of ordinary linear orthogonal projections onto closed linear manifolds (CLMs). The benefit of using POCS for image restoration, or for any other signal-restoration problem, is that one can incorporate nonlinear constraints of a certain type into the POCS method. Since nonlinear constraints on the unknown image do not translate into linear manifolds, linear image-restoration algorithms cannot take advantage of *a priori* information based on nonlinear constraints.

The method of POCS requires *a priori* image information to lie in a well-defined closed convex set. If such properties exist, then **f** is restricted to lie in the region defined by the intersection of all the convex sets, that is,

$$x \in C_0 = \bigcap_{i=1}^{l} C_i \tag{3.110}$$

where C_i denotes the ith closed convex set corresponding to the ith property on **f**, $C_i \in \mathcal{S}$, and $i \in \mathcal{I}$. The unknown image **f** can be restored by using the corresponding projection operators P_i onto each convex set C_i.

The image-restoration problem is once again described from a linear-algebra viewpoint, as was done for the method of alternating orthogonal projections. The original image **f** is known *a priori* to belong to a linear subspace $\mathcal{F}$ of a parent Hilbert space $\mathcal{S}$. All that is available is the measurement of the blurred image **g**, which is the orthogonal projection of **f** onto a linear subspace $\mathcal{H}$ in $\mathcal{S}$. The conditions needed to restore **f** are:

1. **f** can be uniquely determined by **g** if and only if $\mathcal{F}$ and the orthogonal complement of $\mathcal{H}$ have only the zero vector in common,

$$\mathcal{F} \cap \mathcal{H}^{\perp} = 0. \tag{3.111}$$

2. The restoration problem is stable if and only if the angle between $\mathcal{F}$ and $\mathcal{H}^\perp$ is greater than zero,

$$\angle\{\mathcal{F} - \mathcal{H}^\perp\} > 0. \tag{3.112}$$

3. For both 1 and 2 and a consistent set of equations, there exists an iterative algorithm that restores **f** based only on the projections onto $\mathcal{F}$ and the projections onto $\mathcal{H}^\perp$; this is the method of alternating orthogonal projections described previously.

The projection of the point x onto the ith closed convex set $\mathcal{C}_i$ of a Hilbert space $\mathcal{S}$ is defined as

$$\|x - P_i(x)\| = \inf_{a \in \mathcal{C}_i} \|x - a\| = \rho(x, \mathcal{C}_i). \tag{3.113}$$

A property of closed convex sets is that a projection of a point onto a convex set is unique. This is known as the unique-nearest-neighbor property. The method of successive projections onto the convex sets can be expressed as

$$\mathbf{f}^{(k+1)} = \mathbf{f}^{(k)} + \lambda_k(P_{i_k}(\mathbf{f}^{(k)}) - \mathbf{f}^{(k)}), \qquad 0 \leq \lambda_k \leq 2, \tag{3.114}$$

where λ_k denotes the relaxation parameter. If $\lambda_k < 1$, the algorithm is said to be *underrelaxed*; if $\lambda_k > 1$, the algorithm is *overrelaxed*. The order in which the projection operators are chosen can vary, but a common method is *cyclic control*, which was also used in the method of Kaczmarz. For *cyclic control*, $i_k = k \bmod L + 1$, where k is equal to the number of iterations and L is equal to the number of convex sets. The work in [25] suggests using *remotest set control*, where the projection operator to be used is determined by

$$\rho\{\mathbf{f}^{(k)}, \mathcal{C}_{i_k}\} = \sup_{i \in I} \rho\{\mathbf{f}^{(k)}, \mathcal{C}_{i_k}\} = \Phi(\mathbf{f}^{(k)}). \tag{3.115}$$

The most distant set may be difficult to find or inefficient for a real-time application, so the following approximation to the remotest set control can be used:

$$\text{if} \qquad \rho\{^{(k)}, C_{i_k}\} \to \mathbf{0},$$

$$\text{then} \qquad \Phi(^{(k)}) = \sup_{i \in I} \rho\{\mathbf{f}^{(k)}, \mathcal{C}_{i_k}\} \to \mathbf{0}. \tag{3.116}$$

It is important to define some commonly used terms and concepts for POCS. The sequence $\{\mathbf{f}^{(k)}\}_k$ is said to strongly converge to **f** if

$$\lim_{k \to \infty} \|\mathbf{f}^{(k)} - \mathbf{f}\| = 0. \tag{3.117}$$

The sequence $\{f^{(k)}\}_k$ is said to weakly converge to f if

$$\lim_{k \to \infty} (f^{(k)}, g) = (f, g), \qquad g \in \mathcal{H}. \tag{3.118}$$

Strong convergence will always imply weak convergence; in finite-dimensional linear vector space, the converse also holds true. A convex set is *strictly convex* if $(x + y)/2$ is an interior point of C when $x \in C$ and $y \in C$. A convex set is *uniformly convex* if there exists a function $\delta(\tau)$ that is positive for $\tau > 0$ and zero only for $\tau = 0$ such that

$$\left| z - \frac{x+y}{2} \right| \leq \delta(\|x - y\|), \qquad z \in C. \tag{3.119}$$

A convex set is *strongly convex* if one can choose a function $\delta(\tau)$ of the form

$$\delta(\tau) = \mu \tau^2. \tag{3.120}$$

Note that *strong convexity* implies *uniform convexity*, which implies *strict convexity*. Gubin et al. [25] describes the conditions needed for the convergence of POCS and provides extensive proofs for the convergence theorems. Let any one of the following conditions on C_i hold true:

1. $C_{i^*} \cap (\cap C_{i \in I, i \neq i^*})^0$ is nonempty, where R^0 denotes the interior of R.
2. All C_i, with the exception of one, are uniformly convex with the common function $\delta(\tau)$.
3. The Hilbert space $\mathcal{S}$, where $C_i \in \mathcal{S}$, is of finite dimensionality.
4. $I = \{i_1, i_2, \ldots, i_l\}$ is finite and all C_i are halfspaces; that is, $C_i = \{x: (h_i, x) \leq y_i\}$.

Given any one of these four conditions, the general POCS method given in (3.114) for any f^0 and either cyclic control or approximately remotest set control will strongly converge to some point $f^* \in C_0 = \cap_{i \in I} C_i$. For conditions 1 and 4, the rate of convergence is geometric for the POCS algorithm using remotest set control.

Gubin also examines the behavior of the POCS algorithm for the case when the convex sets C_i do not have a common point. This can happen when the data is corrupted with noise. For the case of unity relaxation and cyclic control, the sequence $\{f^{(k)}\}_k$ converges weakly when all C_i are closed convex sets and at least one of them is finite. If condition 2, 3, or 4 is also satisfied, the convergence is strong. If all but one of the convex sets C_i are strongly convex and the intersection of all the convex sets is zero, the rate of convergence is geometric.

The following POCS algorithm includes a modification on how to choose the step size:

$$f^{(i+1)} = P_{i^*}(f^{(i)}) = \gamma_i (\Phi(f^{(i)}))^{-1}(P_i(f^{(i)}) - f^{(i)}), \qquad \gamma_i \geq 0,$$

$$\Phi(\mathbf{f}^{(i)}) = \|P_i(\mathbf{f}^{(i)}) - \mathbf{f}^{(i)}\| = \sup_{i \in I, i \neq i^*} \rho(\mathbf{f}^{(i)}, C_i). \tag{3.121}$$

When condition 1 is satisfied, the method given in (3.121) converges to a solution in a finite number of iterations.

If C is a closed convex set, the following is true for any $\mathbf{x} \in S$:

$$\Re(\mathbf{x} - P_C\mathbf{x}, \mathbf{y} - P_C\mathbf{x}) \leq 0, \qquad \mathbf{y} \in S. \tag{3.122}$$

In a real Hilbert space, the inequality in (3.122) reduces to the following lemma:

Lemma 1

$$(\mathbf{x} - P_C\mathbf{x}, \mathbf{y} - P_C\mathbf{x}) \leq 0, \qquad \mathbf{y} \in S. \tag{3.123}$$

If C is strongly convex,

$$(\mathbf{x} - P_C\mathbf{x}, \mathbf{y} - P_C\mathbf{x}) < 0. \tag{3.124}$$

Then, for a uniformly convex set, we get

$$(\mathbf{x} - P_C\mathbf{x}, \mathbf{y} - P_C\mathbf{x}) \leq -2\delta(\|\mathbf{y} - P_C\mathbf{x}\|)\|\mathbf{x} - P_C\mathbf{x}\|. \tag{3.125}$$

The vector $\mathbf{x} - P_C\mathbf{x}$ is normal to the tangent plane on C at the point $P_C\mathbf{x}$. Note that the angle between the vectors $\mathbf{x} - P_C\mathbf{x}$ and $\mathbf{y} - P_C\mathbf{x}$ is greater than or equal to $90°$. From (3.123) it follows that

$$\|P_C\mathbf{x} - P_C\mathbf{y}\|^2 \leq \Re(\mathbf{x} - \mathbf{y}, P_C\mathbf{x} - P_C\mathbf{y}). \tag{3.126}$$

The property of nonexpansiveness holds for convex projection operators:

Lemma 2

$$\|P_C\mathbf{x} - P_C\mathbf{y}\| \leq \|\mathbf{x} - \mathbf{y}\|. \tag{3.127}$$

If C is strongly convex, the projection operator is compact:

$$\|P_C\mathbf{x} - P_C\mathbf{y}\| < \|\mathbf{x} - \mathbf{y}\|. \tag{3.128}$$

From (3.127), it can be seen that the following convergence property holds true for POCS.

Lemma 3

$$\|\mathbf{f}^{(k+1)} - \mathbf{f}\| \leq \|\mathbf{f}^{(k)} - \mathbf{f}\| \tag{3.129}$$

for every point $\mathbf{f}$ that is in the intersection of all the convex sets at any iteration k. This shows that the actual error between the estimate and any one of the feasible solutions is monotonically decreasing. Therefore the method of projecting onto convex sets gets monotonically closer to the true solution as a function of the number of iterations.

For the projection onto convex sets method given in (3.114) with *cyclic control* or *approximately remotest set control* and $0 < \lambda_k < 2$, the following is true.

Lemma 4

$$\lim_{k \to \infty} \Phi(\mathbf{f}^{(k)}) = 0, \qquad \Phi(\mathbf{f}) = \sup_{i \in I}(\mathbf{f}, P_i). \tag{3.130}$$

If any of the four previously given conditions on the convex sets are satisfied, then

Lemma 5

$$\lim_{k \to \infty} \rho(\mathbf{f}^{(k)}, C) = 0. \tag{3.131}$$

If, for a closed convex set C and sequence $\mathbf{f}^{(k)}$, Lemma 3 and Lemma 5 are satisfied, then

Lemma 6

$$\lim_{k \to \infty} \{\mathbf{f}^{(k)}\} = \mathbf{f}^*, \qquad \mathbf{f}^* \in C. \tag{3.132}$$

The final topic to be covered is the rate of convergence for POCS. The rate of convergence could be slower than any geometrical progression depending on the function describing the convex sets. A typical example where the convergence rate is very slow is alternating projections onto two convex sets represented by two hyperplanes intersecting at a small angle. One way of accelerating the rate of convergence for projection onto two convex sets is

$$\mathbf{f}^{(1)} = P_1 \mathbf{f}^{(0)},$$

$$\mathbf{f}^{(2)} = P_2 \mathbf{f}^{(1)},$$

$$\mathbf{f}^{(3)} = P_1 \mathbf{f}^{(2)},$$

$$\mathbf{f}^{(4)} = \mathbf{f}^{(1)} + \lambda(\mathbf{f}^{(3)} - \mathbf{f}^{(1)}),$$

$$\lambda = \frac{\|\mathbf{f}^{(1)} - \mathbf{f}^{(2)}\|^2}{(\mathbf{f}^{(1)} - \mathbf{f}^{(3)}, \mathbf{f}^{(1)} - \mathbf{f}^{(2)})}. \tag{3.133}$$

In order for the accelerated method given by (3.133) to converge so that $\mathbf{f}^{(k)} \to \mathbf{f}^* \in C$, any one of the following conditions must be met:

1. $C_1 \cap C_2^0$ (or $C_1^0 \cap C_2$) is not empty,
2. C_2 (or C_1) is uniformly convex, or
3. S is finite dimensional.

It has been mentioned that many of the iterative signal-restoration techniques are specific examples of the POCS algorithm. The row-action Kaczmarz algorithm is one version of an algorithm based on POCS. Two tailored versions of the block-action Kaczmarz algorithm—Landweber's iteration and the method of alternating orthogonal projections—are also examples of POCS. It is worth noting that the successful image-restoration technique developed independently by Gerchberg and Saxton [26, 27] and by Papoulis [28] are also versions of the POCS method. The algorithm was developed by Gerchberg for iterative phase retrieval from two images and by Papoulis for superresolution by iterative methods. The Gerchberg–Papoulis (GP) algorithm is based on applying constraints on the estimate in both the signal space and the Fourier space in an iterative fashion until the estimate converges to a solution. For image restoration, the high-frequency components of the image are extrapolated by imposing the constraint that the object be finite in extent in the spatial domain and by imposing the known low-frequency components in the frequency domain. The dual problem solved by the GP algorithm involves spectral estimation, where the signal is extrapolated in the time or space domain. The algorithm consists of imposing the known part of the signal in the time domain and imposing the property of having a limited bandwidth in the frequency domain. The GP algorithm is limited in that a space-invariant (or time-invariant) degradation operator is assumed.

The method of POCS is a very powerful general iterative algorithm. Youla [29] provides a number of convex projection operators that can be applied to many signal-recovery problems. Most of the commonly used constraints for different signal-processing applications fall under the category of convex sets that provide weak convergence. However, in practice, most of the POCS algorithms provide strong convergence. In this chapter, the theory of POCS will be used to create generalized versions of the row-action and block-action iterative recovery algorithms described so far.

3.9 SINGULAR-VALUE DECOMPOSITION

Singular-value decomposition (SVD) is a technique of diagonalizing any arbitrary $M \times N$ matrix. We will use SVD to develop a new generalized iterative algorithm based on the fundamental principles of POCS. Recall that the image-recovery problem can be expressed in vector-space notation as (3.12). The matrix **H** is a discrete model of the degradation effects. Both the vector **f** of length N and the vector **g** of length M are lexicographically ordered vectors of the original two-dimensional images **F** and **G**. The $N_1 \times N_2$ matrix **F** represents the original unknown image, while the $M_1 \times M_2$ matrix **G** represents the measured blurred output image. Since the restoration algorithm depends on the diagonalization of the degradation operator, SVD is applied to the $M \times N$ operator $\mathbf{H}$ to yield

$$\mathbf{H} = \mathbf{U}\Sigma\mathbf{V}^T, \tag{3.134}$$

where

$$\Sigma = \begin{bmatrix} \sigma_1 & 0 & \cdots & \\ 0 & \sigma_2 & 0 & \\ & & \ddots & \\ & & & \sigma_N \end{bmatrix}. \tag{3.135}$$

The unitary matrix $\mathbf{U}$ is $M \times N$ and the unitary matrix $\mathbf{V}$ is $N \times N$. Without loss of generality, we assume $M \geq N$ (the analysis of $\mathbf{H}^T$ will yield the same results for $N \geq M$).

The $M \times N$ matrix $\mathbf{U}$ consists of N orthonormalized eigenvectors associated with the N largest eigenvalues of $\mathbf{HH}^T$ and the $N \times N$ matrix $\mathbf{V}$ consists of the N orthonormalized eigenvectors of $\mathbf{H}^T\mathbf{H}$. The matrix $\mathbf{U}$ is the truncated modal matrix of $\mathbf{HH}^T$, while $\mathbf{V}$ is the modal matrix of $\mathbf{H}^T\mathbf{H}$. Σ is a diagonal matrix whose diagonal terms are composed of the nonnegative square roots of the eigenvalues of $\mathbf{H}^T\mathbf{H}$ or $\mathbf{HH}^T$; these values are called the singular values of $\mathbf{H}$. Because the matrices $\mathbf{U}$ and $\mathbf{V}$ are composed of orthonormal vectors, they are orthogonal matrices[6] that possess the attractive property

$$\mathbf{U}^T\mathbf{U} = \mathbf{VV}^T = \mathbf{V}^T\mathbf{V} = \mathbf{I}_N, \tag{3.136}$$

where

$$\mathbf{I}_N = \begin{pmatrix} 1 & 0 & \cdots & \\ 0 & 1 & 0 & \cdots \\ & & \ddots & \\ & & & 1 \end{pmatrix}. \tag{3.137}$$

In other words, the inverse of a unitary matrix is its transpose.

Assuming

$$\sigma_1 \geq \sigma_2 \geq \cdots \geq \sigma_N, \tag{3.138}$$

if the rank of the $\mathbf{H}$ matrix is r,

$$\sigma_{(r+1)} = \sigma_{(r+2)} = \cdots = \sigma_N = 0. \tag{3.139}$$

---

[6]When $\mathbf{U}$ and $\mathbf{V}$ are complex valued, they are unitary.

There are r nonzero singular values associated with the matrix $\mathbf{H}$ of rank r. When $r < N \le M$, the matrix $\mathbf{H}$ is said to be rank deficient. The inverse of $\mathbf{H}$ can be expressed in terms of its SVD as

$$\mathbf{H}^{-1} = (\mathbf{U}\Sigma\mathbf{V}^T)^{-1} = \mathbf{V}\Sigma^{-1}\mathbf{U}^T, \tag{3.140}$$

where

$$\Sigma^{-1} = \begin{pmatrix} \sigma_1^{-1} & 0 & \cdots & \\ 0 & \sigma_2^{-1} & 0 & \\ & & & \\ & & \sigma_N^{-1} \end{pmatrix}. \tag{3.141}$$

When the matrix $\mathbf{H}$ is rank deficient, the inverse of $\mathbf{H}$ depends on inverting zero-valued singular values. Obviously, this will lead to an unstable solution. An alternative to computing the straightforward inverse for a rank-deficient matrix is the pseudo-inverse. Because the vectors composing $\mathbf{U}$ and $\mathbf{V}$ are orthogonal to each other, one can express $\mathbf{H}$ in series form as

$$\mathbf{H} = \sum_{i=1}^{r} \sigma_i \mathbf{u}_i \mathbf{v}_i^T, \tag{3.142}$$

where the subscript i in $\mathbf{u}$ and $\mathbf{v}$ denotes the ith columns of the respective matrices. The Moore–Penrose generalized inverse, or pseudo-inverse, solution of $\mathbf{H}$ is obtained by inverting only the nonzero singular values of $\mathbf{H}$,

$$H^\dagger = \sum_{i=1}^{r} \sigma_i^{-1} \mathbf{v}_i \mathbf{u}_i^T, \tag{3.143}$$

where the dagger ($\dagger$) denotes the pseudo-inverse. Using the decomposition given by (3.134), we can describe the pseudo-inverse solution as

$$\mathbf{f}^\dagger = \mathbf{H}^\dagger \mathbf{H}\mathbf{f} = \mathbf{V}\Sigma^\dagger\mathbf{U}^T\mathbf{U}\Sigma\mathbf{V}^T\mathbf{f} = \mathbf{V}(\Sigma^\dagger\Sigma)\mathbf{V}^T\mathbf{f}, \tag{3.144}$$

where

$$\Sigma^\dagger = \begin{pmatrix} \sigma_1^{-1} & 0 & \cdots & \\ 0 & \sigma_2^{-1} & 0 & \\ \vdots & & \ddots & \\ & & \sigma_r^{-1} & \\ & & & 0 \end{pmatrix}. \tag{3.145}$$

Therefore,

$$\mathbf{f}^\dagger = \mathbf{V}\Lambda_r\mathbf{V}^T\mathbf{f}, \tag{3.146}$$

where

$$
\Lambda_r = \begin{pmatrix} 1 & 0 & \cdots & & \\ 0 & 1 & 0 & & \\ \vdots & & \ddots & & \\ & & & 1 & \\ & & & & 0 \end{pmatrix} ; \tag{3.147}
$$

the first r terms along the diagonal of Λ_r are equal to one and all other terms are equal to zero. The pseudo-inverse solution overcomes the problem of having a rank-deficient operator $\mathbf{H}$. Recalling the original discrete problem formulation given by (3.12) we can derive the pseudo-inverse solution from the perspective of finding a least squares minimum-norm solution; that is, finding the minimum-norm solution which minimizes the cost function

$$
\|\mathbf{g} - \mathbf{Hf}\| = \|\mathbf{n}\|. \tag{3.148}
$$

The quantity $\|\cdot\|$ denotes the L_2 norm, or Euclidean norm, of a vector. The least squares solution must satisfy the normal equation:

$$
\mathbf{H}^T \mathbf{g} = \mathbf{H}^T \mathbf{Hf}. \tag{3.149}
$$

The pseudo-inverse solution is given by

$$
\mathbf{f}^\dagger = (\mathbf{H}^T \mathbf{H})^\dagger \mathbf{H}^T \mathbf{g}. \tag{3.150}
$$

The condition number of a matrix $\mathbf{H}$ is given by

$$
C_H = \frac{\sigma_{\max}}{\sigma_{\min}}; \tag{3.151}
$$

it is the ratio of the largest singular value to the smallest singular value. For a rank-deficient matrix, the condition number is equal to infinity. Due to the noise term present in the formulation of the problem, a matrix $\mathbf{H}$ with a large condition number (> 100) will lead to a very ill conditioned problem. The ill-conditioned nature of the problem implies that a small change in the input $\mathbf{f}$ can lead to a large change in the output $\mathbf{g}$. Therefore, inverting an ill-conditioned matrix $\mathbf{H}$ in the presence of noise can cause the noise terms to overpower the image terms, leading to a poor restoration of the desired image. When a system is ill conditioned and the output has a relatively low signal-to-noise ratio (SNR), the solution given by (3.146) may result in a poor estimate. One way to overcome the problem of an ill-conditioned operator is to realize that the condition number is an indication of the magnification of the noise power. By taking the reciprocal of the singular values only up to the Kth singular value, one can define the "pseudo-condition number" as

$$C_{\text{pseudo}} = \frac{\sigma_{\max}}{\sigma_K}. \tag{3.152}$$

One can determine the cutoff K for a particular ill-conditioned problem by the pseudo-condition number, which gives an indication of the amount of noise magnification one is willing to tolerate. *A priori* statistics of the noise source are used to determine the cutoff K. The higher the SNR, the larger C_{pseudo} can be without causing intolerable noise magnification. Other techniques, such as smoothing methods and regularization, also attempt to overcome the ill-conditioned nature of many signal-recovery problems. The obvious tradeoff in such techniques is a decrease in noise magnification at the expense of losing high-resolution information.

The image-recovery algorithm presented here uses the SVD algorithm developed by Golub and Reinsch [30]. The algorithm can be divided into two parts. The first part consists of finding two sequences of Householder transformations,

$$\mathbf{P}_k(k = 1, 2, \ldots, N) \tag{3.153}$$

and

$$\mathbf{Q}_k(k = 1, 2, \ldots, N - 2), \tag{3.154}$$

such that

$$\mathbf{P}_N \, \mathbf{P}_{N-1} \, \cdots \, \mathbf{P}_1 \, \mathbf{H} \, \mathbf{Q}_1 \, \cdots \, \mathbf{Q}_{N-3} \, \mathbf{Q}_{N-2} = \begin{pmatrix} j & j & 0 & \cdots \\ & \ddots & \ddots & 0 \\ 0 & \cdots & 0 & j \\ 0 & & \cdots & 0 \\ \vdots & \vdots & & \end{pmatrix} = \mathbf{J}_0. \tag{3.155}$$

The sequential multiplication of $\mathbf{H}$ by $\mathbf{P}_i$ on the left and $\mathbf{Q}_i$ on the right forms an upper bidiagonal matrix $\mathbf{J}_0$. The matrix $\mathbf{P}_i$ zeros out the subdiagonal elements in column i of $\mathbf{H}$, while $\mathbf{Q}_i$ zeros out the appropriate elements in row i.

A Householder transformation is described by the following matrix:

$$\mathbf{HT} = \mathbf{I} - 2\mathbf{w}\mathbf{w}^T, \qquad \|\mathbf{w}\| = 1.0. \tag{3.156}$$

The matrix $\mathbf{HT}$ is simply a reflection that preserves length and angles and is, therefore, a unitary transformation.

Because $\mathbf{H}$ is multiplied by a sequence of Householder transformations (unitary matrices), the singular values of $\mathbf{H}$ are the same as the singular values of $\mathbf{J}_0$.

Thus, if

$$J = R\Sigma S^T \qquad (3.157)$$

is the SVD of J, then

$$H = PJQ^T = PR\Sigma S^T Q^T. \qquad (3.158)$$

Let

$$U = PR \qquad (3.159)$$

and

$$V = QS, \qquad (3.160)$$

where

$$P = P_1 P_2 \cdots P_N \qquad (3.161)$$

and

$$Q = Q_1 Q_2 \cdots Q_{N-2}. \qquad (3.162)$$

After constructing the upper bidiagonal matrix J_0, the second part of the algorithm iteratively diagonalizes J_0 by the QR decomposition method,

$$J_{i+1} = E_i^T J_i F_i, \qquad J_0 \to J_1 \to \cdots \to \Sigma, \qquad (3.163)$$

where E_i and F_i are products of Givens transformations and are therefore orthogonal. The matrices F_i are chosen so that the sequence $J_i^T J_i$ converges to a diagonal matrix, while the matrices E_i are chosen so that all J_i are of bidiagonal form. From (3.162) and (3.163) it follows that

$$R^T = E_1 E_2 \cdots E_i \qquad \text{and} \qquad S^T = F_1 F_2 \cdots F_i. \qquad (3.164)$$

It has been found that the average number of iterations required to diagonalize J are on the order of $2N$. By diagonalizing J, one finds the singular values for H and the unitary matrices U and V.

3.10 GENERALIZED ITERATIVE RESTORATION ALGORITHM

The previous section laid the groundwork for the introduction of a generalized iterative restoration algorithm that can be applied to any type of signal-restoration problem, specifically that of image recovery.

Recall that the least squares minimum-norm, or pseudo-inverse, solution to (3.12) is

$$\mathbf{f}^{\dagger} = \mathbf{H}^{\dagger}\mathbf{H}\mathbf{f} = \mathbf{V}\Lambda\mathbf{V}^{T}\mathbf{f}, \qquad (3.165)$$

where the dagger ($\dagger$) denotes the pseudo-inverse, $\mathbf{V}$ is the unitary matrix found in the diagonalization of $\mathbf{H}$, and Λ is a diagonal matrix for which the first r terms are all equal to one and the rest are zero. By defining

$$\mathbf{P} = \mathbf{V}\Lambda\mathbf{V}^{T}, \qquad (3.166)$$

we can show that $\mathbf{P}$ is a projection operator, since the definition of a projection operator holds:

$$\mathbf{P}^{2} = \mathbf{V}\Lambda\mathbf{V}^{T}\mathbf{V}\Lambda\mathbf{V}^{T} = \mathbf{V}(\Lambda\Lambda)\mathbf{V}^{T} = \mathbf{V}\Lambda\mathbf{V}^{T} = \mathbf{P}. \qquad (3.167)$$

A projection operator is a Hermitian idempotent matrix. The orthogonal complement to the operator $\mathbf{P}$ is defined by the projection operator

$$\mathbf{Q} = \mathbf{I} - \mathbf{P} = \mathbf{V}\Lambda^{C}\mathbf{V}^{T}, \qquad (3.168)$$

where

$$\Lambda^{C} = \begin{pmatrix} 0 & 0 & \cdots & & & & \\ 0 & \ddots & & & & & \\ \vdots & & 1 & & & & \\ & & & 1 & & & \\ & & & & \ddots & & \\ & & & & & 1 \end{pmatrix}. \qquad (3.169)$$

The diagonal matrix Λ^{C} contains ones in the last $N - r$ diagonal positions and zeros elsewhere. The superscript C denotes the complement, signifying that the ones and zeros along the diagonal of Λ have been complemented.

Any arbitrary vector $\mathbf{f}$ can be decomposed as follows:

$$\mathbf{f} = \mathbf{P}\mathbf{f} + \mathbf{Q}\mathbf{f}, \qquad (3.170)$$

where the projection operator $\mathbf{P}$ projects $\mathbf{f}$ onto the range space of the degradation operator $\mathbf{H}^{T}$ and the orthogonal projection operator $\mathbf{Q}$ projects $\mathbf{f}$ onto the null space of the degradation matrix $\mathbf{H}$. The component $\mathbf{P}\mathbf{f}$ will be referred to as the "in-band" term and the component $\mathbf{Q}\mathbf{f}$ will be referred to as the "out-of-band" term.

In general, the least squares family of solutions to the image-restoration problem can be stated as

$$\mathbf{f} = \mathbf{f}_{particular} + \mathbf{f}_{homogeneous}$$

$$= \mathbf{f}^\dagger + K_{r+1}\mathbf{v}_{r+1} + K_{r+2}\mathbf{v}_{r+2} + \cdots + K_N\mathbf{v}_N. \tag{3.171}$$

The vectors $\mathbf{v}_i$ in (3.171) correspond to the eigenvectors of

$$\left\{\sigma_{r+1}^2, \sigma_{r+2}^2, \ldots, \sigma_N^2\right\} \tag{3.172}$$

for $H^T H$; they are the eigenvectors associated with zero-valued eigenvalues. The homogeneous solution $K_{r+1}\mathbf{v}_{r+1} + \cdots + K_N\mathbf{v}_N$ must satisfy

$$\mathbf{H}\mathbf{f}_{homogeneous} = 0. \tag{3.173}$$

Adding the terms $\{K_{r+1}\mathbf{v}_{r+1}, K_{r+2}\mathbf{v}_{r+2}, \ldots, K_N\mathbf{v}_N\}$, to the pseudo-inverse solution $\mathbf{f}^\dagger$ does not change the L_2 norm of the error since

$$\|\mathbf{n}\| = \|\mathbf{g} - \mathbf{H}\mathbf{f}\|$$

$$= \|\mathbf{g} - \mathbf{H}(\mathbf{f}^\dagger + K_{r+1}\mathbf{v}_{r+1} + \cdots + K_N\mathbf{v}_N)\| \tag{3.174}$$

$$= \|\mathbf{g} - \mathbf{H}\mathbf{f}^\dagger - \mathbf{H}K_{r+1}\mathbf{v}_{r+1} - \cdots - \mathbf{H}K_N\mathbf{v}_N\|$$

$$= \|\mathbf{g} - \mathbf{H}\mathbf{f}^\dagger\|,$$

which is the least squares error. The terms $\mathbf{H}K_{r+1}\mathbf{v}_{r+1}, \ldots, \mathbf{H}K_N\mathbf{v}_N$ are all equal to zero because the vectors $\mathbf{v}_{r+1}, \ldots, \mathbf{v}_N$ are in the null space of $\mathbf{H}$. Therefore, any linear combination of $\mathbf{v}_i$ in the null space of $\mathbf{H}$ can be added to the pseudo-inverse solution without affecting the least squares cost function. The pseudo-inverse solution, $\mathbf{f}^\dagger$, provides the unique least squares estimate with minimum norm,

$$\min\|\mathbf{f}_{LS}\| = \|\mathbf{f}^\dagger\|, \tag{3.175}$$

where $\mathbf{f}_{LS}$ denotes the least squares solution. In general, the solution need not satisfy the property of possessing the minimum norm; thus, $\mathbf{f}^\dagger$ is not necessarily the optimum solution. The new algorithm incorporates *a priori* information in the form of signal constraints in order to obtain a better estimate than the least squares minimum-norm solution $\mathbf{f}^\dagger$. The constraint operator will be represented by C and can incorporate a variety of linear and nonlinear *a priori* signal characteristics. In the case of image restoration, the constraint operator C may consist of a nonnegativity constraint and is represented by

$$(Cx)_i = \begin{cases} x_i, & x_i \geq 0, \\ 0, & x_i < 0. \end{cases} \tag{3.176}$$

Concatenating the vectors v_i in (3.171) yields

$$\mathbf{f} = \mathbf{f}^\dagger + V\Lambda^C K, \tag{3.177}$$

where

$$K = \begin{pmatrix} K_1 \\ K_2 \\ \vdots \\ K_N \end{pmatrix} \tag{3.178}$$

and

$$V\Lambda^C = \begin{pmatrix} \mathbf{v}_1 & \mathbf{v}_2 & \cdots & \mathbf{v}_N \end{pmatrix} \begin{pmatrix} 0 & 0 & \cdots & & & \\ 0 & \ddots & & & & \\ \vdots & & 1 & & & \\ & & & 1 & & \\ & & & & \ddots & \\ & & & & & 1 \end{pmatrix}. \tag{3.179}$$

The formulation given in (3.177) calls for the solution of the unknown vector K. A reasonable approach would be to start with the constrained pseudo-inverse solution and solve for K in a least squares manner:

$$\text{minimize} \quad \|C\mathbf{f}^\dagger - \mathbf{f}^\dagger - V\Lambda^C K\| \tag{3.180}$$
$$\text{subject to} \quad C\mathbf{f}^\dagger = \mathbf{f}^\dagger + V\Lambda^C K. \tag{3.181}$$

The least squares solution is found to be

$$C\mathbf{f}^\dagger - \mathbf{f}^\dagger = V\Lambda^C K,$$
$$K = \Lambda^C V^T (C\mathbf{f}^\dagger - \mathbf{f}^\dagger). \tag{3.182}$$

Since $\Lambda^C V^T \mathbf{f}^\dagger = 0.0$,

$$K = \Lambda^C V^T C\mathbf{f}^\dagger. \tag{3.183}$$

Substituting (3.183) into (3.181) and adding a residual vector $\mathbf{n}$ yields

$$C\mathbf{f}^\dagger = \mathbf{f}^\dagger + QC\mathbf{f}^\dagger + \mathbf{n}. \tag{3.184}$$

The solution given by (3.184) minimizes the cost function given by $\|\mathbf{n}\|$. The process of enforcing the overall least squares solution and solving for the homogeneous (out-of-band) component to fit the constraints in a least squares fashion can be implemented iteratively. The resulting recursion is

$$C\mathbf{f}^{(k)} = \mathbf{f}^\dagger + QC\mathbf{f}^{(k)} + \mathbf{n}^{(k)}. \tag{3.185}$$

By defining

$$\mathbf{f}^{(k+1)} \triangleq C\mathbf{f}^{(k)} - \mathbf{n}^{(k)}, \tag{3.186}$$

the final iterative algorithm becomes

$$\mathbf{f}^{(0)} = \mathbf{f}^\dagger,$$
$$\mathbf{f}^{(k+1)} = \mathbf{f}^\dagger + QC\mathbf{f}^{(k)}, \qquad k = 0, 1, 2, \ldots. \tag{3.187}$$

Note that the recursion still yields the least squares solution, with the added property of fitting the *a priori* signal constraints in a least squares fashion through the homogeneous (out-of-band) signal solution. Obviously, this recursion yields a better estimate of $\mathbf{f}$ than the minimum-norm least squares solution $\mathbf{f}^\dagger$. In the case where $\mathbf{f}^\dagger$ satisfies all the constraints exactly, the iterative algorithm's solution reduces to the pseudo-inverse solution.

3.11   ACCELERATION OF THE NEW RESTORATION ALGORITHM

Acceleration of the new general iterative recovery algorithm's convergence speed can be introduced in two ways: relaxation in the spatial domain and relaxation in the transform domain. Relaxation in the spatial domain is incorporated into the algorithm in two different ways. The first way of accelerating in the spatial domain is by realizing that the new algorithm is a form of POCS and using the relaxation parameter introduced in the POCS formulation to speed up convergence. The second way of introducing acceleration is by applying the relaxation suggested by Gubin [25] for projection onto two convex sets. Accelerating the algorithm's convergence speed can also be implemented by introducing relaxation in the transform domain. This type of acceleration is provided by a taper on the projection operators $\mathbf{P}$ and $\mathbf{Q}$ so that they are no longer orthogonal to each other.

3.11.1 Acceleration by Spatial Domain Relaxation

The POCS algorithm is based on incorporating the *a priori* image constraints through projection onto convex sets. A convex function $C(\cdot)$ is defined as

$$C(\lambda x + (1 - \lambda)y) \le \lambda C(x) + (1 - \lambda)C(y), \qquad 0 \le \lambda \le 1, \quad x, y \in \mathcal{S}. \qquad (3.188)$$

The paper on POCS by Youla and Webb [29] shows that many of the commonly used constraints in signal restoration can be viewed as convex functions. The image-restoration algorithm given in (3.187) is one such case. The POCS algorithm can be stated as

$$\mathbf{f}^{(k+1)} = \mathbf{P}\mathbf{f}^{(k)} = \mathbf{P}_l \, \mathbf{P}_{l-1} \, \cdots \, \mathbf{P}_2 \, \mathbf{P}_1 \, \mathbf{f}^{(k)}, \qquad (3.189)$$

where $\mathbf{P}$ is a composition of the projection operators onto all l convex constraint sets. The POCS algorithm can be greatly accelerated by introducing a relaxation parameter λ and replacing $\mathbf{P}$ by

$$\mathbf{T} = 1 + \lambda(\mathbf{P} - 1), \qquad 0 \le \lambda \le 2. \qquad (3.190)$$

By choosing the appropriate relaxation parameter (RP) λ, the algorithm can be greatly accelerated. An RP between 0 and 1 results in the underrelaxed case and an RP between 1 and 2 results in the overrelaxed case. Applying relaxation, as given in (3.190), to the general iterative algorithm in (3.187) results in

$$\mathbf{f}_A^{(k+1)} = (1 - \lambda)\mathbf{f}_A^{(k)} + \lambda\mathbf{f}^{(k+1)}, \qquad k = 0, 1, 2, \ldots, \quad 0 \le \lambda \le 2, \qquad (3.191)$$

where the subscript A denotes the accelerated estimate. Without loss of generality, one can let the vector dimension N equal 1 in (3.191). This equation is a simple first-order difference equation. The solution is stable for $0 \le \lambda \le 2$. This condition holds true for the more general case of an N-dimensional vector, since all the terms act independently of each other. Youla gives a more general proof for the limits on λ. The general accelerated iterative restoration algorithm can be compactly stated as

$$\mathbf{f}_A^{(0)} = \mathbf{f}^\dagger,$$
$$\mathbf{f}^{(k+1)} = \mathbf{f}^\dagger + \mathbf{Q}C\mathbf{f}_A^{(k)}, \qquad k = 0, 1, 2, \ldots,$$
$$\mathbf{f}_A^{(k+1)} = (1 - \lambda)\mathbf{f}_A^{(k)} + \lambda\mathbf{f}^{(k+1)}, \qquad 0 \le \lambda \le 2. \qquad (3.192)$$

Another type of spatial relaxation examined is that proposed by Gubin for projection onto two convex sets. Let the positivity constraint C correspond to the first projection onto a convex set and the concatenation of the projection onto the null space of $\mathbf{H}$ represented by $\mathbf{Q}$ and the addition of the pseudo-inverse solution correspond to the second projection onto a convex set. Recall the method of acceleration for two projection operators proposed by Gubin:

$$\mathbf{a} = \mathbf{P}_1(\mathbf{f}^{(k)}),$$

$$\mathbf{b} = \mathbf{P}_2\mathbf{a},$$

$$\mathbf{c} = \mathbf{P}_1\mathbf{b},$$

$$\mathbf{f}_A^{(k+1)} = \mathbf{a} + \lambda(\mathbf{c} - \mathbf{b}),$$

$$\lambda = \frac{\|\mathbf{a} - \mathbf{b}\|^2}{(\mathbf{a} - \mathbf{c}, \mathbf{a} - \mathbf{b})}. \tag{3.193}$$

Applying this form of acceleration to the new restoration algorithm yields the following step at the kth iteration:

$$\mathbf{a} = C\mathbf{f}^{(k)},$$

$$\mathbf{b} = \mathbf{f}^{(k+1)} = \mathbf{f}^\dagger + \mathbf{Q}C\mathbf{f}^{(k)},$$

$$\mathbf{c} = C\mathbf{b} = C\mathbf{f}^{(k+1)},$$

$$\mathbf{f}_A^{(k+1)} = C\mathbf{f}^{(k)} + \lambda(C\mathbf{f}^{(k+1)} - \mathbf{f}^{(k+1)}),$$

$$\lambda = \frac{\|C\mathbf{f}^{(k)} - \mathbf{f}^{(k+1)}\|^2}{(C\mathbf{f}^{(k)} - C\mathbf{f}^{(k+1)}, C\mathbf{f}^{(k)} - \mathbf{f}^{(k+1)})}. \tag{3.194}$$

3.11.2 Stopping Criterion

An optimum stopping criterion is an important factor in developing an efficient acceleration algorithm. One should be able to monitor the algorithm for error reduction saturation or algorithm instability. A parameter that has been found to monitor the actual error norm precisely in a complementary fashion is the out-of-band energy, defined as

$$\|e_Q^{(k)}\| = \|\mathbf{Q}C\mathbf{f}^{(k)}\|. \tag{3.195}$$

The out-of-band energy is shown to be an effective stopping criterion for both acceleration strategies in [31–34]. When the error norm saturates or starts to become unstable, the out-of-band energy exhibits similar behavior. Therefore, an optimum strategy is to stop when

$$\|\mathbf{Q}C\mathbf{f}^{(k+1)}\| - \|\mathbf{Q}C\mathbf{f}^{(k)}\| \le \epsilon, \tag{3.196}$$

where ϵ is a small positive value determined on the basis of how the error saturation is defined for each application.

Another interesting application of the out-of-band energy deals with the type of constraint operators C one can use in the iterative scheme provided by the algorithm in (3.187). Researchers in the area [35] show that a fixed-point iterative algorithm will converge when the constraint operator is compact:

$$\|Cx - Cy\| < \|x - y\|, \tag{3.197}$$

where x and y are arbitrary vectors. The algorithm may converge when the constraint operator is nonexpansive:

$$\|Cx - Cy\| \leq \|x - y\|. \tag{3.198}$$

However, many constraints for different signal-recovery problems do not obey these properties; or it may be difficult to analyze the algorithms for these properties. Gubin proves that the method of POCS will converge when the data is corrupted with noise under certain limiting circumstances and when $\lambda < 1$. However, choosing $\lambda < 1$ will often lead to very slow convergence speed, impractical for many applications. Therefore, an iterative scheme as given in (3.187) is not guaranteed to converge for certain constraint operators and $\lambda > 1$. By monitoring the out-of-band energy, one can determine whether a given constraint operator and RP value results in a converging scheme. By monitoring all the *a priori* known constraints, one can determine which ones can be used in the algorithm. Therefore, any constraint operator can be tested for convergence, as monitored by the out-of-band energy. A large λ may also be used for fast convergence with the stopping criterion monitoring the algorithm's stability.

3.11.3 Acceleration by Transform Domain Relaxation

It has been shown how the new iterative algorithm can be accelerated through relaxation in the spatial domain. Relaxation in the transform domain is now introduced to provide further improvement in the overall algorithm performance. Recall the original algorithm formulation for the new image restoration technique,

$$f^{(0)} = f^{\dagger},$$
$$f^{(k+1)} = f^{\dagger} + QCf^{(k)}, \tag{3.199}$$

where

$$f^{\dagger} = Pf = V\Lambda V^{T}f \tag{3.200}$$

and

$$Q = V\Lambda^{C}V^{T}. \tag{3.201}$$

We found that introducing a taper on Λ results in faster acceleration and an improved final estimate. The reason for introducing a taper is the noise term, which corrupts the pseudo-inverse $\mathbf{f}^\dagger = \mathbf{P}\mathbf{f} + \mathbf{H}^\dagger \mathbf{n}$. The taper on Λ establishes the uncertainty in the estimate of the in-band component $\mathbf{f}^\dagger$ due to noise-corrupted data. The taper allows some of the in-band component to contribute to the positivity constraint on the solution. The algorithm in (3.187) is replaced with

$$\mathbf{f}^{(0)} = \mathbf{f}^\dagger,$$

$$\mathbf{f}^{(k+1)} = \mathbf{f}_t^\dagger + \mathbf{Q}_t C \mathbf{f}^{(k)}, \tag{3.202}$$

where

$$\mathbf{f}_t^\dagger = \mathbf{P}_t \mathbf{f} = \mathbf{V}\Lambda_t \mathbf{V}^T \mathbf{f} \tag{3.203}$$

and

$$\mathbf{Q}_t = \mathbf{V}\Lambda_t^C \mathbf{V}^T. \tag{3.204}$$

The diagonal matrices Λ_t and Λ_t^C are defined as

$$\Lambda_t = \begin{pmatrix} t_1 & 0 & \cdots & & & \\ & t_2 & & & & \\ & & \ddots & & & \\ & & & t_r & & \\ & & & & 0 & \\ & & & & & \ddots \end{pmatrix} \tag{3.205}$$

and

$$\Lambda_t^C = \begin{pmatrix} 1 - t_1 & 0 & \cdots & & & \\ & 1 - t_2 & & & & \\ & & \ddots & & & \\ & & & 1 - t_r & & \\ & & & & 1 & \\ & & & & & \ddots \end{pmatrix}. \tag{3.206}$$

The new algorithm with the taper introduces relaxation in the transform domain, which results in acceleration of the original iterative algorithm.

The taper, which was found to produce excellent results for image recovery, is described as

$$
\Lambda_t = \begin{pmatrix} (\frac{\lambda_1}{\lambda_1})^\alpha & & & & & \\ & (\frac{\lambda_2}{\lambda_1})^\alpha & & & & \\ & & \ddots & & & \\ & & & (\frac{\lambda_r}{\lambda_1})^\alpha & & \\ & & & & 0 & \\ & & & & & \ddots \end{pmatrix}, \tag{3.207}
$$

so that

$$
\Lambda_t^C = \begin{pmatrix} 1-(\frac{\lambda_1}{\lambda_1})^\alpha & & & & & \\ & 1-(\frac{\lambda_2}{\lambda_1})^\alpha & & & & \\ & & \ddots & & & \\ & & & 1-(\frac{\lambda_r}{\lambda_1})^\alpha & & \\ & & & & 1 & \\ & & & & & \ddots \end{pmatrix}, \tag{3.208}
$$

where $0 \leq \alpha \leq 1$. Note that such a taper always results in $t_i \leq 1$. The scalar α is chosen on the basis of how well the "untapered" pseudo-inverse $\mathbf{f}^\dagger$ conforms to the constraint given by C. The norm of the constraint error for the pseudo-inverse is defined as

$$
\|\mathbf{e}_C^\dagger\| = \|C\mathbf{f}^\dagger - \mathbf{f}^\dagger\|. \tag{3.209}
$$

The larger the quantity $\|\mathbf{e}_C^\dagger\|$ is, the larger $\acute{\alpha}$ is chosen. When α is large—that is, $\alpha \to 1$—Λ_t becomes more tapered: the terms $\lambda_i/\lambda_1 \leq 1$ raised to the power α become very small. Because the pseudo-inverse does not comply with the *a priori* constraint information, the taper allows a variable amount of the in-band component to satisfy the constraint.

A combination of the taper in the transform domain with spatial relaxation and the out-of-band energy as a stopping criterion proves to be a very powerful iterative recovery algorithm. Generalized versions of row-action and block-action recovery algorithms are shown to solve many different signal-recovery problems, as shown in subsequent chapters.

3.12 CYCLIC AND SHIFT-INVARIANT DEGRADATION OPERATORS

When the degradation operator under consideration is a cyclic matrix,

$$
H = \begin{pmatrix}
h_1 & h_2 & \cdots & & h_N \\
h_N & h_1 & & & h_{N-1} \\
\vdots & \vdots & & &
\end{pmatrix},
\tag{3.210}
$$

it can be diagonalized using the discrete Fourier transform (DFT) matrix. Therefore, the SVD of **H** as given by (3.134) is reduced to

$$
\mathbf{H} = \mathbf{W}^{-1}\Sigma\mathbf{W},
\tag{3.211}
$$

where **W** is the (N-point) DFT matrix given by

$$
\mathbf{W} = \begin{pmatrix}
w_0^0 & w_0^0 & \cdots & \cdots & w_0^0 \\
w_0^0 & w_0^1 & w_0^2 & \cdots & w_0^{N-1} \\
w_0^0 & w_0^2 & w_0^4 & \cdots & w_0^{2(N-1)} \\
\vdots & \vdots & \vdots & & \vdots \\
w_0^0 & w_0^{N-1} & w_0^{2(N-1)} & \cdots & w_0^{(N-1)(N-1)}
\end{pmatrix},
\tag{3.212}
$$

where

$$
w_0 = e^{j2\pi/N} = \cos(2\pi/N) + j\sin(2\pi/N)
\tag{3.213}
$$

and $\mathbf{W}^{-1}$ is the inverse DFT matrix, which is related to **W** by

$$
\mathbf{W}^{-1} = \mathbf{W}^H = \bar{\mathbf{W}}^T.
\tag{3.214}
$$

By taking advantage of the symmetry found in the DFT, one can use FFTs for computational efficiency. The diagonal elements of the matrix Σ can be obtained by taking the DFT of the shifted version of the first row of the degradation operator **H**:

$$
\text{Diagonal}\{\Sigma\} = \mathbf{W} \begin{pmatrix}
h_1 \\
h_N \\
h_{N-1} \\
\vdots \\
h_2
\end{pmatrix}.
\tag{3.215}
$$

In the case of a cyclic matrix **H**, one can reduce the computational complexity of the algorithm by using FFTs instead of performing the computationally complex

SVD. In the case where the degradation operator is a shift-invariant operator, $\mathbf{H}$ is a Toeplitz matrix,

$$
\mathbf{H} = \begin{pmatrix} h_0 & h_1 & \cdots & h_{N-1} \\ h_{-1} & h_0 & & h_{N-2} \\ h_{-2} & h_{-1} & & \\ \vdots & \vdots & & \end{pmatrix}. \tag{3.216}
$$

The terms along each subdiagonal are the same. For such a matrix, one can approximate the linear convolution with a cyclic convolution by appending $\mathbf{H}$ and $\mathbf{f}$ as follows:

$$
\mathbf{g} = \begin{pmatrix} & h_{-M} & h_{-M+1} & \cdots & h_{-1} \\ \mathbf{H} & h_{N-1} & h_{-M} & \cdots & h_{-2} \\ & h_{N-2} & h_{N-1} & \cdots & h_{-3} \\ & \vdots & & & \end{pmatrix} \begin{pmatrix} \mathbf{f} \\ 0 \end{pmatrix}. \tag{3.217}
$$

Note that $\mathbf{g}$ is the same and $\mathbf{f}$ is appended with zeros, but $\mathbf{H}$ has been replaced with a cyclic matrix. Therefore, FFTs can be used to approximate the SVD of a Toeplitz matrix $\mathbf{H}$.

In summary, for cyclic degradation operators, the IFFT and FFT can be used for the unitary operators $\mathbf{U}$ and $\mathbf{V^H}$, respectively.[7] In general, when the operator $\mathbf{H}$ is symmetric, $\mathbf{U} = \mathbf{V}$, as is the case for the cyclic operator, since $\mathbf{W}^{-1} = \mathbf{W}^H$ with a scaling factor $1/N$. For Toeplitz matrices $\mathbf{H}$, the diagonalization of $\mathbf{H}$ can be approximated with the IFFT and FFT. The loss in accuracy is compensated by the computational ease and efficiency of the FFT. The algorithm given by (3.187) can be applied to cyclic and Toeplitz degradation operators using only FFTs, IFFTs, and the constraint operators.

3.13   SEPARABLE DEGRADATION OPERATOR

In the case where the degradation operator is separable—that is, when the operator can be decomposed into a horizontal and vertical component—the iterative algorithm can be reformulated for a dramatic reduction in the computational load. This can be applied to the algorithm of the previous section by realizing that the two-dimensional DFT matrix is separable. A separable operator $\mathbf{H}$ can be expressed in the outer product form,

$$
\mathbf{H} = \mathbf{H}_C \otimes \mathbf{H}_R, \tag{3.218}
$$

---

[7]For complex matrices, the Hermitian transpose H must be used in place of the transpose T.

where $\mathbf{H}_C$ and $\mathbf{H}_R$ are column- and row-degradation operators, respectively. The two-dimensional blurred output $\mathbf{G}$ can be represented by a sequential row and column convolution of the two-dimensional image $\mathbf{F}$ with the row and column-degradation kernels, that is,

$$\mathbf{G} = \mathbf{H}_C \mathbf{F} \mathbf{H}_R^T. \tag{3.219}$$

The iterative algorithm will now be based on the singular-value decomposition (SVD) of the column- and row-degradation operators $\mathbf{H}_C$ and $\mathbf{H}_R$, separately. The SVDs of the two operators are expressed as

$$\mathbf{H}_C = \mathbf{U}_C \mathbf{\Sigma}_C \mathbf{V}_C^T,$$

and

$$\mathbf{H}_R = \mathbf{U}_R \mathbf{\Sigma}_R \mathbf{V}_R^T. \tag{3.220}$$

As in the nonseparable case, $\mathbf{U}_{C,R}$ and $\mathbf{V}_{C,R}$ are unitary matrices composed of the orthonormal eigenvectors associated with $\mathbf{H}_{C,R}\mathbf{H}_{C,R}^T$ and $\mathbf{H}_{C,R}^T\mathbf{H}_{C,R}$, respectively. The term $\mathbf{\Sigma}_{C,R}$ is a diagonal matrix whose diagonal terms are the singular values of $\mathbf{H}_{C,R}$. The pseudo-inverse is derived as follows:

$$\mathbf{F}^\dagger = \mathbf{H}_C^\dagger \mathbf{G}(\mathbf{H}_R^T)^\dagger = \mathbf{H}_C^\dagger \mathbf{H}_C \mathbf{F} \mathbf{H}_R^T (\mathbf{H}_R^T)^\dagger = \mathbf{P}_C \mathbf{F} \mathbf{P}_R^T, \tag{3.221}$$

where

$$\mathbf{P}_C = \mathbf{V}_C \mathbf{\Lambda}_C \mathbf{V}_C^T \tag{3.222}$$

and

$$\mathbf{P}_R = \mathbf{V}_R \mathbf{\Lambda}_R \mathbf{V}_R^T. \tag{3.223}$$

Recall that, as in the case of the nonseparable degradation operator, $\mathbf{\Lambda}_C$ and $\mathbf{\Lambda}_R$ are diagonal matrices whose diagonal terms are composed of ones and zeros. Note that if the original image $\mathbf{F}$ is $N_1 \times N_2$ and the blurred image $\mathbf{G}$ is $M_1 \times M_2$, the column-degradation matrix $\mathbf{H}_C$ will be $M_1 \times N_1$, while the row-degradation matrix $\mathbf{H}_R$ will be $N_2 \times M_2$. The SVD must be performed on two matrices, one of size $M_1 \times N_1$, the other of size $N_2 \times M_2$. Compare this to the nonseparable case, where $\mathbf{F}$ is concatenated by rows or columns into an $N = N_1 \times N_2$ vector $\mathbf{f}$, $\mathbf{G}$ is an $M = M_1 \times M_2$ vector $\mathbf{g}$, and the degradation matrix $\mathbf{H}$ is $M \times N$. Performing the SVD on the $M \times N$ matrix $\mathbf{H}$ is computationally much more costly than performing two SVDs on the much smaller row and column matrices $\mathbf{H}_R$ and $\mathbf{H}_C$. In general, the column- and row-degradation operators $\mathbf{H}_C$ and $\mathbf{H}_R$ can be decomposed into the projection operators $\mathbf{P}_{C,R}$ and orthogonal projection operators $\mathbf{Q}_{C,R}$ as follows:

$$\mathbf{F} = (\mathbf{P}_C + \mathbf{Q}_C)\mathbf{F}(\mathbf{P}_R + \mathbf{Q}_R)^T. \tag{3.224}$$

It is easy to show that this equality holds by substituting $\mathbf{I} - \mathbf{Q}_{C,R}$ for $\mathbf{P}_{C,R}$,

$$\mathbf{F} = (\mathbf{I} - \mathbf{Q}_C + \mathbf{Q}_C)\mathbf{F}(\mathbf{I} - \mathbf{Q}_R + \mathbf{Q}_R)^T = \mathbf{F}. \tag{3.225}$$

Applying the same arguments as those used in the formulation of the algorithm for the nonseparable degradation operator, the iterative algorithm in (3.187) can be replaced by the following for the separable case:

$$\mathbf{F}^{(0)} = \mathbf{F}^\dagger,$$
$$\mathbf{F}^{(k+1)} = \mathbf{F}^\dagger + \{\mathbf{Q}_C C \mathbf{F}^{(k)} \mathbf{Q}_R^T + \mathbf{Q}_C C \mathbf{F}^{(k)} \mathbf{P}_R^T + \mathbf{P}_C C \mathbf{F}^{(k)} \mathbf{Q}_R^T\}. \tag{3.226}$$

The out-of-band term is represented by all the terms in the bracket. The first term in the bracket is very similar to the out-of-band term for the nonseparable case. However, two additional terms are found in the out-of-band term due to cross-multiplications of the projection operators. The Frobenius norm of a matrix is very similar to the Euclidean norm of a vector and is defined as

$$\|\mathbf{H}\|_2 = \sqrt{\left\{\sum_{i,j} h_{i,j}^2\right\}}. \tag{3.227}$$

The Frobenius norm of the out-of-band term given within the brackets in (3.226) is used as a stopping criterion in the same manner as the out-of-band energy for the nonseparable degradation operator. The algorithm for the separable degradation operator should be used whenever applicable due to the significant decrease in computational complexity and memory space required. The next chapter applies the concepts introduced in this chapter to some image-recovery problems.

REFERENCES

1. M. M. Sondhi. Image restoration: The removal of spatially invariant degradation. *Proc. IEEE*, 1972.

2. C. W. Helstrom. Image restoration by the method of least squares. *J. Opt. Soc. Am.*, **57**:297–303, 1967.

3. D. Slepian. Restoration of photographs blurred by image motion. *Bell Sys. Tech. J.*, **46**:2353–2362, 1967.

4. W. K. Pratt. *Digital Image Processing*. New York: Wiley, 1978.

5. N. D. A. Mascarenhas and W. K. Pratt. Digital image restoration under a regression model. *IEEE Trans. Circ. Sys.*, **22**:252–266, 1975.

6. S. Twomey. On the numerical solution of fredholm integral equations of the first kind by the inversion of the linear sustem produced by quadrature. *J. ACM*, **10**:97–101, 1963.

7. D. L. Phillips. A technique for the numerical solution of certain integral equations of the first kind. *J. ACM*, **9**:84–97, 1964.

8. A. N. Tikonov. Regularization of incorrectly posed problems. *Sov. Math.*, **4**:1624–1627, 1963.

9. S. Kacmarz. Angenaherte au flosung von systemen linearer gleichungen. *Bull. Acad. Polon. Sci. Lett.*, **A**:355–357, 1937.

10. A. Lent, G. T. Herman, and P. H. Lutz. Relaxation methods for image reconstruction. *Commun. ACM*, **21**:152–158, 1978.

11. A. Lent, G. T. Herman, and S. W. Rowland. Art: Mathematics and applications. *J. Theoret. Biol.*, **42**:1–32, 1973.

12. K. Tanabe. Projection method for solving a singular system of linear equations and its applications. *Numer. Math.*, **17**:203–214, 1971.

13. P. P. B. Eggermont. Iterative algorithms for large partitioned linear systems with applications to image reconstruction. *Linear Alg. Appl.*, **40**:37–67, 1981.

14. S. Kuo and R. J. Mammone. Resolution enhancement of tomographic images using the row action projection method. *IEEE Trans. Med. Imaging*, 1990.

15. B. Widrow and J. M. McCool. A comparison of adaptive algorithms based on the methods of steepest descent and random search. *IEEE Trans. Antennas Propag.*, **24**:615–637, 1976.

16. S. Agmon. The relaxation method for linear inequalities. *Canad. J. Math.*, **6**:382–392, 1954.

17. T. S. Motzkin and I. J. Schoenberg. The relaxation method for linear inequalities. *Canad. J. Math.*, **6**:393–404, 1954.

18. M. Minsky and S. Papert. *Perceptrons: An Introduction to Computational Geometry.* Cambridge, Mass.: MIT Press, 1969.

19. C. Hildreth. A quadratic programming procedure. *Naval Res. Log. Quart.*, **4**:79–85, 1957.

20. G. T. Herman and A. Lent. A family of iterative quadratic optimization algorithms for pairs of inequalities with applications in diagnostic radiology. *Math. Programming Stud.*, **9**:15–29, 1978.

21. R. Bender, R. Gordon, and G. T. Herman. Algebraic reconstruction techniques (art) for three-dimensional electron microscopy and x-ray photography. *J. Theoret. Biol.*, **29**:471–481, 1970.

22. O. N. Strand. Theory and methods related to the singular-function expansion and landweber's iteration for integral equations of the first kind. *SIAM J. Numer. Anal.*, **11**:798–825, 1974.

23. D. C. Youla. Generalized image retsoration by the method of alternating orthogonal projections. *IEEE Trans. Circ. Sys.*, **25**:694–702, 1978.

24. J. von Neumann. *The Geometry of Orthogonal Spaces.* Princeton, N.J.: Princeton University Press, 1950.

25. B. T. Polyak L. G. Gubin and E. V. Raik. The method of projections for finding the common point of convex sets. *USSR Comp. Math. Math. Phy.*, **7**:1–24, 1967.

26. R. W. Gerchberg and W. O. Saxton. A practical algorithm for the determination of phase from image and diffraction plane pictures. *Optik*, **35**:237–246, 1972.

27. R. W. Gerchberg. Super resolution through error energy reduction. *Opt. Acta*, **21**:709–720, 1974.

28. A. Papoulis. A new algorithm in spectral analysis and band-limited extrapolation. *IEEE Trans. Circ. Sys.*, **22**:735–742, 1975.

29. D. C. Youla and H. Webb. Image restoration by the method of convex projections: Part ii—theory. *IEEE Trans. Med. Imaging*, **1**:81–94, 1982.

30. G. M. Golub and C. Reinsch. Singular value decomposition and least squares solutions. *Numer. Math.*, **14**:403–420, 1970.

31. C. I. Podilchuk and R. J. Mammone. Row and block action projection techniques for image restoration. In *Signal Recovery and Synthesis III*, volume 15 of *1989 Technical Digest*, pp. 18–21. Cape Cod, Mass., Optical Society of America, North Falmouth, June 1989.

32. C. I. Podilchuk and R. J. Mammone. A comparison of projection techniques for image restoration. In Y.-w. Lin and R. Srinivasan, eds., *Digital Image Processing Applications*, volume 1075 of *Proceedings*, pages 303–310. SPIE, 1989.

33. R. J. Mammone and C. I. Podilchuk. Acceleration of the generalized projection onto convex sets (gpocs) algorithm. In *Applied Control, Filtering and Signal Processing*. Geneva, 1987.

34. C. I. Podilchuk and R. J. Mammone. Image recovery by convex projections using a least-squares constraint. *J. Opt. Soc. Am. A*, 1990.

35. M. H. Hayes V. T. Tom, T. F. Quatieri, and J. H. McClellan. Convergence of iterative nonexpansive signal reconstruction algorithms. *IEEE Acoust. Speech Signal Process.*, **29**:1052–1058, 1981.

4

IMAGE RECOVERY
USING ROW-ACTION
PROJECTION METHODS

Shyh-shiaw Kuo[1]

AT&T Bell Laboratories
600 Mountain Avenue
Murray Hill, New Jersey 07947
email: skuo@research.att.com

4.1 INTRODUCTION

A recorded image is generally a version of the original scene that is degraded by
the imperfections of any real imaging system. In some situations, the quality of
the imaging system can be improved only at very high cost (e.g., a astronomical
imaging system) or by physical requirements that are unacceptable and unrealizable (e.g., in medical imaging). On the other hand, in many cases the degradation
cannot be avoided and the image must be accepted as is (e.g., a blurred picture of
a unique event). For these reasons, the image-recovery problem has been widely
studied [1]. The applications of image-recovery techniques cover a broad range,
including space and astronomical research, consumer and commercial imaging,
documentary and forensic science, and medical imaging. Interest in image recovery is currently increasing due to the widespread use of imaging systems in
modern society. An image-degradation system can be modeled by Figure 4.1. The
image-recovery problem can be viewed as a process that attempts to recover an
image that has been degraded by a blur function $h(x, y; a, b)$ and by noise. For
a linear imaging system, the blurred image in Figure 4.1 can be expressed by a

---

[1] This research was done when Dr. Kuo was a Ph.D. candidate at Rutgers University.

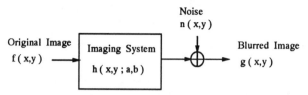

Figure 4.1 A model of an image-degradation system. The image-recovery problem is to find the estimate of the original, $f(x,y)$, given the blurred image $g(x,y)$ and the degradation operator $h(x,y; \alpha, \beta)$.

Fredholm integral equation of the first kind [2],

$$g(x,y) = \int_{-\infty}^{\infty} \int_{-\infty}^{\infty} h(x,y;a,b)f(a,b)\,da\,db + n(x,y) \qquad (4.1)$$

where x and y denote the image coordinates; $f(a,b)$ and $g(x,y)$ denote the original and observed blurred images, respectively; $n(x,y)$ denotes the additive noise function; and $h(x,y;a,b)$ denotes the response of the imaging system to an impulse at coordinate (a,b). $h(\cdot)$ is commonly called the point-spread function (PSF) because the impulse function represents a point of light in optics and $h(\cdot)$ is the response to a particular point. The superposition integral in (4.1) becomes a convolution integral when the kernel of (4.1), $h(\cdot)$, is shift invariant.

In this chapter, we address the discrete linear image-recovery problem. The discrete formulation of (4.1) can be obtained by replacing the continuous arguments f, g, h, and n by arrays of samples in two dimensions [2, Chap. 3]. For ease of representation, $N \times N$ rectangular lattices of equispaced samples will be used in this chapter. The discrete formulation can be written as

$$g(i,j) = \sum_{m} \sum_{n} h(i,j;m,n)f(m,n) + n(i,j). \qquad (4.2)$$

It is also convenient to write (4.2) as a system of linear equations [2] given by

$$\mathbf{g} = \mathbf{Hf} + \mathbf{n}, \qquad (4.3)$$

where $\mathbf{g}$, $\mathbf{f}$, and $\mathbf{n}$ are the lexicographic row-stacked vectors of the discretized versions of g, f, and n in (4.1), respectively. $\mathbf{H}$ is the degradation matrix composed of the PSF.

4.2 THE POSSIBILITIES AND DIFFICULTIES OF IMAGE RECOVERY

We demonstrate the possibilities and difficulties of restoring an image by considering a one-dimensional, linear, time-invariant imaging system:

$$g(x) = \int_{-T}^{T} f(\alpha)h(x - \alpha)\,d\alpha, \tag{4.4}$$

where $g(x)$ and $f(\alpha)$ are the observed and original images, respectively ($f(\alpha)$ is of finite support on the interval $(-T, T)$), and $h(x)$ denotes the PSF which corresponds to a diffraction-limited imaging system; i.e., the only limitation is that due to the finite size of the aperture. In this case, the PSF is given by

$$h(x) = \frac{\sin(x)}{x}. \tag{4.5}$$

Theoretically, it is possible to solve (4.4) in this special case by using the prolate spheroidal wavefunctions [3] to get the exact image $f(x)$. The eigenfunctions of the integral equation (4.4) with the kernel function given by (4.5) are the prolate spheroidal wavefunctions $\psi_n(x)$,

$$\int_{-T}^{T} \psi_n(\alpha)h(x - \alpha)\,d\alpha = \lambda_n\psi_n(x), n = 0, 1, 2, \ldots, \tag{4.6}$$

where the λ_n are the eigenvalues of (4.4).

The prolate spheroidal wavefunctions are complete orthogonal bases in both intervals, $(-\infty, \infty)$ and $(-T, T)$; i.e.,

$$\int_{-\infty}^{\infty} \psi_n(x)\psi_m(x)\,dx = \begin{cases} 1, & \text{if } n = m, \\ 0, & \text{if } n \neq m, \end{cases} \tag{4.7}$$

and

$$\int_{-T}^{T} \psi_n(x)\psi_m(x)\,dx = \begin{cases} \lambda_n, & \text{if } n = m, \\ 0, & \text{if } n \neq m. \end{cases} \tag{4.8}$$

Thus, the functions $g(x)$ and $f(x)$ can be represented as the series expansion,

$$g(x) = \sum_{n=0}^{\infty} c_n\psi_n(x), \tag{4.9}$$

$$f(x) = \sum_{n=0}^{\infty} d_n\psi_{Ln}(x), \tag{4.10}$$

where the $\psi_{Ln}(x)$ are the prolate spheroidal functions truncated to the interval $(-T, T)$. The coefficients are obtained by

$$c_n = \int_{-\infty}^{\infty} g(x)\psi_n(x)\,dx \tag{4.11}$$

and

$$d_n = \frac{1}{\lambda_n} \int_{-T}^{T} f(x)\psi_n(x)\,dx. \tag{4.12}$$

We substitute the series expansion of $g(x)$ and $f(x)$, (4.9) and (4.10), into the integral equation given by (4.4) to get

$$\sum_{n=0}^{\infty} c_n\psi_n(x) = \int_{-T}^{T} \left[\sum_{n=0}^{\infty} d_n\psi_{Ln}(\alpha)\right] h(x - \alpha)\,d\alpha \tag{4.13}$$

$$= \sum_{n=0}^{\infty} d_n \left[\int_{-T}^{T} \psi_n(\alpha)h(x - \alpha)\,d\alpha\right].$$

From (4.6) and (4.13), we have

$$\sum_{n=0}^{\infty} c_n\psi_n(x) = \sum_{n=0}^{\infty} \lambda_n d_n\psi_n(x). \tag{4.14}$$

Thus,

$$c_n = \lambda_n d_n, \tag{4.15}$$

or

$$d_n = \frac{c_n}{\lambda_n}. \tag{4.16}$$

Therefore, the exact solution of (4.4) can be expressed as

$$f(x) = \sum_{n=0}^{\infty} \frac{c_n}{\lambda_n}\psi_{Ln}(x). \tag{4.17}$$

The above expression demonstrates how the exact image $f(x)$ can be obtained from the diffraction-limited image $g(x)$. This requires the coefficients c_n, which are projections of the observed image $g(x)$ onto the set of prolate spheroidal wavefunctions ψ_n, the eigenfunctions of the integral equation (4.4). Similarly, we can analyze the discrete formulation of this problem by replacing the orthogonal prolate spheroidal wavefunctions with a set of orthonormal vectors.

However, the eigenvalues λ_n behave such that the first several eigenvalues are approximately equal to one but thereafter fall to zero very quickly. This fact makes (4.17) very sensitive to noise because of the division by small eigenvalues. The same problem occurs in the discrete imaging system given by (4.3). The unit-step behavior of the eigenvalues indicates that the degradation operator **H** is an ill-conditioned, or rank-deficient, matrix. If the direct inversion method is applied, the solution of the system is found to be unstable because it will be extremely sensitive to noise. Therefore, many methods have been proposed to seek a compromise between exact deblurring and noise amplification due to division by small eigenvalues. Methods such as least squares, Wiener filtering, pseudo-inverse filtering, and Kalman filtering have been used. The central idea of these methods is to find a solution that satisfies a predetermined optimality criterion. A different approach is to use regularized deconvolution methods. These methods optimize some measure of the error subject to constraints on the original image. Constraints can be obtained from *a priori* information about the degradation operator, the noise, and the original image itself. The most widely used regularization methods are usually associated with the names Tikhonov [4] and Miller [5]. Another family of image-recovery methods is based on restricting the space of feasible solutions, e.g., the method of projection onto convex sets (POCS). The new projection technique proposed in this chapter falls into this latter category.

The POCS methods have found many applications in the field of signal processing, such as bandlimited extrapolation, space-limited extrapolation, phase and magnitude retrieval [1, 6–11] image restoration [12–17], image reconstruction [18, 19], and adaptive filtering [20]. Note that the algorithm developed independently by Gerchberg and Saxton [9, 21] and by Papoulis [10] are also versions of the POCS method. In fact, many iterative signal-restoration techniques are specific examples of the POCS algorithm. For example, Tikhonov–Miller regularization can also be used within the POCS framework [16, 22, 23] since their solutions are selected from feasible solution sets that define convex sets. The POCS method is a very powerful general iterative algorithm for signal recovery. Youla [24] provides a number of convex projection operators that can be applied to many signal-recovery problems. Most of the commonly used constraints for different signal-processing applications fall in the category of convex sets. Although POCS has only been shown to provide weak convergence in theory, most of POCS algorithms provide strong convergence in practice.

This chapter presents an iterative *row-action projection* (RAP) technique based on the Kaczmarz [25] algorithm. We shall extend the use of the RAP algorithm for image restoration by recognizing that it is a subset of POCS. The first modification is to implement the RAP algorithm by forming local projection operators based on the 2-D image and the 2-D PSF directly, as opposed to the usual lexicographically stacked global approach. The resulting RAP algorithm can be implemented in parallel, with a significant decrease in processing time and can be extended to solve space-varying problems. The new RAP algorithm also facilitates the use of adaptive updates of the constraints. One such constraint is the minimum L_1-norm projection operator which projects onto the set of the minimum L_1-norm solutions. In addition, a local adaptive smoothness constraint based on the local average value is used. The use of such adaptive constraints has been found to significantly improve the performance of RAP.

4.3 THE ROW-ACTION PROJECTION (RAP) ALGORITHM

The RAP algorithm originally developed by Kaczmarz [25] forms a solution to a set of linear equations,

$$\mathbf{g} = \mathbf{Hf}, \tag{4.18}$$

by iterative orthogonal projections onto the hyperplanes specified by each equation. The method converges to the intersection of all the hyperplanes. The RAP update equation is given by

$$\mathbf{f}^{(k+1)} = \mathbf{f}^{(k)} + \lambda \frac{g_p - \mathbf{h}_p^T \mathbf{f}^{(k)}}{\|\mathbf{h}_p\|^2} \mathbf{h}_p = \mathbf{f}^{(k)} + \lambda \frac{\epsilon_p^{(k)}}{\|\mathbf{h}_p\|^2} \mathbf{h}_p, \tag{4.19}$$

where λ is the relaxation factor, g_p is the pth element of the vector $\mathbf{g}_{M \times 1}$, $\mathbf{h}_p^T$ is the pth row of the matrix $\mathbf{H}_{M \times M}$, $\epsilon_p^{(k)}$ is the residual error for the pth equation in (4.18) after k iterations, and $\| \cdot \|$ is the L_2 norm of a vector, defined as $\|X\|^2 = \sum_{i=1}^{N} x_i^2$. The iteration index is related to the equation index by $p = k(\bmod M)$, indicating that each row is used multiple times in the estimation process. The Kaczmarz algorithm can be interpreted geometrically as Figure 4.2. In the kth iteration, the pth equation of (4.18) determines the hyperplane $HP_p = \{\mathbf{f} \in R^n \mid \langle \mathbf{h}_p, \mathbf{f} \rangle = g_p\}$, where $\langle \cdot, \cdot \rangle$ denotes the standard inner product between two vectors. For unity relaxation, $\lambda = 1$, $f^{(k+1)}$ is the orthogonal projection of $f^{(k)}$ onto HP_p. Since $\langle h_p, f^{(k+1)} \rangle = g_p$, the distance between $f^{(k)}$ and $f^{(k+1)}$ can be expressed as the inner product,

$$\left\langle \left(\frac{h_p}{\|h_p\|} \right), (f^{(k+1)} - f^{(k)}) \right\rangle = \frac{\langle h_p, (f^{(k+1)} - f^{(k)}) \rangle}{\|h_p\|}$$

$$= \frac{g_p - \langle h_p, f^{(k)} \rangle}{\|h_p\|}, \tag{4.20}$$

where $h_p/\|h_p\|$ is the unit normal vector of the hyperplane HP_p. Then $f^{(k+1)}$ is the result of $f^{(k)}$ after moving along the normal direction of the hyperplane a distance given by (4.20); i.e.,

$$f^{(k+1)} = f^{(k)} + \frac{g_p - \langle h_p, f^{(k)} \rangle}{\|h_p\|} \times \frac{h_p}{\|h_p\|}, \tag{4.21}$$

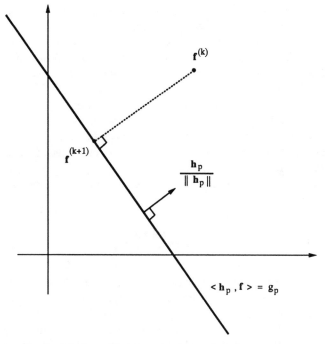

Figure 4.2 Geometrical interpretation of Kaczmarz's method.

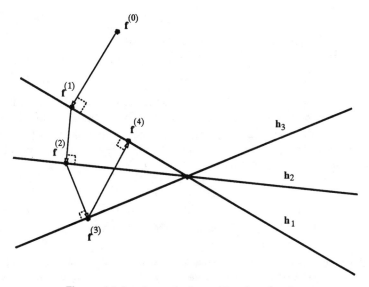

Figure 4.3 Iterative projections with unity relaxation.

which is the same as (4.19) for unity relaxation. Figure 4.3 illustrates the iterative projections onto the hyperplanes that represent the set of system equations (4.18). Successive projections converge to the intersection of these equations. The relaxation parameter, λ, actually allows $f^{(k+1)}$ to be a point anywhere along the line segment connecting $f^{(k)}$ and its orthogonal reflection with respect to the hyperplane.

Kaczmarz [25] has shown that the iterative process given by (4.19) converges to the solution of (4.18) when $\mathbf{H}$ is a square nonsingular matrix—i.e., $\mathbf{H}$ has an inverse—and the relaxation parameter is unity. Proof of the convergence of RAP for the consistent [26] and inconsistent [27] cases has also been demonstrated. The conventional Kaczmarz algorithm expressed by (4.19) converges to the pseudo-inverse solution when the initial estimate $\mathbf{f}^{(0)}$ is in the range space of $\mathbf{H}^T$ [28], as given by

$$\lim_{k \to \infty} \mathbf{f}^{(k)} = \mathbf{H}^\dagger \mathbf{g}, \tag{4.22}$$

where $\mathbf{H}^\dagger$ denotes the pseudo-inverse of $\mathbf{H}$. It has been shown that this iterative approach of obtaining the pseudo-inverse is computationally more attractive than alternative methods [17].

The RAP algorithm is a form of the POCS method, since each hyperplane is a linear manifold that is closed and convex. Therefore, the RAP algorithm can be generalized to include nonlinear constraints in each iteration and (4.19) can be rewritten as a projection operator of POCS,

$$P_{C_R}\mathbf{f} = \mathbf{P}_C\mathbf{f} + \lambda \frac{g_p - \mathbf{h}_p^T(\mathbf{P}_C\mathbf{f})}{\|\mathbf{h}_p\|^2}\mathbf{h}_p, \tag{4.23}$$

where $\mathbf{P}_C = P_{C_m}P_{C_{m-1}} \cdots P_{C_1}$ is the concatenated projection operator which describes any *a priori* constraints. This generalized row-action method is an enhancement of the conventional Kaczmarz algorithm expressed by (4.19), because the Kaczmarz algorithm converges to the pseudo-inverse solution, which does not necessarily satisfy the constraints.

4.4   TWO-DIMENSIONAL FORMULATION OF RAP

Imaging systems are generally designed so that the degradation matrix $\mathbf{H}$ in (4.3) is sparse and compact. The sparseness of the matrix $\mathbf{H}$ is due to the fact that the size of the PSF is generally much smaller than the size of the image. This is desirable, in that an ideal imaging system should provide a point-to-point mapping. This characteristic is typical of well-designed imaging systems. In addition, $\mathbf{H}$ is a block Toeplitz matrix in the shift-invariant case and (4.2) will become a 2-D linear convolution given by

$$g(i,j) = \sum_m \sum_n h(i-m, j-n) f(m,n). \tag{4.24}$$

The sparseness of the matrix $\mathbf{H}$ is due to the fact that the size of the PSF is generally much smaller than the size of the image.

The RAP algorithm given by (4.19) can be implemented by considering only a subregion of the 2-D image $\hat{\mathbf{f}}$ that is determined by the size of the 2-D support of the PSF. This can be seen in the following way. Consider the case where the support of the PSF matrix is a 3×3 window, as indicated by the solid outline in Figure 4.4(a). In this case, every row of the matrix $\mathbf{H}$ in (4.18) contains only 9 nonzero elements. Each pixel $g(i,j)$ of the blurred image $\mathbf{g}$ corresponds to a specific equation of the set given by (4.18). The RAP algorithm, (4.19), requires only those 9 pixel values in $\hat{f}$ which correspond to the 9 nonzero elements in $\mathbf{h}_p$ that cover the support of the PSF. Hence, the 2-D formulation of the RAP algorithm can be written as

$$\hat{\mathbf{f}}^{(k+1)} = \begin{cases} \hat{f}^{(k)}(m,n) + \\ \qquad \lambda \dfrac{\epsilon(i,j)}{\|\mathbf{h}(i,j)\|^2} h(i-m, j-n; i,j), & \text{if } \hat{f}^{(k)}(m,n) \in S_{h(i,j)}, \\ \hat{f}^{(k)}(m,n), & \text{otherwise,} \end{cases} \tag{4.25}$$

where

$$\epsilon(i,j) = g(i,j) - \sum_{m,n \in S_{h(i,j)}} h(i-m, j-n; i,j) \hat{f}^{(k)}(m,n),$$

$$\|\mathbf{h}(i,j)\|^2 = \sum_{m,n \in S_{h(i,j)}} h(m,n; i,j)^2, \tag{4.26}$$

and $S_{h(i,j)}$ is the support of the PSF centered at pixel (i,j). For example, $S_{h(i,j)}$ is a 3×3 window for the case shown in Figure 4.4. The equation selected next for the projection operation can be obtained by shifting the PSF matrix one pixel (down, for example) as indicated in Figure 4.4(b). This shift corresponds to a specific ordering (cycling strategy) of the projections onto the equations given in (4.18). Hence, the RAP algorithm can be implemented in a manner similar to a 2-D convolution. That is, each projection operator is local, requiring only the neighborhood $S_{h(i,j)}$ of the image $\hat{\mathbf{f}}$, at each iteration. In addition, the next local operator can be obtained by simply shifting the window by one pixel. The computational savings for a sparse degradation matrix $\mathbf{H}$ is apparent for this convolution-type implementation of RAP. The 2-D implementation of the RAP operators provides the advantage of efficient parallel implementation and facilitates the use both of local adaptive constraints imposed on the 2-D

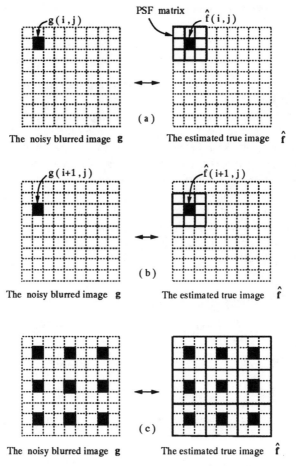

Figure 4.4 A 2-D implementation of the RAP algorithm: (a) a projection onto one of the equations in the set $\mathbf{g} = \mathbf{Hf}$, (b) a projection onto the next equation in the set of equations $\mathbf{g} = \mathbf{Hf}$, and (c) a parallel implementation of RAP.

neighborhood $S_{h(i,j)}$ and of cycling strategies based on 2-D properties of the image, such as edges and textures.

The RAP method can be implemented in parallel by separating the image into nonoverlapping regions determined by $S_{h(i,j)}$, as depicted in Figure 4.4(c). Each region can be processed by a separate processor. The image can be shifted through the processors to effect a complete cycle of the RAP algorithm after $\mathcal{L}$ shifts, where $\mathcal{L}$ is the total number of pixels in the PSF matrix. The computation time for the parallel implementation requires only $O(\mathcal{L})$ multiplications and additions. However, the computation time for the conventional RAP algorithm requires $O(\mathcal{N}\mathcal{L})$ multiplications and additions [17], where $\mathcal{N}$ is the total number of pixels in the image. In addition, this parallel implementation is not possible for the Gerchberg–Papoulis POCS algorithms [9, 13, 16, 18, 29], which require global operations, such as 2-D FFT, to be performed on the entire image. The multiply-add complexity of 2-D FFT is approximately $O(\mathcal{N})$. Thus, the processing time of the new method is approximately $\mathcal{N}/\mathcal{L}$ times faster than the methods that use 2-D FFT operations. In addition, the new method is able to handle shift-variant degradation systems, while the FFT-based methods cannot.

The 2-D implementation of RAP can readily be extended to include local adaptive constraints based on 2-D features similar to those given by Sezan and Tekalp [16], such as local variance bounds. However, unlike the algorithm proposed in [16], the algorithm presented here applies to shift-variant degradation systems. The 2-D implementation of RAP also facilitates the use of new cycling strategies. These extensions will be described in the following sections. In addition, we shall introduce the new constraint set of the minimum L_1-norm solutions and also a novel noise-reduction technique.

4.5 SMOOTHNESS CONSTRAINT AND CYCLING

Since RAP makes use of local projection operators, the update strategy can be made adaptive based on local 2-D features of the image. Sezan and Tekalp [16] have shown that scene-dependent adaptive processes can be facilitated by segmenting the image into regions of comparable gray levels according to the local variance. Uniform-, texture-, and edge-like regions—denoted by R_1, R_2, and R_3, respectively—are defined as

$$R_1 = \left\{ (i,j) \mid 0 \le \sigma_g(i,j) \le T_a \right\},$$
$$R_2 = \left\{ (i,j) \mid T_a \le \sigma_g(i,j) \le T_b \right\},$$
$$R_3 = \left\{ (i,j) \mid T_b \le \sigma_g(i,j) \right\}, \tag{4.27}$$

where $\sigma_g(i,j)$ denotes the local variance of the blurred image $g(i,j)$. This can be computed over some rectangular window of pixels centered at (i,j). The thresholds T_a and T_b are determined experimentally [16].

There are two possible ways to apply the smoothness constraint to the semi-uniform region R_1. For a fairly smooth R_1, we simply apply the neighborhood averaging,

$$f(i,j) = \begin{cases} \bar{f}(i,j), & \text{if } f(i,j) \in R_1, \\ f(i,j), & \text{otherwise}, \end{cases} \tag{4.28}$$

where

$$\bar{f}(i,j) = \frac{1}{N} \sum_{(m,n) \in S} f(m,n), \tag{4.29}$$

S is some rectangular window of pixels centered at (i, j), and N is the total number of pixels of $f(m, n)$ in that window. This is the simplest smoothing method in terms of computational complexity. Neighborhood averaging shall be shown to be a nonexpansive mapping in the following.

The smoothness operation described in (4.28) and (4.29) can be rewritten in matrix form,

$$\bar{\mathbf{f}} = A\mathbf{f}, \tag{4.30}$$

where $\bar{\mathbf{f}}$ and $\mathbf{f}$ are stacked vectors composed of all the pixels in R_1, and each row and column in the matrix A has at most N nonzero elements all equal to $1/N$. From the definition of the norm of a matrix in linear algebra [30, p. 266], the norm of A is smaller than unity. For example, the L_2 norm of a matrix is defined as

$$\|A\|_2 = (\text{maximum eigenvalue of } A^T A)^{1/2} = \lambda_{\max}^{1/2} \leq 1. \tag{4.31}$$

It can easily be shown from the structure of A that each row of $A^T A$ has at most N nonzero elements and that their largest possible value is $1/N$. According to the Gerschgorin circles theorem [30, p. 317],

$$\lambda_{\max} \leq \frac{1}{N} + \sum_{1}^{N-1} \frac{1}{N} = 1, \tag{4.32}$$

i.e., $\|A\|_2 \leq 1$. Therefore, we can show that neighborhood averaging is a nonexpansive mapping [1, Def. 2.3-3] as follows:

If $A: \mathfrak{I} \rightarrow \mathfrak{I}$ denotes the mapping described in (4.30), then

$$\|Ax - Ay\| = \|A(x - y)\| \leq \|A\|\|x - y\| \leq \|x - y\|, \qquad \text{for all } x, y \in \mathfrak{I}. \tag{4.33}$$

Thus, since it is a nonexpansive mapping, the neighborhood averaging operator can be used in cooperation with POCS [1, Chap. 2].

There is one equation in the set given by $\mathbf{g} = H\mathbf{f}$ for each pixel in the blurred image $\mathbf{g}$. The selection of which equation to project onto next is called the cycling strategy. One example of a cycling strategy is when projections are made only along the edge-like region of an image via RAP. Thus, the edges can be deblurred as much as possible. This can be expressed as a projection operator P_{C_e} of POCS:

$$P_{C_e}\mathbf{f} = \begin{cases} \mathbf{f} + \lambda \dfrac{g(i,j) - \mathbf{h}^T\mathbf{f}}{\|\mathbf{h}\|^2}\mathbf{h}, & \forall\, g(i,j) \in R_3, \\[2mm] \mathbf{f}, & \text{otherwise,} \end{cases} \tag{4.34}$$

where $\mathbf{h}$ denotes the normal vector of the hyperplane which associates with the (i,j)th pixel of the blurred image g. Similarly, the projection operator of which projections are made along texture- and edge-like regions can be defined as

$$
P_{C_i}\mathbf{f} = \begin{cases} \mathbf{f} + \lambda \dfrac{g(i,j) - \mathbf{d}^T\mathbf{f}}{\|\mathbf{d}\|^2}\mathbf{d}, & \forall\ g(i,j) \in R_2 \text{ or } R_3, \\[2mm] \mathbf{f}, & \text{otherwise.} \end{cases} \tag{4.35}
$$

4.6 PROJECTION ONTO THE SET OF MINIMUM L_1-NORM SOLUTIONS

The RAP algorithm yields the least squares solution, or the minimum L_2-norm solution in the linear unconstrained case. However, there is no such optimality condition when inconsistent or nonlinear constraints are used. In this section, we show that it is possible to find the minimum L_1-norm solution via the RAP algorithm when both nonlinear and linear constraints are imposed. It should be mentioned that the different minimization criterion will result in slightly different statistical properties of the residual [1, Chap. 4]. The minimum L_1-norm formulation of the problem is

$$
\text{minimize} \quad \sum_{i=1}^{M} |r_i| \tag{4.36}
$$

$$
\text{subject to} \quad \mathbf{g} = \mathbf{Hf} + \mathbf{r}, \tag{4.37}
$$

where $r_i = g_i - \mathbf{h}_i\mathbf{f}$ is the residual of the ith equation. According to the current estimate $\hat{\mathbf{f}}$, we can express each residual of the estimate as

$$
|r_1| = |g_1 - h_{1,1}\hat{f}_1 - h_{1,2}\hat{f}_2 - \cdots - h_{1,j}\hat{f}_j - \cdots - h_{1,M}\hat{f}_M|,
$$
$$
|r_2| = |g_2 - h_{2,1}\hat{f}_1 - h_{2,2}\hat{f}_2 - \cdots - h_{2,j}\hat{f}_j - \cdots - h_{2,M}\hat{f}_M|,
$$
$$
\vdots
$$
$$
|r_i| = |g_i - h_{i,1}\hat{f}_1 - h_{i,2}\hat{f}_2 - \cdots - h_{i,j}\hat{f}_j - \cdots - h_{i,M}\hat{f}_M|,
$$
$$
\vdots
$$
$$
|r_M| = |g_M - h_{M,1}\hat{f}_1 - h_{M,2}\hat{f}_2 - \cdots - h_{M,j}\hat{f}_j - \cdots - h_{M,M}\hat{f}_M|,
$$

where $h_{i,j}$ indicates the element of $\mathbf{H}$ at location (i,j). The absolute-value operator can be removed by checking the value of each residual. If it is negative, then that equation is multiplied by -1. Otherwise it is left unchanged. The L_1 norm of the error can now be calculated by summing the above expressions to obtain

$$C = \sum_{i=1}^{M} |r_i| = \tilde{g} - \tilde{h}_1 \hat{f}_1 - \tilde{h}_2 \hat{f}_2 - \cdots - \tilde{h}_j \hat{f}_j - \cdots - \tilde{h}_M \hat{f}_M$$

$$= \tilde{g} - \tilde{\mathbf{h}}^T \hat{\mathbf{f}},$$

where $\tilde{g}$ and $\tilde{h}_j$ are the summations over the sign-corrected values of g_i and the jth column of $h_{i,j}$, respectively, and $\tilde{\mathbf{h}}$ is the vector which consists of the $\tilde{h}_j$. The above equation can be restated as

$$\tilde{g} - C = \tilde{\mathbf{h}}^T \hat{\mathbf{f}}. \tag{4.38}$$

Since this is a linear equation, we can immediately define a projection operator onto the set that satisfies (4.38). Note that the particular estimate changes with time. This is similar to the use of a piecewise linear approximation of the true cost function, as is frequently used in convex programming [31]. The L_1 norm of the error will converge to zero for a noise-free environment. The current L_1 norm of the error, C, may not be zero for the current estimate $\hat{\mathbf{f}}$. Hence, we set C to zero at that time and project onto this new L_1-norm hyperplane in order to reduce the current L_1-norm value. The L_1-norm projection is performed alternately with the RAP algorithm. This is shown in Figure 4.5. The procedure can be repeated until the value of the L_1 norm converges to its expected value. The expected value of the L_1 norm will not converge to zero when noise is present. Consider the case when the noise is zero-mean Gaussian with standard deviation σ; the expected value for the absolute value of the error can then easily

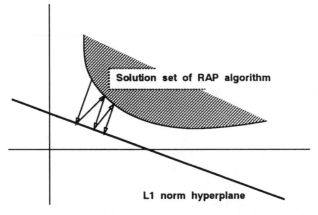

Figure 4.5 Alternating projections between the L_1-norm hyperplane and the solution set of the RAP algorithm.

be found by [32]:

$$E(|n_i|) = \sqrt{\frac{2}{\pi}}\sigma. \tag{4.39}$$

The expected value of the L_1 norm of the error is

$$E\left(\sum_{i=1}^{M}|n_i|\right) = M\sqrt{\frac{2}{\pi}}\sigma, \tag{4.40}$$

where M is the total number of random variables. The seminorm of the residual is defined by

$$\sum_{i=1}^{M}|r_i| = \sum_{i=1}^{M}|g_i - \langle \mathbf{h}_i, \hat{\mathbf{f}} \rangle|. \tag{4.41}$$

Note that a seminorm is similar to a norm except that $|g_i - \langle \mathbf{h}_i, \hat{\mathbf{f}} \rangle| = 0$ does not necessarily imply that $\hat{\mathbf{f}}$ is equal to its true value $\mathbf{f}$. Thus, in a noisy environment, the constant C in (4.38) should be set to its expected value as calculated by (4.40). The process can be terminated when the seminorm of the residual is equal to the expected value of the L_1 norm of the error. This stopping criterion is based on the desire not to *overfit* the data.

4.7   NOISE-REDUCTION PROJECTION OPERATOR

For a rank-deficient system of equations, the vector $\mathbf{f}$ can be decomposed into two components [13, 14]:

$$\mathbf{f} = P\mathbf{f} + Q\mathbf{f} = \mathbf{f}^\dagger + Q\mathbf{f}, \tag{4.42}$$

where $P = \mathbf{H}^\dagger \mathbf{H}$ and $Q = I - P$ are the range and null-space projection operators of $\mathbf{H}$, respectively. For an ill-conditioned problem, a suitable pseudo-rank can be defined [33] to transform the problem into a rank-deficient system. The null space of $\mathbf{H}$ can be precisely defined as

$$\{\mathbf{x} \mid \mathbf{H}\mathbf{x} = \mathbf{0}\}. \tag{4.43}$$

According to (4.42), we can decompose the noisy blurred image $\mathbf{g}$ as

$$\mathbf{g} = \mathbf{g}_{\text{blur}} + \mathbf{n}_{\text{white}}$$

$$= \mathbf{g}_{\text{blur}} + P\mathbf{n}_{\text{white}} + Q\mathbf{n}_{\text{white}},$$

where $\mathbf{g}_{blur}$ = $\mathbf{Hf}$ and $\mathbf{n}_{white}$ denotes the white Gaussian noise. Since $\mathbf{g}_{blur}$ only contains the range-space information, the noise in the null space, $Q\mathbf{n}_{white}$, is irrelevant [34, Chap. 4]. Thus the noisy blurred image $\mathbf{g}$ can be preprocessed by projecting onto the range space of $\mathbf{H}$; i.e.,

$$\mathbf{g}_{range} = P\mathbf{g}. \tag{4.44}$$

For the image-restoration problem we proposed in this chapter, the null-space consists of the high-frequency component. When $\mathbf{H}$ is cyclic, it can be decomposed by using discrete Fourier transform (DFT). A Toeplitz matrix $\mathbf{H}$ can be augmented to be similar to a cyclic matrix [2]. Thus, for a shift-invariant $\mathbf{H}$, the noise-reduction step can be implemented by using the FFT to form the ideal low-pass filtered version of $\mathbf{g}$,

$$\mathbf{g}_{range} = \text{IFFT}[G^{\ddagger}], \tag{4.45}$$

where $G^{\ddagger}$ is the FFT of $\mathbf{g}$ modified by setting the high-frequency components to zero.

4.8 NUMERICAL RESULTS

The original test image is the 256×256 image shown in Figure 4.6. The degradation operator is a two-dimensional digital FIR low-pass filter with a symmetric separable 11×11 kernel constructed from the outer vector product of a one-dimensional low-pass filter that taps with itself. The taps of the one-dimensional low-pass filter are computed by using the program provided by McClellan and Parks [35]. In order to eliminate the high spatial frequency information effectively, a low-pass filter with more than 43 dB stopband attenuation over 50% of the usable bandwidth is specified. Consequently, the resulting two-dimensional

Figure 4.6 The original image.

degradation PSF exhibits more than 86 dB stopband attenuation in the spatial frequency domain, as shown in Figure 4.7. The original image has been convolved with this 11×11 PSF. The resulting blurred image shows a loss of 75% of the high-frequency information. The performance of the new algorithm is indicated by the relative error, ϵ, given by

$$\epsilon = 100 \times \frac{\|\mathbf{f} - \hat{\mathbf{f}}\|}{\|\mathbf{f}\|}. \tag{4.46}$$

We present the following two examples to illustrate the performance of the new algorithm.

Experiment I: Noise-Dominant Environment

White Gaussian noise is added to the blurred image to obtain an SNR of 25 dB. The relative error of the resulting image shown in Figure 4.8(a) is 23.4%. The best restored image found by using the conventional Kaczmarz algorithm, (4.25), is shown in Figure 4.8(b). The relative error of this image is as high as 9.8%. However, the relative error is reduced to 7.2% when the new algorithm is used. The new method is implemented by the following steps:

1. Eliminate the irrelevant noise by projecting the noisy blurred image $\mathbf{g}$ onto the range space of $\mathbf{H}$. The relative error of the postprocessed image $\mathbf{g}_{range}$ is still as high as 22.7%.
2. Set the initial estimate $\mathbf{f}^{(0)} = \mathbf{g}_{range}$.

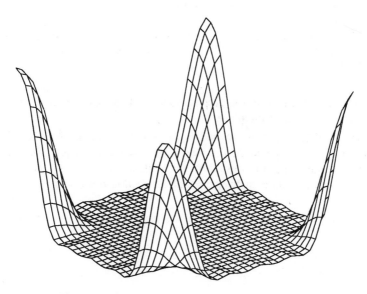

Figure 4.7 The spatial frequency response of the PSF.

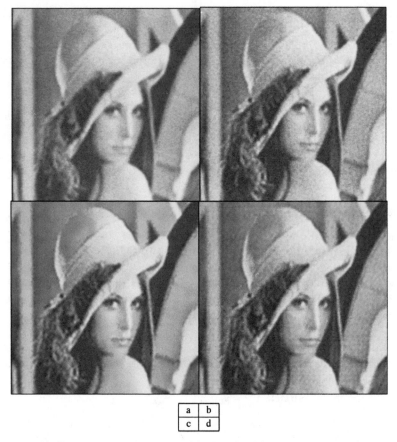

a	b
c	d

Figure 4.8 (a) The noisy blurred image at 25 dB SNR. (b) The best restored image of the regular RAP algorithm. (c) The restored image of the new algorithm. (d) The restored image of the PWAS algorithm.

3. Form the new L_1-norm hyperplane and find the projection of the current estimate onto the hyperplane.

4. Use the RAP algorithm to update the estimated image $\hat{\mathbf{f}}$. The relaxation factor λ is selected to be 0.1 due to the high noise level. The minimum and maximum bound constraints are applied after each projection. These constraints are imposed by the POCS projection operator given by

$$
P_{C_b}f = \begin{cases} f_{\max}, & \text{if } f(i,j) > f_{\max}, \\ f_{\min}, & \text{if } f(i,j) < f_{\min}, \\ f(i,j) & \text{otherwise}, \end{cases} \tag{4.47}
$$

5. Apply the smoothness constraint described in (4.28).

6. Go to step 3.

The final restored image after 3 iterations is shown in Figure 4.8(c). Figures 4.9(a), 4.9(b), and 4.9(c) show the spatial frequency response of the original, blurred, and restored images, respectively. It is clear that most of the high-frequency information lost in the blurred image, Figure 4.9(b), has been restored. Figure 4.8(d) illustrates the restored image obtained by using the partial Wiener solution with adaptive smoothness constraints (PWAS) [16], which is also a subset of the POCS method. The relative error obtained by the PWAS method is 8.5%, which is larger than the 7.2% obtained by the new algorithm. Also, the processing time of the new method's parallel implementation would be approximately 500 ($\approx 256^2/11^2$) times faster than that of the PWAS algorithm.

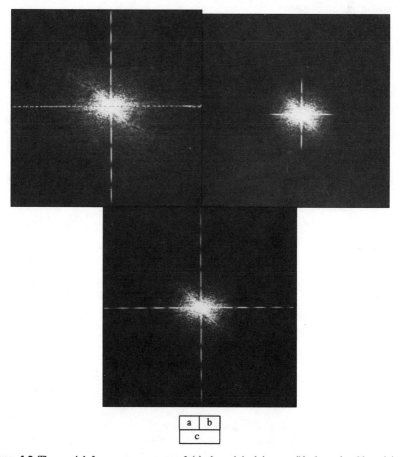

a	b
c	

Figure 4.9 The spatial frequency response of (a) the original image, (b) the noisy blurred image at 25 dB, and (c) the restored image of the new algorithm.

TABLE 4.1 The relative errors of the high noise environment.

Images	Blurred	Conventional RAP	PWAS Algorithm	New Algorithm
Errors	23.4%	9.8%	8.5%	7.2%

The relative error of the blurred and restored images are summarized in Table 4.1.

Experiment II: Low-Noise Environment

White Gaussian noise is added to the blurred image to obtain an SNR of 50 dB. The resulting image is shown in Figure 4.10(a). Since the noise level is low, there is no need for the noise-reduction step. The smoothness constraint is also weak in this situation. Thus, these two steps are not used in the high-SNR case in order

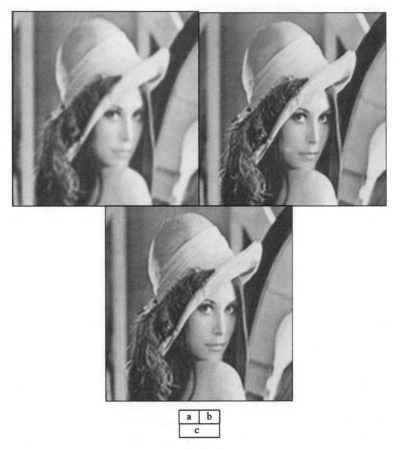

Figure 4.10 (a) The noisy blurred image at 50 dB. (b) The restored image of the new algorithm. (c) The restored image of the PWAS algorithm.

Figure 4.11 The restored image (a) with and (b) without the L_1-norm projection operator.

to reduce the computational complexity. Figure 4.10(b) illustrates the restored image. It is clear that the new algorithm restores the high-frequency information successfully. Figure 4.10(c) illustrates the restored image obtained by the PWAS algorithm. The relative error of the noisy blurred image, the restored image of PWAS algorithm, and the restored image of the new algorithm are 22.5%, 6.1%, and 5.6%, respectively. Even though the relative error of the image restored by the new algorithm is only 0.5% better than that obtained by PWAS algorithm, the new algorithm has a significant improvement (approximately 500 times faster) in processing time. The L_1-norm projection operator accelerates the first few iterations significantly. Figures 4.11(a) and (b) are the restored images with and without the minimum L_1 norm projection operator after one iteration. The relative errors of Figure 4.11 (a) and (b) are 7.5% and 12.3%, respectively. The relative error of the blurred and restored images are summarized in Table 4.2.

4.9 SUMMARY AND CONCLUSION

A new projection method of image restoration has been introduced with a number of practical advantages. The new algorithm does not require the storage of the entire degradation matrix **H** due to the use of 2-D local projection operators. This reduces not only the memory space requirements but also the processing time, since it can be implemented in parallel. The RAP algorithm is shown

TABLE 4.2 The relative errors of the low-noise environment.

Images	Blurred	PWAS Algorithm	New Algorithm
Errors	22.5%	6.1%	5.6%

to be a subset of the POCS method and therefore allows the addition of convex constraints to reduce ambiguity in the solution and accelerate the speed of convergence. Since POCS is also a subset of nonexpansive mappings, the new algorithm can be incorporated with other nonexpansive mapping strategies, such as the smoothness constraint [36, 37]. Because the new algorithm is implemented by local operations, it provides the ability to deconvolve shift-variant degradation operators and also allows the use of local adaptive constraints, as well as cycling strategies. In addition, the restored image can be monitored during the process because the new algorithm is an iterative process. The new algorithm can also be incorporated with the expectation maximization (EM) algorithm to restore a degraded noisy image when the PSF is unknown [38].

ACKNOWLEDGMENTS

This research reported here was made possible through the support of the New Jersey Commission on Science and Technology and the Center for Computer Aids for Industrial Productivity at Rutgers University.

REFERENCES

1. H. Stark, editor. *Image Recovery, Theory and Application*. Orlando, Fla.: Academic Press, 1987.

2. H. C. Andrews and B. R. Hunt. *Digital Image Restoration*. Englewood Cliffs, N.J.: Prentice-Hall, 1977.

3. D. Slepian and H. O. Pollak. Prolate spheroidal wave functions, Fourier analysis and uncertainty—I. *Bell Sys. Tech. J.*, **40**:43–63, 1961.

4. A. N. Tikhonov and V. Y. Arsenin. *Solutions of Ill-Posed Problems*. New York: Wiley, 1977.

5. K. Miller. Least squares methods for ill-posed problems with a prescribed bound. *SIAM J. Math. Anal.*, **1**:52–74, 1970.

6. R. W. Schafer, R. M. Mersereau, and M. A. Richards. Constrained iterative signal restoration algorithms. *Proc. IEEE*, **69**:432–450, 1981.

7. V. T. Tom, T. F. Quatieri, M. H. Hayes, and J. H. McClellan. Convergence of iterative non-expansive signal restoration algorithms. *IEEE Trans. Acoust. Speech Signal Process.*, **ASSP-29**:1052–1058, 1981.

8. A. Levi and H. Stark. Image restoration by the method of generalized projections with applications to restoration from magnitude. *J. Opt. Soc. Am.*, **74**:932–943, 1984.

9. R. W. Gerchberg. Super-resolution through error reduction. *Opt. Acta*, **21**:709–720, 1974.

10. A. Papoulis. A new algorithm in spectral analysis and bandlimited extrapolation. *IEEE Trans. Circ. Sys.*, **CAS-22**:735–742, 1975.

11. M. H. Hayes, J. S. Lim, and A. V. Oppenheim. Signal reconstruction from phase or magnitude. *IEEE Trans. Acoust. Speech Signal Process.*, **ASSP-28**:672–680, 1980.

12. C. I. Podilchuk and R. J. Mammone. Image recovery by convex projections using a least squares constraint. *J. Opt. Soc. Am.*, **7**:517–521, March 1990.

13. R. J. Mammone and R. J. Rothacker. General iterative method of restoring linearly degraded images. *J. Opt. Soc. Am.*, **A4**:208–215, 1987.

14. C. I. Podilchuk and R. J. Mammone. Step size for the general iterative image recovery algorithm. *Opt. Eng.*, **27**(9):806–811, 1988.

15. H. J. Trussel and M. R. Civanlar. Feasible solution in signal restoration. *IEEE Trans. Acoust. Speech Signal Process.*, **ASSP-32**(2):201–212, 1984.

16. M. I. Sezan and A. M. Tekalp. Adaptive image restoration with artifact suppression using the theory of convex projections. *IEEE Trans. Acoust. Speech Signal Process.*, **ASSP-38**(1):181–185, 1990.

17. T. S. Huang. Restoring images with shift-varying degradation. In J. C. Urbach, ed., *Proceedings of SPIE/OSA, Image Processing*, pp. 149–151, 1976.

18. M. I. Sezan and H. Stark. Tomographic image reconstruction from incomplete view data by convex projections and Fourier inversion. *IEEE Trans. Med. Imaging*, **3**:91–98, 1984.

19. G. T. Herman. *Image Reconstruction from Projections*. New York: Academic Press, 1980.

20. J. F. Doherty and R. J. Mammone. A row-action projection algorithm for adaptive filtering. *Proceedings of the IEEE International Conference on Acoustics, Speech, and Signal Processing*, 1990.

21. R. W. Gerchberg and W. O. Saxton. A practical algorithm for the determination of phase from image and diffraction plane pictures. *Optik*, **35**:237–246, 1972.

22. M. I. Sezan, A. M. Tekalp, and C. T. Chen. Regularized signal restoration using the theory of convex projections. *Proceedings of the IEEE International Conference on Acoustics, Speech, and Signal Processing*, pp. 1565–1568, 1987.

23. M. I. Sezan and A. M. Tekalp. Iterative image restoration with ringing suppression using pocs. *Proceedings of the IEEE International Conference on Acoustics, Speech, and Signal Processing*, pp. 1300–1303, 1988.

24. D. C. Youla and H. Webb. Image restoration by the method of convex projections: Part I—theory. *IEEE Trans. Med. Imaging*, **MI-1**(1):81–94, 1982.

25. S. Kaczmarz. Angenaherta auflosung von systemen linearer gleichungen. *Bull. Acad. Polon. Sci. Lett.*, **A35**:355–357, 1937.

26. G. T. Herman, A. Lent, and P. Lutz. Relaxation methods for image reconstruction. *Commun. ACM*, **21**:152–158, 1978.

27. G. T. Herman, A. Lent, and S. Rowland. Art: Mathematics and applications. *J. Theoret. Biol.*, **42**:1–32, 1973.

28. P. Eggermont, G. Herman, and A. Lent. Iterative algorithm for large partitioned linear systems with applications to image reconstruction. *Linear Alg. Appl.*, **40**, 1981.

29. A. Papoulis. A new algorithm in spectral analysis and band-limited extrapolation. *IEEE Trans. Circ. Sys.*, **CAS-22**(9):735–741, Sept. 1975.

30. Ben Noble and James W. Daniel. *Applied Linear Algebra*, 3rd ed. Englewood Cliffs, N.J.: Prentice-Hall, 1988.

31. T. S. Arthanari and Y. dodge. *Mathematical Programming in statistics*, Chap. 2. New York: Wiley, 1981.

32. A. Papoulis. *Probability, Random Variables, and Stochastic Processes,.* New York: McGraw-Hill, 1965.

33. C. Eckhart and G. Young. The approximation of one matrix by another of lower rank. *Psychometrika*, 1:211–218, 1936.

34. J. M. Wozencraft and I. M. Jacobs. *Principles of Communication Engineering*, Chap. 4. New York: Wiley, 1962.

35. J. H. McClellan, T. W. Parks, and L. R. Rabiner. FIR linear phase filter design. In *Programs for Digital Signal Processing*, Chapter 5. New York: IEEE Press, 1979.

36. S. Kuo and R. J. Mammone. Image restoration by convex projections using adaptive constraints and the l_1 norm. *IEEE Trans. Signal Process.*, 40(1):159–168, 1992.

37. S. Kuo and R. J. Mammone. Resolution enhancement of tomographic images using the row action projection method. *IEEE Trans. Med. Imaging*, 10(4):593–601, 1991.

38. S. Kuo and R. Mammone. Refinement of EM restored images using adaptive POCS. In *Proceedings of the IEEE International Conference on Acoustics, Speech, and Signal Processing*, 1991.

5

THE RESTORATION OF THREE-DIMENSIONAL SURFACES FROM TWO-DIMENSIONAL IMAGES

K. Venkatesh Prasad[1]

Laboratory of Vision Research
Rutgers University
Piscataway, New Jersey 08854–5930
email: kprasad@caip.rutgers.edu

5.1 INTRODUCTION

The purpose of this chapter is to describe a computational scheme that uses two-dimensional (2-D) picture element (pixel) data, to restore the three-dimensional (3-D) coordinates of volume-elements (voxels). For didactic purposes, the 3-D coordinate restoration task is limited to only those voxels, referred to as visible voxels, which constitute visible opaque surfaces. The framework developed in this chapter demonstrates the extendibility of the vector-space formulation [1, 2]:

$$g = Hf, \tag{5.1}$$

where g and f are vectors and H is a matrix, to applications where the unknown vector f represents 3-D information. Previous descriptions of f in this book have related this vector to 2-D information. In the class of problems addressed in this chapter, each element $g(i)$ of g represents the intensity (an integer in the range 0–255) of the ith pixel of a 2-D image of a 3-D surface. This vector g is known, or given, to the computational scheme since the image is observable. The image is

---

[1] In Spring 1992, at the Laboratory of Vision Research, California Institute of Technology, Pasadena, CA 91125.

observed through an image-formation system that is characterized by a matrix ($\mathbf{H}$ in the vector-space equation (5.1)). The columns of $\mathbf{H}$ represent the point-spread function of the optical system for a unit source located at the position of each voxel. The matrix $\mathbf{H}$ is also a given or known entity in the computational process. What is not known from the 2-D image is the set of coordinates associated with the visible voxels that describes the observable surface. An ordered subset of the indices of $\mathbf{f}$ can, however, be uniquely and linearly mapped to yield the desired coordinate information. The restoration problem therefore translates to finding the most suitable subset of indices of $\mathbf{f}$ to associate with the visible voxels. It is this re-posed problem that will form the crux of the computational formulation to be described in this chapter. The initial estimates of $\mathbf{f}$ and, hence, of the visible voxel coordinates, is made using the row-action projection (RAP) algorithm first stated by Kaczmarz and described in detail in Chapter 2. These initial estimates are refined using a simulated annealing algorithm [3].

Illustrative Example 5.1 *To introduce some of the terminology and notation to be used in this chapter, consider the following illustrative example: It is given that a simple opaque object composed of four voxels creates a two-pixel defocused image* $\mathbf{g} = (100, 50)^T$, *where the superscript T denotes the vector transpose operation. It is also given that this simple two-pixel image is created by an imaging system,*

$$\mathbf{H} = \begin{bmatrix} 1.0 & 0.5 & 0.0 & 0.0 \\ 0.0 & 0.0 & 1.0 & 0.5 \end{bmatrix}. \tag{5.2}$$

Each column of $\mathbf{H}$ *represents the response of the system to each of the four voxels with unit radiant intensity. The problem is to estimate the four elements* $\{f(1), f(2), f(3), f(4)\}$ *of* $\mathbf{f}$ *and, from this, to determine the indices of the elements which correspond to the visible voxels. The indices of the visible voxels can then be used to determine the 3-D coordinates of the visible voxels uniquely. Since it is known that the object is opaque, we will further assume that in each of the sets* (f_1, f_2) *and* (f_3, f_4) *that correspond to voxels arranged along a line of sight, at most one element can have a nonzero value. This assumption is justified by the fact that for opaque objects at most one voxel can be visible (that is, have a nonzero radiance) down a line of sight. Invisible or occluded voxels are assumed to have zero radiance. In addition to the defocused image, a well-focused image, denoted by* $\mathbf{g}^w$, *is given to be such that* $\mathbf{g}^w = (100, 100)^T$. *This implies that one of the elements in the set* (f_1, f_2) *is equal to 100 and that one of the elements in the set* (f_3, f_4) *is also equal to 100. Possible solutions to this problem are of the form*

$$\begin{bmatrix} f_1 \\ 0.0 \\ f_3 \\ 0.0 \end{bmatrix}, \quad \begin{bmatrix} f_1 \\ 0.0 \\ 0.0 \\ f_4 \end{bmatrix}, \quad \begin{bmatrix} 0.0 \\ f_2 \\ f_3 \\ 0.0 \end{bmatrix}, \quad or \quad \begin{bmatrix} 0.0 \\ f_2 \\ 0.0 \\ f_4 \end{bmatrix}. \tag{5.3}$$

From the above solutions of **f**, *the corresponding indices of the visible voxels (nonzero elements of* **f**) *are seen to be* $(1,3)$, $(1,4)$, $(2,3)$, *or* $(2,4)$. *The
3-D coordinates are easily obtained from the index pairs using a simple linear
transformation. Figure 5.1 illustrates the interpretation of the solution using the
coordinate pair* $(1,3)$ *as an example.*

The restoration of the 3-D coordinates of visible voxels from their 2-D images
has numerous applications in areas such as robot vision [4], medical imaging,
crystallography, and surface science [5]. In addition, coordinate information
forms an important primary input to surface interpolation algorithms [6, 7]. The
restoration of 3-D coordinate information can be made using several of the
numerous methods that are available for recovering depth information from 2-D
images [8]. The method presented in this chapter to recover depth and then 3-D
coordinate information is called the "depth-from-defocus" (DFD) method [3]. The
motivation for the DFD method, as pointed out by Pentland [9], is that it can
provide depth information comparable to the stereo disparity or motion-parallax
methods, while avoiding image-to-image point-matching problems.

Pentland [9] has also observed that, unlike some of the earlier methods [4,
10, 11], which extracted depth by searching for a well-focused point in a scene,

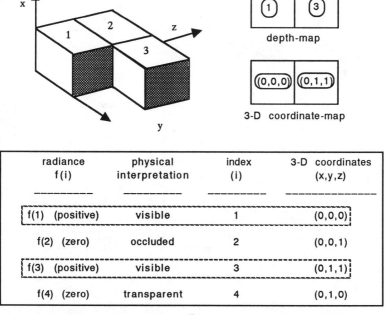

radiance f(i)	physical interpretation	index (i)	3-D coordinates (x,y,z)
f(1) (positive)	visible	1	(0,0,0)
f(2) (zero)	occluded	2	(0,0,1)
f(3) (positive)	visible	3	(0,1,1)
f(4) (zero)	transparent	4	(0,1,0)

Figure 5.1 For the solution $\mathbf{f} = [f(1),0,f(3),0]^T$ of the example problem, the schematic sketch
above shows the voxel arrangement in a 3-D coordinate reference frame. The visible voxels are
shown with shaded front surfaces. The table lists the physical interpretation, index value, and
3-D coordinate information obtained from each case of voxel radiance shown in the first column.

the DFD method requires no search at all and hence is significantly more efficient. Independently, Grossman [12] has demonstrated an experimental scheme for the recovery of depth from defocused images. Further, Subbarao [13, 14] has presented a comprehensive analysis of different DFD techniques. More recently, Pentland et al. [15] have designed a range camera based on the DFD method. This camera is simple and economical to implement, but its performance is limited to coarse estimates of depth [15]. Several other DFD cameras have also been proposed recently [16, 17]. These cameras also recover only coarse depth information. We shall refer to all these coarse depth-recovery approaches [9, 12–17] as *coarse DFD methods*.

In this chapter we introduce a coarse-to-fine coordinate restoration algorithm. The algorithm estimates a fine-grain coordinate map from a coarse-grain coordinate-map obtained by a coarse DFD method. We shall refer to the new approach taken here as "3-D coordinate restoration," since it extends the classical 2-D image-restoration formulation to the 3-D coordinate-restoration problem. The 3-D coordinate-restoration process is schematically outlined in Figure 5.2. A coarse-grain coordinate map of the discrete visible surface (DVS) is obtained

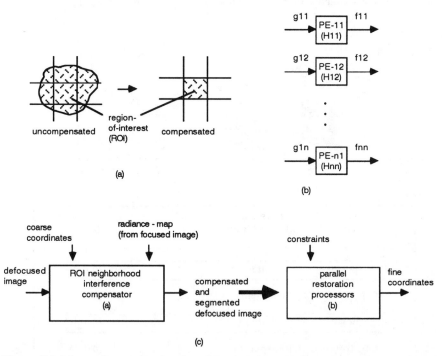

Figure 5.2 3-D coordinate restoration from 2-D pixel data. (a) Pixel data within a region of interest (ROI) is influenced by the radiant energy of its neighboring pixels. Before the pixel data can be used, the influence of its neighbors must first be subtracted out. (b) Each segmented ROI may then be processed independently. (c) ROI pixel data compensation is made using data from a coarse depth map and a well-focused image that provides a radiance map.

using a coarse DFD method. This coordinate map labels large blocks of pixels, referred to as *coarse pixels*. In addition a relatively well-focused image is obtained using a beamsplitter. This image is used to obtain coarse bounds on the radiance distribution of the visible surface. This distribution is called a coarse radiance map. Figure 5.3 illustrates the use of a well-focused image and a defocused image in the coordinate restoration process and schematically outlines the 3-D restoration problem. The intensity distribution of each coarse pixel is influenced by the intensity spread of all its neighboring pixels. These interpixel intensity blurs first have to be compensated for before each coarse pixel can be processed independently. An estimate of the interpixel spread is made by using the intensity map and the coarse depth map. For a coarse pixel of interest, the estimated interpixel spread of its neighboring coarse pixels is first subtracted out. Therefore, the unknown 3-D object can be partitioned into roughly independent 3-D tessellation blocks, and the 3-D coordinates of the visible voxels of each block can be estimated in parallel (Figure 5.2(c)).

The coarse DFD methods make the assumption that the optical point-spread-function (PSF) has a Gaussian distribution [9] and is laterally shift-invariant. In this chapter, an accurate representation of the optical PSFs and the use of a model that is shift varying with respect to depth is presented in order to significantly enhance the accuracy of the depth estimates over that obtained by the coarse DFD method alone. *A priori* knowledge of the 3-D object—such as its opacity property—a coarse-grain depth map, and a fine-grain intensity map provide constraints that are used to reduce the ambiguity caused by the many possible solutions that could exist. A constrained optimization problem is then formulated, where the outcome is a fine-grain coordinate map. We investigate the use of three algorithms to solve the constrained optimization problem: simulated annealing (SA), linear programming (LP), and Kaczmarz's row-action projections (KM). In general, the combinatorial complexity of the algorithms restricts their application to small images. However, this restriction is circumvented by presegmenting the large image into a set of small images. This is accomplished by compensating the neighboring segments using estimates of the overlapping intensity spread based on the coarse coordinate map and coarse radiance map, as illustrated in Figure 5.2.

5.2 3-D COORDINATE RESTORATION

To help understand the formulation of the restoration problem, some basic image-formation relationships will be reviewed. A conventional imaging system consists of an illuminated 3-D object, an imaging system, and a 2-D image of the object. An image-formation system can be characterized by representing the object as a set of axially stacked slices and by considering a simple convex lens to model the optical system. If the focal length z_f of the lens and the lens-sensor distance z_s are known, the homogeneous Gaussian lens equation $1/z_s + 1/z_o - 1/z_f = \epsilon = 0$ can be used to solve for $z_o = z_{\text{in-focus}}$, the distance of the slice that appears

Image Formation:

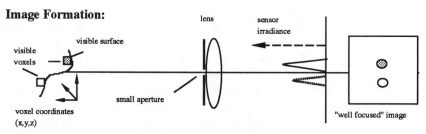

A "well focused" image , recorded using a small aperture lens, contains point spread functions (psfs) that vary in magnitude as a function of voxel radiance (or object "intensity"). A larger psf magnitude implies a "brighter" voxel. The white and hatched shades are used to illustrate two different radiance levels of the visible voxels. The well focused image can be used as a coarse radiance map, but contains little depth information.

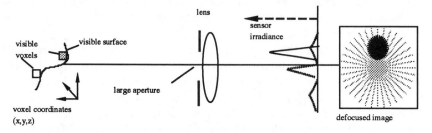

A defocused image, recorded using a large aperture lens, contains point spread functions that vary, in spatial extent and distribution, as a function of depth: Radiance from a relatively well focused (hatched) voxel generates a smaller spatial spread than that from a defocused (white) voxel. The defocused image is rich in depth information.

The 3-D Coordinate Restoration Problem:

Given: A well focused and a defocused image of a visible surface and other constraints based on prior knowledge of the object.

Obtain: The voxel coordinates of the discrete visible surface (DVS).

"well focused" image ⟶ ┌─────────────┐
defocused image ⟶ │ 3-D   │
A priori ⟶ │ Coordinate │ ⟶ voxel coordinates
constraints │ restoration │   (x,y,z)
│ process   │
└─────────────┘

Figure 5.3 Image formation and the restoration problem. During the process of image formation, the 3-D coordinates of the visible surface are mapped onto the 2-D coordinates of its image. A well-focused image is a useful source of radiance information; a defocused image is a rich source of depth information. By utilizing the information in the two types of images and by making use of constraints to eliminate infeasible solutions, the 3-D coordinates can be restored.

focused on the sensor-plane. The slices at all other distances $z_o \neq z_{\text{in-focus}}$ appear defocused on the sensor plane. The extent of defocusing associated with these slices is related to their distance from the lens plane. For the purposes of this paper, *depth* is defined to be the distance from the lens plane. Defocused image formation, can be thought of as a depth-weighted mapping of the radiances of the 3-D object points onto a 2-D sensor plane. A detailed analysis of *continuous* image formation is contained in Born and Wolf [18, p. 435]. An ideal imaging system with a 3-D object can be represented by the Fredholm integral equation of the first kind,

$$g(x, y) = \int_{-\infty}^{\infty} \int_{-\infty}^{\infty} \int_{-\infty}^{\infty} h(x, y; X, Y, D) f(X, Y, D) \, dX \, dY \, dD, \qquad (5.4)$$

where x, y are image coordinates and X, Y, D are object coordinates. The X–Y plane contains the lateral planes of the object and the D axis is aligned along the normal to the x–y plane; $g(x, y)$ is the irradiance at point (x, y) on the image-plane $\mathcal{G}$; $f(\cdot)$ is the radiance at the known lateral coordinates (X, Y) but the *unknown* depth coordinate (D) on the 3-D object $\mathcal{F}$; and $h(\cdot)$ is the point-spread function (PSF) of the imaging system with respect to an object point located at (X, Y, D) on $\mathcal{F}$ and an image point located at (x, y) on $\mathcal{G}$. In general, the PSFs are not Gaussian distributions and the imaging process is not shift invariant for planes that are at out-of-focus depths, as assumed by the DFD methods, although these approximations yield good coarse-depth estimates [9, 13]. In the case of monochromatic light, the in-focus PSF is the Airy pattern, which consists of Bessel functions of the first order and first kind; the out-of-focus PSFs are approximated by Lommel functions, which are summations of higher-order Bessel functions of the first kind [18, p. 435]. These PSFs can be evaluated using the asymptotic recurrence relationships given in expressions found in Gray et al. [19] (see Figure 5.4). For the polychromatic, or "white" light, case, the PSFs have usually been approximated by Gaussian distributions. Recently, Dhadwal and Hantgan [20] have presented an approximate closed-form expression for poly-chromatic PSFs.

In this chapter, the polychromatic PSFs are experimentally determined. The line-spread function (LSF) associated with an edge at each voxel position is first measured; the relationship $\sigma = \sqrt{2}\sigma_l$, where σ and σ_l are, respectively, the spread parameters of a PSF and its corresponding LSF, is used with an imaging system having a circular aperture. The PSF is therefore assumed to be circularly symmetric. A two-dimensional PSF is evaluated by interpolating the 1-D PSF.

For the purpose of numerical computation, (5.4) is discretized. The image $\mathcal{G}$ is stacked as a lexicographic-vector $\mathbf{g}$ in a Euclidean space of dimension n^2, that is, $\mathbf{g} \in E^{n^2}$, and the optical system is represented by a *space-varying point-spread function* (SVPSF) matrix $\mathbf{H} \in E^{n^2 \times n^3}$ whose columns represent lexicographically ordered and discretized PSFs. The object $\mathcal{F}$ is discretized into a set of volume elements or *voxels* $\{F\}$. The radiance values of the voxels are stacked as a lexicographic vector $\mathbf{f} \in E^{n^3}$. The discrete defocused imaging process is expressed

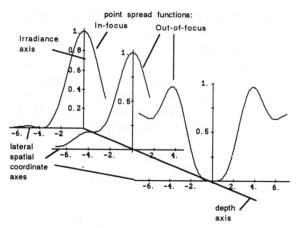

Figure 5.4 The in-focus and out-of-focus point-spread functions, computed for monochromatic light.

as

$$
\begin{bmatrix} g(1) \\ \vdots \\ g(n^2) \end{bmatrix} = \begin{bmatrix} h(1,1) & \cdots & h(1,n^2) & \cdots & h(1,n^3) \\ \vdots & \ddots & \vdots & \ddots & \vdots \\ h(n^2,1) & \cdots & h(n^2,n^2) & \cdots & h(n^2,n^3) \end{bmatrix} \begin{bmatrix} f(1) \\ \vdots \\ f(n^2) \\ \vdots \\ f(n^3) \end{bmatrix}, \quad (5.5)
$$

or, in matrix form,

$$\mathbf{g} = \mathbf{Hf}. \qquad (5.6)$$

In general, the two dimensions of $\mathcal{G}$ could be discretized into n and m intervals. Also, the three dimensions of $\mathcal{F}$ could be discretized into p, q, and r interval. In this chapter, without loss of generality, it is assumed that $m = p = q = r = n$, as shown in (5.5). The SVPSF matrix $\mathbf{H}$ is computed by discretizing the PSF associated with each voxel and concatenating these PSFs in lexicographic order. This process is computationally complex. However, for a given object space, which could enclose many different shapes of objects, and a given imaging system, the matrix needs to be computed only once. If an element $f(i)$ of $\mathbf{f}$ is greater than zero, that is, $f(i) > 0$, then its corresponding voxel $F(i)$ is referred to as a *feasible* voxel. If $f(i) = 0$, then $F(i)$ is referred to as an infeasible voxel. It is assumed that $f(i)$ is nonnegative. The DVS is therefore the set of all feasible voxels. The problem of 3-D coordinate restoration is to first estimate the DVS. The coordinate map is obtained by translating the indices of the feasible voxels into 3-D coordinates, as shown in the illustrative example above. The set of voxels $\{F(1), F(2), \ldots, F(n^3)\} \in \mathcal{F}$ that forms the 3-D object space can be considered a set of n^2 laterally bundled *tubes* T of n voxels each, where

$$T_i \overset{\text{def}}{=} \begin{bmatrix} F(i + 0 \cdot n^2) \\ F(i + 1 \cdot n^2) \\ \vdots \\ F(i + (n-1)n^2) \end{bmatrix}, \qquad \text{for} \quad 1 \leq i \leq n^2. \tag{5.7}$$

That is, $\mathcal{F} = \{T_1, T_2, \cdots, T_{n^2}\}$.

An object is defined to be *opaque* if there exists only one feasible voxel in each tube of the object. To reduce the complexity of the depth-restoration problem, the objects of interest are assumed to be opaque. The opacity restriction forces a unique value of depth $D(i)$ to be associated with every tube T_i. This also implies that the vector $\mathbf{f}$ in (5.6) for an opaque object is very sparse—having at most n^2 nonzero elements out of a total of n^3 elements. To further reduce the complexity of the restoration, a large depth-of-field image is obtained using a small-aperture lens system. This image is used to evaluate the radiance map of the object, as illustrated in Figure 5.3.

This restricts the values of the nonzero elements of $\mathbf{f}$. The possible diffraction effects of the small aperture in altering the true radiance values of the object are accounted for by using a radiance-constraint relaxation factor. Let $I(i)$ represent the index of the feasible voxel $F(I(i))$ in tube T_i and $\mathcal{F}_{\text{feasible}}$ and $\mathcal{I}_{\text{feasible}}$ represent the set of all feasible voxels $\{F(I(1)), F(I(2)), \ldots, F(I(n^2))\}$, and their indices $\{I(1), I(2), \ldots, I(n^2)\}$, respectively. The computational problem of depth restoration involves estimating the set of feasible voxels $\mathcal{F}_{\text{feasible}}$, and then decoding the lexicographic relationship between the elements of $\mathcal{I}_{\text{feasible}}$ and the 3-D spatial coordinates of each voxel in $\mathcal{F}_{\text{feasible}}$. The depth map of the DVS is the set $\{\mathcal{X}_{\text{feasible}}, \mathcal{Y}_{\text{feasible}}, \mathcal{D}_{\text{feasible}}\}$ of lateral (X, Y) and depth coordinates D, respectively, which correspond to $\mathcal{I}_{\text{feasible}}$.

The system of equations given by (5.5) is underdetermined, since there are n^3 unknowns but only n^2 equations. The matrix $\mathbf{H}$ is, therefore, rank deficient. In terms of a vector-space description of the imaging process, the observed image $\mathbf{g}$ is the projection of the voxel radiances ($\mathbf{f}$) onto the range space G_r of $\mathbf{H}$. The null space of $\mathbf{H}$ is the linear subspace $F_n \in F$, such that for any $\mathbf{f}_n \in F_n$, $\mathbf{H} \cdot \mathbf{f}_n = \mathbf{0}$. The smaller the number of independent equations (as compared to the number of unknowns), the larger the space F_n and the greater will be the number of possible solutions for $\mathbf{f}$. This set of equations is ill posed in the sense of the definition given by Hadamard [21] and its relationship to early vision by Bertero et al. [22, 23]. In addition, the matrix $\mathbf{H}^T\mathbf{H}$, where $\mathbf{H}^T$ denotes the transpose of $\mathbf{H}$, is usually ill conditioned or rank deficient [22]. Therefore, the conventional least squares solution to the inverse problem, that is, $\mathbf{f} = [\mathbf{H}^T\mathbf{H}]^{-1}\mathbf{H}^T\mathbf{g}$, is not applicable. Problems of the type represented by (5.5) have a large null space $(E^{(n^3-n^2)})$ and, hence, there is an infinite set of least squared solutions given by $\mathbf{f} = \mathbf{f}^\dagger + [\mathbf{I} - \mathbf{H}^T\mathbf{H}]\mathbf{f}$, where $\mathbf{f}^\dagger$ is the pseudo-inverse solution of $\mathbf{f}$ and $\mathbf{H}^\dagger$ is the pseudo-inverse matrix of $\mathbf{H}$. The reduction of volume of the null space component $[\mathbf{I} - \mathbf{H}^T\mathbf{H}]\mathbf{f}$ is implemented by enforcing constraints in the restoration process.

5.3 CONSTRAINED OPTIMIZATION

The problem of coordinate restoration, which involves estimating the set of feasible voxels $\mathcal{F}_{\text{feasible}}$, may now be formulated as a constrained optimization problem. The *cost function* $e(\mathcal{F}_{\text{feasible}})$ to be minimized is defined as

$$e(\mathcal{F}_{\text{feasible}}) = e_{\text{image}} + \sum_{k=1}^{N_c} \beta_k \cdot e_k, \tag{5.8}$$

where $e_{\text{image}} \in E^1$ is an image-error cost factor and is a measure of the disparity between the given defocused image and the image computed using (5.5); $e_k \in E^1$ denotes the kth constraint-error cost factor. The second term on the right-hand side of (5.8) is a weighted summation of all the constraint-error cost factors. It represents the total penalty for not satisfying one or more of the N_c constraints. The parameter β_k controls the importance of the kth constraint-error cost factor, relative to the image-error cost factor. N_c is the total number of constraints used. The values of β_k are selected by trying a range of values and selecting the set that results in the lowest cost function.

The *optimum* depth map $\{\mathcal{X}_{\text{feasible}}, \mathcal{Y}_{\text{feasible}}, \mathcal{D}_{\text{feasible}}\}$ of the DVS is defined to be that estimate of the DVS, or $\mathcal{F}_{\text{feasible}}$, which has the lowest value of the cost function $e(\mathcal{F}_{\text{feasible}})$. This minimum-cost estimate is obtained by (i) making an *initial estimate*; (ii) using an *active-set strategy*, that is, the set of constraints to be satisfied with an equality; (iii) choosing a *stopping criterion*; (iv) perturbing the old, or initial, estimate to make a new estimate; (v) enforcing constraints prescribed by the *active-set strategy*; (vi) computing the change in the cost function between two successive estimates; (vii) deciding whether to accept the new estimate or retain the old estimate; and (viii) repeating steps (iv)–(vii) until the *stopping criterion* is satisfied.

The image-error cost factor e_{image} is defined as

$$\begin{aligned}
e_{\text{image}} &= \|\mathbf{g} - \hat{\mathbf{g}}\|^2 \\
&= \|\mathbf{g} - \mathbf{H}_s \cdot \hat{\mathbf{f}}\|^2 = \|\mathbf{g} - \mathbf{H}_s \cdot \mathbf{H}_s^{\dagger} \cdot \mathbf{g}\|^2 \\
&= \|(\mathbf{I} - \mathbf{H}_s \cdot \mathbf{H}_s^{\dagger})\mathbf{g}\|^2 = \|\mathbf{Q}_s\mathbf{f}\|^2,
\end{aligned} \tag{5.9}$$

where the subscript s in $\mathbf{H}$ denotes the fact that $\mathbf{H}_s$ is a reduced form of the SVPSF matrix $\mathbf{H}$; $\mathbf{H}_s^{\dagger}$ denotes the pseudo-inverse of $\mathbf{H}_s$; $\mathbf{Q}_s$ is a projection operator of the image-error space; and $\mathbf{I}$ is the identity matrix. The notation $\|\cdot\|$ is used to denote the L_2 norm of a vector $(\cdot)$. The least squares error is measured by the energy of $\mathbf{g}$ projected on the kernel of $\mathbf{H}$.

The constraint-error cost factors are given by

$$
e_1 = \begin{cases} 0, & \text{if } \dfrac{1}{n^2} \sum_{i=1}^{n^2} (\bar{D} - D(i))^2 \leq \delta_s, \\[2em] \beta_1 \left[\dfrac{1}{n^2} \sum_{i=1}^{n^2} (\bar{D} - D(i))^2 - \delta_s \right], & \text{otherwise}, \end{cases} \tag{5.10}
$$

for a depth-smoothness constraint, where $\bar{D}$ is the average depth of the DVS and δ_s is a constraint-relaxation parameter;

$$
e_2 = \begin{cases} 0, & \text{if } [\hat{D}_S(i) - D_S(i)]^2 \leq \delta_c, \\[1em] \beta_2[(\hat{D}_S(i) - D_S(i))^2 - \delta_c], & \text{otherwise}, \end{cases} \tag{5.11}
$$

for a coarse depth-consistency constraint, where $\hat{D}_S(i)$ is the estimated depth of the ith coarse voxel and δ_c is a constraint-relaxation parameter;

$$
e_3 = \begin{cases} 0, & \text{if } \dfrac{1}{n^2} \sum_{i=1}^{n^2} (\bar{f} - f(i))^2 \leq i_s, \\[2em] \beta_3 \left[\dfrac{1}{n^2} \sum_{i=1}^{n^2} (\bar{f} - f(i))^2 - i_s \right], & \text{otherwise}, \end{cases} \tag{5.12}
$$

for a radiance-smoothness constraint, where $\bar{f}$ is the average radiance of the DVS and i_s is a constraint-relaxation parameter; and

$$
e_4 = \begin{cases} 0, & \text{if } [\hat{f}_S(i) - f_S(i)]^2 \leq i_c, \\[1em] \beta_4 \left[(\hat{f}_S(i) - f_S(i))^2 - i_c \right], & \text{otherwise}, \end{cases} \tag{5.13}
$$

for a coarse radiance-consistency constraint, where $\hat{f}_S(i)$ is the radiance of the ith coarse voxel and i_c is a constraint-relaxation factor.

In practice, the values of relaxation parameters are based on the nature of the *a priori* information, including local shading information [24], that is available. If an object is perfectly smooth in both its depth and radiance distributions, for example, then their corresponding relaxation parameters—namely, δ_s and i_s— would both be equal to zero. For the experimental results shown in this chapter, the constraint weights β_1 and β_3 were selected to be 10^5 and 10^6 respectively.

The constraint weights β_2 and β_4 were chosen to be 1 and 0, respectively. The choice of these values was based on experimental trials.

5.3.1 Active-Set Strategies

Let $\mathcal{C}_{\text{all}}$ denote the set of all possible constraints that apply to the problem being considered. A constraint is defined to be *active* if it is satisfied with equality. An *active set* $\mathcal{C}_{\text{active}}^k \in \mathcal{C}_{\text{all}}$ is the set of all constraints that are satisfied by the kth iterative estimate. The active set is enforced by using the active-set operator $\mathbf{C}_{\text{active}}$. An example of the use of $\mathbf{C}_{\text{active}}^k$ is

$$\hat{\mathbf{f}}^k = \mathbf{C}_{\text{active}}^k \cdot \hat{\mathbf{f}}_u^k$$

$$\mathbf{C}_{\text{active}}^k = C_1 \cdot C_3, \tag{5.14}$$

where C_1 is the depth-smoothness constraint, C_3 is the radiance-smoothness constraint, $\hat{\mathbf{f}}_u^k$ is the kth vector unconstrained by the kth active set, and $\hat{\mathbf{f}}$ is the result of imposing the kth active set on $\hat{\mathbf{f}}_u^k$.

An active-set strategy is an ordered sequence

$$\left\{ C_{\text{active}}^1, C_{\text{active}}^2, \dots, C_{\text{active}}^N \right\} \tag{5.15}$$

that specifies which active-set operator should be used for each iteration of $\mathbf{f}_u^k$. It is necessary to order the sequence of active-set operators because some constraints are more important than others at different stages in the restoration process. For example, in the simulated annealing method, discussed later in this section, it was necessary to enforce the coarse constraints in the early iterations of the algorithm and the fine constraints in the later stages.

5.3.2 Coarse Image Segmentation

The segmented image-restoration problem can be represented as

$$\begin{bmatrix} \mathbf{g}_1 \\ \mathbf{g}_2 \\ \vdots \\ \mathbf{g}_{n_{ps}} \end{bmatrix} = \begin{bmatrix} \mathbf{H}_{1,1} & \mathbf{H}_{1,2} & \cdots & \mathbf{H}_{1,n_{vs}} \\ \mathbf{H}_{2,1} & \mathbf{H}_{2,2} & \cdots & \mathbf{H}_{2,n_{vs}} \\ \vdots & \vdots & \ddots & \vdots \\ \mathbf{H}_{n_{ps},1} & \mathbf{H}_{n_{ps},2} & \cdots & \mathbf{H}_{n_{ps},n_{vs}} \end{bmatrix} \begin{bmatrix} \mathbf{f}_1 \\ \mathbf{f}_2 \\ \vdots \\ \mathbf{f}_{n_{vs}} \end{bmatrix}, \tag{5.16}$$

where $\mathbf{g}_i$ denotes the ith image segment and is a result of the contributions from n_{vs} voxel segments. The submatrices $\mathbf{H}_{i,j}$ denote the PSF matrices that relate the intensity of the ith image segment $\mathbf{g}_i$ to the jth voxel segment $\mathbf{f}_j$.

From (5.16), the ith image segment is given by

$$\mathbf{g}_i = [\mathbf{H}_{i,1}\mathbf{f}_1 + \cdots + \mathbf{H}_{i,i}\mathbf{f}_i + \cdots + \mathbf{H}_{i,nv}\mathbf{f}_{nv}]. \tag{5.17}$$

Using coarse depth values, the image g_i can be segmented into the compensated form by

$$g_i^c = g_i - [H_{i,1}f_1 + \cdots + H_{i,1}f_{nv}] = H_{i,i}f_i, \qquad (5.18)$$

where the interference of the coarse-voxel regions neighboring the image g_i has been coarsely compensated for by using the coarse coordinate and radiance maps. The system of equations that forms the compensated image segment has the form given by (5.5). The resulting problem is significantly smaller than the system of equations that represents the whole unsegmented image. In addition, the segmented system of equations is now amenable to independent parallel processing of each region of interest (ROI).

5.3.3 Simulated Annealing

Under the assumption that the unknown DVS has a bounded and discrete range of depth, image formation could be considered a combinatorial process where the observed image is a blurred snapshot of one particular combination of feasible voxels. The problem is to determine which combination could have produced a given image. Each combination is referred to as a *configuration* and has a *cost* associated with it, as computed by the cost function (5.8). A *perturbation* is defined as the process where a configuration c^k is altered to a configuration c^{k+1} by changing the depth of either one feasible voxel or several feasible voxels simultaneously. That is, $c^k \xrightarrow{\text{perturbation}} c^{k+1}$. By performing a sequence of perturbations and saving only the *acceptable* configurations, a minimum-cost configuration or "solution," is obtained. Any configuration c_j that has a lower cost than a previously obtained minimum-cost configuration c_{min} is an acceptable configuration and is saved as the new minimum-cost configuration, that is, $c_{min} \leftarrow c_j$. In order to assist the optimization process in getting out of possible local minima on the configuration cost versus perturbation curve, certain configurations that result in positive cost transitions with respect to c_{min} are also accepted. A control parameter is used to weight the probability of accepting those configurations which result in increased cost functions. This parameter, referred to as the *temperature* is initially assigned a high value and then is decreased according to a *cooling schedule* until a final value. A multilevel perturbation strategy is used to update the configurations. The method consists of initially perturbing a combination of feasible voxels over large neighborhoods. Such perturbations include several coarse voxels and occur at high temperatures. Then the perturbation regions are gradually reduced to single coarse voxels. Finally, just single voxels are perturbed at low temperatures. This approach is particularly amenable to parallel implementation.

The SA algorithm is outlined in Figure 5.5. In order to use the algorithm, the following specifications have to be made: (i) a starting configuration; (ii) a perturbation strategy; (iii) a criterion for accepting a configuration with a positive cost transition; (iv) a cooling schedule; and (v) a stopping criterion. The choice

```
/* SIMULATED ANNEALING */
/* Set initial conditions: */
/* Temperature*k_B  T * k_B = t_0 */
/* Configuration c_0 = c_initial */
/* Minimum-cost configuration c_min = c_0 */
/* Active-Set=C_active^0 */
/* Perturbation-level=neighborhood */
/* k=0; */
repeat
repeat
PERTURB(c_i -> c_j);
COST-DIFFERENCE(Δe_ij = e(c_j) - e(c_i));
if Δe_ij ≤ 0 then accept else
if exp(-Δe_ij/k_BT) > random[0,1) then accept;
if accept then UPDATE(c_min <- c_j);
c_i <- c_min
until equilibrium is approached sufficiently closely;
COOL(T * k_B <- t_k)
k = k + 1
until stopping criterion is satisfied
end
```

Figure 5.5 Outline of the annealing algorithm.

of the starting configuration is based on the available coarse-depth information. For example, a coarse depth map might indicate that, in an 8×8 voxel DVS, the top 4×8 block of voxels is in the foreground, say at slice 1, and that the 4×8 voxels are in the background, say at slice 8. A starting configuration will assign a depth value of 1 to each of the top 4×8 voxels and a depth value of 8 to each of the bottom 4×8 voxels. The perturbation strategy is chosen such that at high temperature values large-voxel regions are perturbed and at low temperature values only single voxels are perturbed. The criterion used to accept configurations with positive cost transitions is the Metropolis criterion, which is

$$\text{if } \exp\left(-\frac{\Delta e_{ij}}{k_B T}\right) > \text{random}[0,1) \text{ then accept the perturbation.} \qquad (5.19)$$

The function $\text{random}[0,1)$ represents a uniformly distributed random number generator, in the range $[0,1)$. The quantity k_B, conventionally called the Boltzmann constant, is not explicitly determined, but has been incorporated into the temperature parameter indirectly. The cooling schedule, or the rate at which the temperature parameter is decreased, is a function of the duration for which the temperature is kept a constant and the amount by which the temperature is decreased. Each constant temperature period is referred to as an *era*. A popular schedule is the one where the temperature is exponentially lowered, that is, $T_{k+1} = \nu T_k$, where ν is a factor between 0 and 1 and T_{k+1} and T_k are the new and old temperatures, respectively. For the schedule used in the example in this chapter, ν was set at 0.9. The initial temperature was set at a value that ensured over 80% reconfigurations, as suggested by Kirkpatrick et al. [25]. An initial temperature of 15000, for example, was chosen for the cases considered in this chapter. The stopping criterion is based on the rate of decrease of the cost associated with the c_{min} configurations. If the cost of the c_{min} configuration of two successive eras is the same, then the algorithm is stopped. To be able to deal with cases where the cost decrease between successive c_{min} configurations is very small, there is an upper bound on the number of eras that the algorithm will permit. The algorithm will stop if this upper bound is exceeded. A upper bound of 50 was selected for the examples considered in this chapter. This number was chosen based on the estimated time complexity of the algorithms. For all the cases considered, however, the algorithm stopped within 20 eras, since the first stopping criterion was satisfied. As can be seen, there are many parameters that have to be chosen before the simulated annealing algorithm can be put to use. For the results obtained in this chapter, the parameters were arrived at by repeated trials. For these cases it was found that both the cooling schedule and the initial temperature were not scene dependent.

5.3.4 Other Optimization Methods

Linear Programming Linear Programming (LP) is probably the most commonly used method of constrained optimization. A lucid textbook-style description of the application of LP to image restoration was given by Mammone [26]. The LP application can easily be adapted to the problem of surfac-coordinate restoration for several reasons. The object-radiance vector is a sparse vector. This is due to the opacity constraint. LP implicitly uses this sparseness constraint [26, 27]. Since LP is a linear optimization technique, linear equality, and inequality constraints are easily implemented. Certain nonlinear constraints are also imposed via pivot selection strategies. For the range-estimation problem, we use the L_1 norm of the error as the objective function.

Considering additive noise, we express the observed image vector as

$$\mathbf{g} = \mathbf{Hf} + \mathbf{n}, \tag{5.20}$$

where $\mathbf{n}$ is the vector of random errors. The equation may be rewritten by decomposing $\mathbf{n}$ into two nonnegative components as

$$\mathbf{g} = \mathbf{Hf} + \mathbf{n}^+ - \mathbf{n}^-. \tag{5.21}$$

The LP formulation can then be written as

$$\text{minimize} \quad \mathbf{n}^+ + \mathbf{n}^- \tag{5.22}$$

$$\text{subject to} \quad \mathbf{g} = \mathbf{Hf} + \mathbf{n}^+ - \mathbf{n}^-, \qquad \mathbf{f}, \mathbf{n}^+, \mathbf{n}^- \geq 0.$$

Direct implementation of the above formulation by the simplex method is inefficient. More efficient algorithms have been developed [28]. A modified

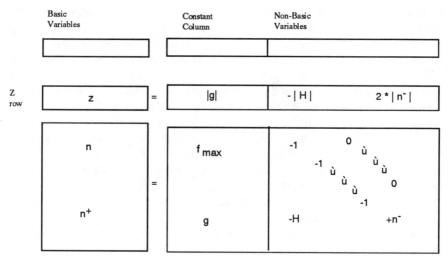

Figure 5.6 Tableau for the LP algorithm.

minimum-$\mathcal{L}_1$ formulation [27] is pursued with a modified pivot strategy to accommodate certain nonlinear constraints. The equality constraints are represented by (5.21). Radiance bounds are imposed using inequality constraints. These constraints are built into the LP tableau, as shown in Figure 5.6.

The opacity constraint is included in the simplex algorithm by modifying the ratio test. The pivot operation in a simplex algorithm exchanges a basic variable for a nonbasic variable. The choice of a pivot column is in the order of decreasing gradients within the objective function row. The choice of a pivot row is preceded by the ratio test. This test determines the largest permissible increase in the pivot column variable without rendering any basic variable negative. The row corresponding to the smallest ratio of a constant column element to a pivot column element is chosen as the pivot row. A pivot is allowed only if the right- and left-hand-side variables corresponding to the pivot row and column belong to the same tube of voxels along the line of sight. The smoothness constraint is enforced by limiting the range of indices searched for each pivot. All the depth and radiance constraints listed in Figure 5.7 are enforced in the above algorithm.

Depth Constraint (Symbol)	Radiance Constraint (Symbol) Rank
Depth uniqueness (D0)	Radiance opacity (I0)
Depth nonnegativity (D1)	Radiance nonnegativity (I1)
Depth boundedness and discreteness (D2)	Radiance boundedness and discreteness (I2)
Depth smoothness (D3)	Radiance smoothness (I3)
Coarse depth consistency (D4)	Radiance consistency (I4)
—	Coarse radiance consistency (I5)

Figure 5.7 Ranking of the effectiveness of constraints.

The radiance-boundedness constraint (I2) is used to set an upper bound (f_{max}) on the inequalities. The lower bound cannot be used, since it is valid only for voxels on the visible surface. Uniqueness, opacity, and smoothness constraints (D0, I0, D3, and I3) are implemented via the modified pivot strategy described above. The nonnegativity constraints (D1 and I1) are implicit in the simplex algorithm.

Kaczmarz Row-Action Projections [29] A comprehensive treatment of the application of Kaczmarz's row-action projection method to image restoration can be found in Chapters 2, 3 and 4. The general form of the Kaczmarz algorithm is

$$\hat{\mathbf{f}}^{k+1} = \mathbf{f}^k + \lambda \cdot \frac{g(i_k) - \langle \mathbf{h}^{i_k}, \mathbf{f}^k \rangle}{\|\mathbf{h}^{i_k}\|^2} \cdot \mathbf{h}^{i_k}$$

$$\mathbf{f}^{k+1} = \mathbf{C}_{active}^j \cdot \hat{\mathbf{f}}^{k+1} \tag{5.23}$$

where k is the $\mathbf{f}$ iteration index, λ is a relaxation parameter that is defined between $(0, 2)$ [30], $i_k = \mathrm{mod}(k, n^2)$, $\mathbf{h}^{i_k}$ is the i_kth row of the $n^2 \times n^3$ matrix $\mathbf{H}$. $\mathbf{C}_{active}^j$ denotes the jth composite active-set operator that $\hat{\mathbf{f}}^k$ is projected onto to generate the $(k+1)$th constrained projection, $\mathbf{f}^{k+1}$. This method allows for all the radiance and depth constraints. In the simulation presented in this chapter, the relaxation parameter $\lambda = 1$.

5.4 COMPUTATIONAL RESULTS

The optimization algorithms were applied to several synthetic objects.

The Results of the SA Algorithm The use of the SA algorithm was investigated for several synthetic objects, using two cases of the optical system matrix $\mathbf{H}$. In one case, an ideal convex-lens camera model was used and the $\mathbf{H}$ matrix was formed using the PSF expressions for monochromatic light found in Born and Wolf [18]. In the other case, the $\mathbf{H}$ matrix was formed using the experimentally measured PSFs of an RCA TV camera (Model TC2511U) with a Nikkor f/2.8 55 mm lens. In this case, a uniform white-light source was used. The algorithms all used two images: one highly defocused image, which was used to form the $\mathbf{g}$ vector, and one large depth-of-field image, which was used to restrict the radiance values of the DVS. In addition, a coarse depth map was first obtained, the images were segmented into 8×8 regions of interest, and a central segment was used. The output of the SA algorithm was the depth map $\{\mathcal{X}_{feasible}, \mathcal{Y}_{feasible}, \mathcal{D}_{feasible}\}$. Two types of error metrics have been used here: The image error (e_{image}) and the depth-error (e_d). Only the image error is observable, in practice, since the true object is unknown. However, the true depth value was used to form the depth error in our computer simulations to accurately indicate the performance of the method.

1. Image error:

$$e_{\text{image}} \stackrel{\text{def}}{=} \|g - \hat{g}\|^2, \tag{5.24}$$

where g and $\hat{g}$ are vectors that represent the observed defocused image and the estimated image, respectively, and $\| \cdot \|^2$ represents the square of the L_2 norm.

2. Depth error:

$$e_d \stackrel{\text{def}}{=} \frac{\text{magnitude of wrong indices}}{\text{total number of voxels}} \times 100. \tag{5.25}$$

The SA algorithm was used to restore depth from a single segmented defocused image of a $8 \times 8 \times 8$ voxel staircase object in a $24 \times 24 \times 8$ object space. Both an ideal camera model and a real TV camera were considered. The RCA TV camera PSF matrix was noisy since it had been experimentally obtained. In the case of the convex-lens model the lateral dimensions of each voxel were about 50×50 μm and the axial dimension was about 1 mm. For the RCA TV camera model, the lateral dimensions were about 0.5 mm $\times$ 0.5 mm and the axial dimension was about 4 mm. Figure 5.7 is a qualitative ranking of the depth and radiance constraints, where the topmost depth and radiance constraints are the most effective in the terms of reducing the search space. The ranking was obtained by physical considerations and computer simulations. A coarse depth map of the top four steps in the foreground and the bottom four steps in the background was assumed. Figure 5.8 shows the highly defocused image of the on-axis top view of the $8 \times 8 \times 8$ voxel staircase object also shown in the same figure. This image was obtained using a simple convex-lens model with monochromatic light. The object had uniform reflectivity, as was measured from its large depth-of-field image. Figure 5.9 was an imperfect reconstruction that resulted from underweighting the penalty function. The plots of e_{image} and e_d versus the index of successful reconfigurations are shown in Figure 5.10(a) and

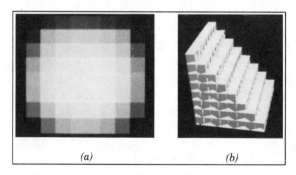

(a)   *(b)*

Figure 5.8 An 8×8 pixel defocused overhead image (a) of an $8 \times 8 \times 8$ voxel staircase object (b) The light source was assumed to be directly above the object and the image was computed based on an ideal convex-lens camera model.

Figure 5.9 The effect of the depth smoothness control parameter (β_{D3}) is shown above. An imperfect restoration results when $\beta_{D3} = 10^5$. At $\beta_{D3} = 10^6$, the restoration is perfect.

a plot of the original and restored depth maps is shown in Figure 5.10(b). At small values of the index, the annealing temperature is high; hence, large upward transitions in the cost function appear in the plots. The highly defocused uncompensated and compensated images obtained using the RCA TV camera PSF are shown in Figure 5.11, along with the original isolated staircase object and the coarse depth map of a 24×24 pixel neighborhood. In the coarse depth map, the top three coarse voxels were assumed to be at slice 1, or closest to the camera. The center three coarse voxels were assumed to be at slice 5 and the bottom three coarse voxels were assumed to be at slice 7. The depth-restored objects are shown in Figure 5.12. The depth restoration using the uncompensated image is very poor. For the case where the compensated image was used, only the voxels at the top- and bottommost levels are improperly restored but at the remaining six levels the restoration is perfect. Figure 5.13 shows plots of e_{image} and e_d versus the index of successful reconfigurations. This figure also shows a plot of the voxel depth map.

The Results of the LP and KM Algorithms Figure 5.14 shows the original object and a reconstruction using the LP algorithm. In the $9 \times 9 \times 9$ reconstruction shown, all but two voxels were reconstructed to within one-voxel accuracy in depth. The mean error in depth was found to be 0.66 voxels, which corresponds to better than two bits of enhancement over the coarse depth estimate given by the DFD method. LP does not show the flexibility of the SA technique. Nonlinear constraints are not easily implemented, and the size of the problem grows very rapidly with increasing size of the object to be reconstructed. The performance of LP was also found to be dependent on the rank of the PSF matrix **H**. The use of pivoting strategies to enforce a nonlinear constraint was found not necessarily to result in convergence to the optimal solution. For these

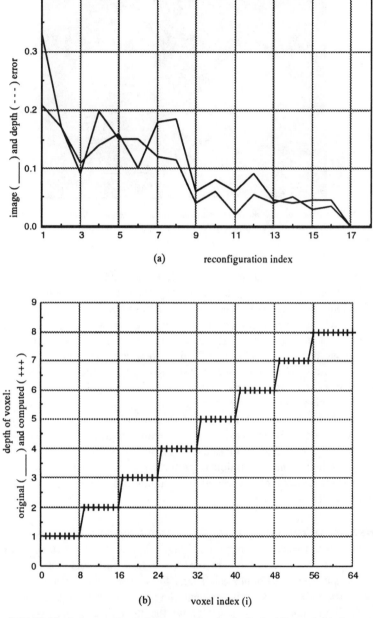

Figure 5.10 (a) The image error and depth error for the $8 \times 8 \times 8$ object and (b) the voxel-depth plot.

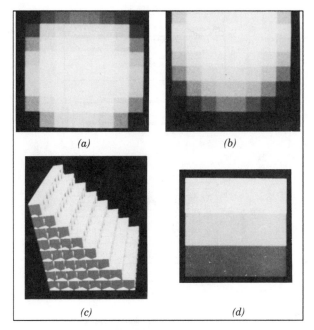

Figure 5.11 RCA camera case: (a) the uncompensated image, (b) the compensated image, (c) the original object, and (d) the 24 × 24 × 24 coarse depth map.

reasons, the results obtained from LP were not found to compare well with the SA technique.

A partially restored 4×4×4 voxel staircase object is shown in Figure 5.15. This was the result of applying the KM algorithm, with a relaxation parameter $\lambda = 1$, on a compensated segment of the blurred image obtained by the convex-lens

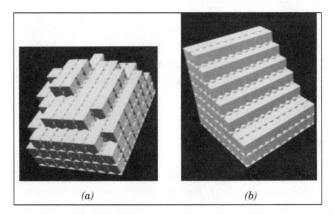

Figure 5.12 The restored objects: (a) uncompensated restoration and (b) compensated restoration.

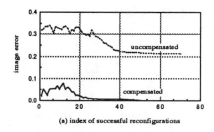

(a) index of successful reconfigurations

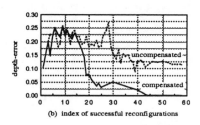

(b) index of successful reconfigurations

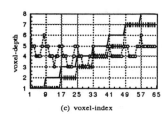

(c) voxel-index

Figure 5.13 For uncompensated and compensated cases: (a) image-error, (b) depth-error, and (c) voxel-depth plots. (—, the original; +++, the compensated, and ooo, the uncompensated cases.)

camera. Two voxels out of sixteen on the visible surface were incorrectly restored. While each step of the KM algorithm is computationally simple, it is not possible to partition the **f** vector so as to use only a set of feasible voxels at each iteration. The result is that the radiance values of all voxels are evaluated at each iteration; this leads to a significant increase in the total complexity. For these reasons the Kaczmarz algorithm was found to be unsuitable for the depth-restoration problem.

5.5 SUMMARY AND CONCLUSIONS

An approach to 3-D coordinate restoration has been developed to obtain high-resolution coordinate maps from coarse maps. An image segmentation strategy has been used to isolate regions of interest over which the coarse-to-fine processing can take place. This method is highly amenable to parallel and distributed

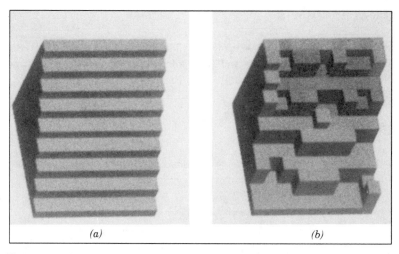

Figure 5.14 (a) The original object used for the LP case and (b) the restored object.

processing. Three algorithms—simulated annealing, linear programming, and Kaczmarz's method of row-action projections (RAP)—have been investigated to solve the depth-restoration problem for noise-free 3-D objects. A combination of first using RAP to make initial estimates followed by the simulated annealing algorithm was found to perform most efficiently, in terms of both the accurate restoration of the 3-D coordinates and the implementation complexity.

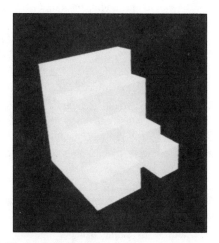

Figure 5.15 For small (4 × 4 × 4) voxel objects, the RAP algorithm estimates the coordinates with only a small error: In the 16-visible-voxel surface shown, only one voxel is incorrectly placed—the original object was a 4-step staircase.

ACKNOWLEDGMENTS

The research reported here was made possible through the support of the New Jersey Commission on Science and Technology and the Center for Computer Aids for Industrial Productivity at Rutgers University.

REFERENCES

1. A. N. Tikhonov and V. Y. Arsenin. *Solutions of Ill-Posed Problems*. New York: Wiley, 1977.

2. H. C. Andrews and B. R. Hunt. *Digital Image Restoration* (Signal Processing Series). Englewood Cliffs, N.J.: Prentice-Hall, 1977.

3. K. Venkatesh Prasad, R. J. Mammone, and J. Yogeshwar. 3D image restoration using constrained optimization techniques. *Opt. Eng.*, **29**(4):279–288, 1990.

4. B. K. P. Horn. *Robot Vision*. Cambridge, Mass.: MIT Press, 1986.

5. G. Binnig and H. Rohrer. Scanning tunneling microscopy— Part I: From birth to adolescence. *Rev. Mod. Phys.*, **59**(3):615–625, 1987.

6. W. E. L. Grimson. *From Images to Surfaces: A Computational Study of the Human Early Visual System*. Cambridge, Mass: MIT Press, 1981.

7. D. Terzopoulos. Multilevel computational processes for visual surface reconstruction. *Comput. Vision Graphics Image Process.*, **14**:52–96, 1983.

8. P. Besl. Active optical range imaging sensors. In J. Sanz, ed., *Advances in Machine Vision: Applications and Architectures*, pp. 1–63, New York: Springer Verlag, 1989.

9. A. P. Pentland. A new sense for depth of field. *IEEE Trans Pattern Recogn. Machine Intell.*, **PAMI-9**(4):523–531, 1987.

10. R. A. Jarvis. A perspective of range finding techniques for computer vision. *IEEE Trans. Pattern Recogn. Machine Intell.*, **PAMI-5**(2):122–139, 1983.

11. E. P. Krotkov and J-P. Martin. Range from focus. In *Proceedings of the IEEE International Conference on Robotics and Automation*, IEEE-CS, pp. 1093–1098. April 7–10, 1986.

12. P. Grossman. Depth from focus. *Pattern Recogn. Lett.*, **5**:63–69, 1987.

13. M. Subbarao and N. Gurumoorthy. Depth recovery from blurred edges. In *Proceedings of the IEEE Computer Society Conference on Computer Vision and Pattern Recognition*, IEEE-CS, pp. 498–503. Ann Arbor, Michigan, June 5–9, 1988.

14. M. Subbarao. Parallel depth recovery by changing camera parameters. In *The Second ICCV*, 1988.

15. A. Pentland, T. Darrell, M. Turk, and W. Huang. A simple real-time range camera. In *Proceedngs of the IEEE Computer Society Conference on Computer Vision and Pattern Recognition*, IEEE-CS, pp. 256–261, San Diego, CA, June 4, 1989.

16. T.-l. Hwang, J. J. Clark, and A. L. Yuille. A depth recovery algorithm using defocus information. In *Proceedings of the IEEE Computer Society Conference on Computer Vision and Pattern Recognition*, IEEE-CS, pp. 476–482, San Diego, California, June 4, 1989.

17. V. M. Bove, Jr. Discrete fourier transform based depth-from-focus. In *Image Understanding and Machine Vision*, volume 14 of *1989 Technical Digest*, pp. 118–121. North Falmouth, Cape Cod, Mass.: Optical Society of America, 1989.

18. M. Born and E. Wolf. *Principles of Optics*, 5th ed. Oxford: Pergamon Press, 1975.

19. A. Gray, G. B. Matthews, and T. M. MacRobert. *A Treatise on Bessel Functions and Their Applications to Physics*, 2nd ed., Chapter 14. London: Macmillan, 1922.

20. H. S. Dhadwal and J. Hantgan. Generalized point spread function for a diffraction limited abberation free system under polychromatic illumination. *Opt. Eng.*, **28**(11):1237–1240, 1989.

21. J. Hadamard. Sur les problèmes aux dérivés partilles et leur signification physique. Princeton University: Bulletin 13, 1902.

22. M. Bertero, T. A. Poggio, and V. Torre. Ill-posed problems in early vision. *Proc. IEEE*, **76**(8):869–889, 1988.

23. T. Poggio, V. Torre, and C. Koch. Computational vision and regularization theory. *Nature*, **317**:314–319, 1985.

24. A. P. Pentland. Local shading analysis. *IEEE Trans. Pattern Recogn. Machine Intell.*, **PAMI-6**(2):170–187, 1984.

25. S. Kirkpatrick, C. D. Gelatt, Jr., and M. P. Vecchi. Optimization by simulated annealing. IBM Research Report RC-9355, 1982.

26. R. J. Mammone. Image restoration using linear programming. In Henry Stark, ed., *Image Recovery: Theory and Recovery*, pp. 127–156. London: Academic Press, 1987.

27. R. J. Mammone. Spectral extrapolation of constrained signals. *J. Opt. Soc. Am.*, **73**(11):1476–1480, 1983.

28. I. Barrodale and F. Roberts. Algorithms 478: Solution of the overdetermined system of equations in the l_1 norm. *Commun. ACM*, **17**(319), 1973.

29. C. I. Podilchuk and R. J. Mammone. Row and block action projection techniques for image restoration. In *Signal Recovery and Synthesis III*, volume 15 of *1989 Technical Digest*, pp. 18–21. North Falmouth, Cape Cod, Mass.: Optical Society of America, 1989.

30. D. C. Youla. Generalized image restoration by the method of alternating orthogonal projections. *IEEE Trans. Circ. Sys.*, **25**(9):694–702, 1978.

PART 3

ADAPTIVE SIGNAL RECOVERY

6

A VECTOR-SPACE APPROACH
TO ADAPTIVE FILTERING

John F. Doherty[1]

Department of Electrical and Computer Engineering
Iowa State University
Ames, Iowa 50011
email: doherty@isuee1.ee.iastate.edu

6.1 INTRODUCTION

The origins of discrete-time adaptive filtering can be traced back to the time of Gauss and the genesis of linear estimation theory. The problem Gauss addressed was the determination of planetary orbits from a limited set of telescopic observations [1]. The study of matrix inversion has had a virtually independent development in the mathematics literature. These two disciplines can be linked through the rather innocuous equation

$$\mathbf{d} = \mathbf{Xc}. \qquad (6.1)$$

The adaptive-filter problem is cast in the form of (6.1) if we allow $\mathbf{X}$ to represent the set of measurement data; $\mathbf{d}$, the adaptive filter output; and $\mathbf{c}$, the adaptive-filter coefficients. The goal of the adaptive filter can be interpreted as the inversion of (6.1). Although the analysis of adaptive algorithms has recently taken this viewpoint [2], the full panoply of the mathematics literature on linear systems inversion has not been used for the understanding of adaptive filtering techniques. For example, the analysis of the stability of an adaptive filter is directly related to the condition number of the measurement matrix $\mathbf{X}$. The development of the

---

[1]This research was done when Dr. Doherty was a Ph.D. candidate at Rutgers University.

adaptive algorithm presented in this chapter will borrow freely from the concepts of matrix theory to perform adaptive filter analysis.

The inversion of (6.1) is written in generic form as

$$c = X^{-1}d, \tag{6.2}$$

where X^{-1} represents the *inverse* of the matrix X. The solution of (6.2), although simple in form, is fraught with theoretical concerns about its existence. The techniques for inverting a square matrix are well established. Perhaps the one most familiar is the method of Gaussian elimination, which relies on the fact that scaling both sides of an equation does not change its solution. The Gaussian elimination method, although straightforward, is computationally expensive. The required calculations[2] are $O(N^3)$ for an $N \times N$ matrix with a required storage capacity that is $O(N^2)$.

It is often the objective in engineering to obtain a limited set of measurements of an unknown system and then estimate the system's behavior outside the field of observation. The goal is often the attainment of a model that best fits the observed data and is capable of producing the best estimate of the missing data. The model often consists of a finite set of parameters that must be determined from the observed data. The estimation process is termed *interpolation* when the estimate occurs between the end points of the observed data; for example, curve fitting. *Extrapolation*, or *prediction*, represents a process whose estimate is outside the boundary of the observation set. Two examples of this are weather prediction and stock market forecasting.

Autoregressive models are used in many areas of signal processing—for example, spectral analysis and speech LPC—where block-processing methods have typically been used to estimate the unknown coefficients. These methods were discussed in Chapter 2. Currently, iterative methods for adaptive estimation fall into two broad categories: the least mean square (LMS) algorithm and the recursive least squares (RLS) algorithm. The LMS algorithm offers low complexity and stable operation at the expense of convergence speed. The RLS algorithm offers improved convergence performance at the expense of possible stability problems. Low computational complexity means that the number of multiplies and adds per update is proportional to the number of adaptive coefficients.

The LMS algorithm is the preferred algorithm for applications where low complexity is essential. The LMS algorithm produces an approximation to the minimum mean square error estimate; that is, the expected value of the output error approaches zero. The major drawback of the LMS algorithm is that its rate of convergence is dependent upon the eigenvalue spread of the input data correlation matrix, usually excluding its use in high-speed real-time signal-processing applications [3]. However, it is widely used where convergence speed is not a problem [4]. The convergence rate of the LMS algorithm can be increased

---

[2]$O(N)$ should be interpreted as "on the order of N."

by introducing methods that attempt to orthogonalize the input data and reduce the eigenvalue spread of the correlation matrix [5, 6]. The convergence rate of any adaptive algorithm can be specified by either the coefficient error or the estimation error, or both. It is well known that, for the LMS algorithm, the mean square error in the output converges faster than the mean square error in the coefficients [7, 8]. Thus, the LMS algorithm is useful in situations where the primary function of the adaptive system is to estimate a signal in the presence of noise. This is in contrast to system identification—that is, the estimation of the model of an unknown system from its output—for which the RLS algorithm usually performs better. In the specific case of an adaptive equalizer, minimizing the equalizer output error (signal estimation) is more important than minimizing the coefficient error (coefficient estimation) because the output error determines the bit-error rate of the communications system.

The class of RLS adaptive algorithms converges faster than the LMS algorithm, given the same number of input samples. The RLS algorithm forms a deterministic least squares estimate based solely upon solving the deterministic normal equations. The convergence rate of the RLS algorithm can be an order of magnitude faster than the LMS algorithm with much less dependence on the eigenvalue spread of the data-correlation matrix [9]. However, severely ill-conditioned—that is, high eigenvalue spread—data may still deleteriously affect the convergence of the RLS algorithm [10]. The generic RLS algorithm has $O(N^2)$ complexity because it involves a recursive matrix inverse. The special case of a transversal filter structure results in fast RLS (FRLS) implementations with $O(N)$ complexity [11]. The RLS and FRLS algorithms are also potentially unstable and usually employ rescue devices that detect the onset of instability [12]. The unstable behavior has been attributed to finite precision implementation and the effects of ill-conditioned input data [13]. Much analysis has been devoted to numerical round-off error propagation in the FRLS algorithm—e.g., [14–17], however, ill-conditioned correlation matrices still pose a stability problem. Ill-conditioned data naturally occur in a variety of signal-processing problems, such as fractionally spaced channel equalization [18], equalization for channels with spectral nulls [19], narrowband jammer excision [20], autoregressive spectral estimation [21], maximum entropy linear prediction [22], and linear predictive coding of speech [23]. The RLS algorithm implicitly computes the *direct* inverse of the data-correlation matrix that arises in the deterministic normal equations. The inverse may not exist analytically and the inversion process can be numerically unstable for correlation matrices with high eigenvalue spread. Thus, it is desirable to use a matrix inversion method that is insensitive to the eigenvalue spread of the matrix.

This chapter presents an adaptive method that combines the simplicity of the LMS algorithm with the performance of the FRLS algorithm and is also stable when operating with ill-conditioned data. The iterative *row-action projection* (RAP) algorithm performs updates of the unknown coefficients by using the rows of the data matrix more than once. The RAP algorithm offers $O(N)$ complexity, easy implementation, facile multiprocessor implementation, stable operation, and comparable convergence performance to the RLS algorithm. This chapter will also address the fact that the least squares estimate is not robust either in the presence of statistical outliers in the data or when the data matrix is ill conditioned. Outliers in the data can be detected by the RAP algorithm and subsequently ignored in the estimation process. The RAP algorithm is stable for ill-conditioned data matrices due to the fact that it performs an implicit pseudo-inverse of the data matrix to obtain an estimate of the unknown coefficients. Unlike the direct inverse, the pseudo-inverse is well defined and bounded for rank-deficient matrices. This property is achieved by noninversion of the zero-valued eigenvalues of the correlation matrix. Spectral tapering of the inverse eigenvalues—that is, regularization—avoids noise enhancement for ill-conditioned data corrupted by noise. We will present a method of controlling the regularization inherent in the RAP algorithm.

6.2 DISCRETE-TIME ADAPTIVE FILTERING

The advantages of describing adaptive filtering in terms of matrix-vector operations are numerous. First, it allows the process of adaptation to be unified with the field of linear algebra dealing with inverting linear algebraic systems. The mathematics associated with vector spaces and matrix inversion is well understood and is very strong. Second, the mechanism of adaptation can be given an intuitive description based upon a search for an optimal solution in a particular vector space. Third, implementations of matrix-vector operations are becoming available in both semiconductor and optical computing devices. Thus, recasting the adaptive signal problem as a sequence of matrix-vector operations offers conceptual and practical advantages.

A word about semantics is in order before we proceed. The term "adaptive signal processing" has a rather broad scope and can have numerous interpretations. The adaptive filtering problem can be decomposed into three distinct categories [24]:

- Interpolation: past and future observations are used for the current estimate.
- Filtering: past and present, but no future, observations are used for the current estimate.
- Prediction: past observations only are used for the current estimate.

Note that interpolation is noncausal, while both filtering and prediction are causal operations. The mathematical procedures used are essentially the same for each type of estimation. Thus, the term adaptive filtering will be used for the remainder of this chapter; the actual type of estimation problem must be inferred from the context.

The goal of adaptive filtering is to form an optimal estimate of some unknown quantity based upon a set of observations. The optimality criterion for the estimate may vary [see Chapter 2]; however, the minimum mean square error

(MMSE) criterion is often used. Minimizing the sum of squared errors is an optimal approach when the probability distribution of the observation noise is a Gaussian distribution. Although this is hardly ever true for any particular set of observations, the central limit theorem of probability theory indicates that this Gaussian assumption is a valid one for many cases.

6.2.1   The Transversal Filter

The discrete-time transversal filter is a good starting point for understanding the connection between adaptive filtering and vector spaces. A typical transversal filter is shown in Figure 6.1. The inputs are normally considered to be sampled values of a continuous waveform which are cascaded through the delay line at the sample rate. The output of the filter at time k is given by the standard

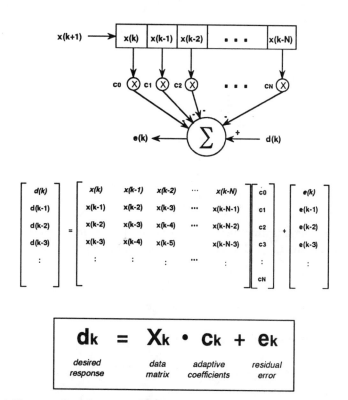

MATRIX FORMULATION OF ADAPTIVE FILTER

$$\mathbf{d_k} = \mathbf{X_k} \cdot \mathbf{c_k} + \mathbf{e_k}$$

| desired response | data matrix | adaptive coefficients | residual error |

Figure 6.1 The generation of the data matrix for a transversal filter structure. The operation of the adaptive filter is equivalent to the solution of the system of linear equations.

convolution formula

$$y_k = \sum_{n=0}^{N} c_k x_{k-n}, \tag{6.3}$$

where c_k is the impulse response of the filter and x_k is the filter input. The operation in (6.3) has an equivalent vector-space representation:

$$y_k = \mathbf{c}_k^T \mathbf{x}_k, \tag{6.4}$$

where the $(N + 1) \times 1$ column vectors are defined as $\mathbf{x}_k = [x_k x_{k-1} \cdots x_{k-N}]^T$ and $\mathbf{c}^T = [c_0 c_1 \cdots c_N]$. The transpose operator is defined by $(\cdot)^T$. By comparing (6.3) and (6.4), we can see that the convolution of the input and the output at time k is equivalent to the scalar product of the impulse-response vector with the tapped delay line vector at time k. This may be extended to the multiple output case by

$$\mathbf{y}_k = \mathbf{X}_k \mathbf{c}, \tag{6.5}$$

where

$$\mathbf{X}_k = \begin{bmatrix} x_k & x_{k-1} & \cdots & x_{k-N} \\ x_{k-1} & x_{k-2} & \cdots & x_{k-1-N} \\ \vdots & & \ddots & \vdots \\ x_{k-L} & x_{k-L-1} & \cdots & x_{k-L-N} \end{bmatrix}, \tag{6.6}$$

$$\mathbf{y}_k = \begin{bmatrix} y_k \\ y_{k-1} \\ \vdots \\ y_{k-L} \end{bmatrix}. \tag{6.7}$$

The output of the adaptive filter, y_k, can be viewed as an estimate of a desired response of the filter, d_k. The error of estimation e_k can be used to form

$$d_k = y_k + e_k, \tag{6.8a}$$

$$\mathbf{d}_k = \mathbf{X}_k \mathbf{c} + \mathbf{e}_k. \tag{6.8b}$$

The mean square error of (6.8b) is defined as

$$\epsilon = E\left\{\mathbf{e}_k^T \mathbf{e}_k\right\} = E\left\{\left[\mathbf{d}_k^T - \mathbf{c}^T \mathbf{X}_k^T\right] [\mathbf{d}_k - \mathbf{X}_k \mathbf{c}]\right\}. \tag{6.9}$$

The optimal coefficients $\mathbf{c}_{opt}$ that satisfy (6.9) are constant in time for the stationary random processes x_k and d_k. Thus, the expectation operation only applies to the data matrices

$$\epsilon = E\left\{\mathbf{d}_k^T\mathbf{d}_k\right\} - 2E\left\{\mathbf{X}_k^T\mathbf{d}_k\right\}\mathbf{c} + \mathbf{c}^T E\left\{\mathbf{X}_k^T\mathbf{X}_k\right\}\mathbf{c}. \tag{6.10}$$

The cross-correlation between the input and the output is given by the cross-correlation vector

$$\mathbf{r} = E\left\{\mathbf{X}_k^T\mathbf{d}_k\right\}. \tag{6.11}$$

The autocorrelation of the input is given by the statistical correlation matrix

$$\mathbf{R} = E\left\{\mathbf{X}_k^T\mathbf{X}_k\right\}. \tag{6.12}$$

Finally, the variance of the output is given by

$$\sigma_d^2 = E\left\{\mathbf{d}_k^T\mathbf{d}_k\right\}. \tag{6.13}$$

Thus, (6.9) can be written as

$$\epsilon = \sigma_d^2 - 2\mathbf{r}^T\mathbf{c} + \mathbf{c}^T\mathbf{R}\mathbf{c}. \tag{6.14}$$

The MMSE is obtained from setting $\partial\epsilon/\partial\mathbf{c} = 0$:

$$\nabla = \frac{\partial\epsilon}{\partial\mathbf{c}} = -2\mathbf{r} + 2\mathbf{R}\mathbf{c} = 0. \tag{6.15}$$

This leads to the well-known Wiener–Hopf equations,

$$\mathbf{r} = \mathbf{R}\mathbf{c}_{\text{opt}}. \tag{6.16}$$

The optimal coefficients become

$$\mathbf{c}_{\text{opt}} = \mathbf{R}^{-1}\mathbf{r}. \tag{6.17}$$

The solution of (6.17) forms the basis of most least squares adaptive filtering. It should be noted that the preceding analysis assumes that the signals involved are stationary. It will be seen that solutions for nonstationary signals follow a similar procedure, except that the statistical quantities become time dependent. A variety of different filtering applications can be put into the form of (6.17). The method of calculation of $\mathbf{c}_{\text{opt}}$ depends primarily on the data matrix characteristics and the computational resources available. The techniques of most importance in adaptive filtering involve iterative solutions that allow tracking of nonstationary signals and also require minimal computational resources. We will now discuss these techniques and how they relate to matrix-vector operations.

6.3   ADAPTIVE SOLUTION TECHNIQUES

6.3.1   Gradient Descent Technique

The MMSE in (6.14) can be obtained from the inversion of the statistical correlation matrix $\mathbf{R}$ or it can be calculated iteratively. The iterative technique of gradient descent uses the fact that the MSE is a quadratic function of the filter coefficients. The implication of this is that there exists an unique minimum $\epsilon_{\min}$ corresponding to the optimal filter coefficients $\mathbf{c}_{\text{opt}}$. In $N + 1$ dimensions, this implies an $(N + 1)$-dimensional "bowl" or parabolic error surface. The bottom of the bowl corresponds to the unique minimum for the error function ϵ. Searching the error surface for the unique minimum has been called gradient descent. That is, at any arbitrary point on the surface the gradient points away from the minimum. This translates into a conceptually pleasing iterative technique for error reduction:

$$\mathbf{c}_{k+1} = \mathbf{c}_k + [\text{a step in the direction opposite the local gradient}],$$

where $\mathbf{c}_{k+1}$ is a better estimate than $\mathbf{c}_k$ in terms of reducing the MSE. The filter coefficients are now adaptive and are usually updated with each sample. The iterative objective can be stated formally as

$$\mathbf{c}_{k+1} = \mathbf{c}_k + \frac{\mu}{2}[-\nabla_k], \tag{6.18a}$$

$$\mathbf{c}_{k+1} = \mathbf{c}_k + \mu[\mathbf{r} - \mathbf{R}\mathbf{c}_k], \tag{6.18b}$$

where ∇_k is obtained from (6.15) by replacing $\mathbf{c}$ with $\mathbf{c}_k$. The scalar μ is known as the step size and controls the length of the step in the direction of the negative gradient. The convergence of the iterative process in (6.18a) can be seen by observing the deviation from the optimal coefficients [9]:

$$\Delta_{k+1} = [\mathbf{I} - \mu\mathbf{R}]\Delta_k, \tag{6.19a}$$

$$\Delta_k = [\mathbf{I} - \mu\mathbf{R}]^k \Delta_0, \tag{6.19b}$$

where $\Delta_k = \mathbf{h}_k - \mathbf{h}_{\text{opt}}$. It can also be shown that (6.19) converges to zero if the eigenvalues of the statistical correlation matrix $\mathbf{R}$ satisfy $|1 - \mu\lambda_l| < 1$ [24]. Thus, the convergence properties of the steepest descent method depend primarily on the eigenvalues of the statistical correlation matrix and the step size μ. The problem with the gradient descent technique is that it requires the product of a matrix-vector product, which is usually too computationally expensive.

6.3.2   The Least Mean Squares Algorithm

The calculation and storage of the statistical correlation matrix and the cross-correlation vector often present a problem to the practical implementation of the gradient descent algorithm. This leads to the technique of stochastic approximation of the statistical quantities, which was first proposed by Robbins and Monro [25] and later used by Widrow and Hoff [26] in deriving the widely used least mean squares (LMS) algorithm. The basis of the LMS algorithm is the approximation of the statistical quantities by their instantaneous values,

$$\mathbf{R} = E\left\{\mathbf{x}_k \mathbf{x}_k^T\right\} \rightarrow \mathbf{x}_k \mathbf{x}_k^T, \tag{6.20a}$$

$$\mathbf{r} = E\left\{\mathbf{x}_k d_k\right\} \rightarrow \mathbf{x}_k d_k, \tag{6.20b}$$

Inserting (6.20) into (6.18b), we get the standard formulation of the LMS algorithm update equation:

$$\mathbf{c}_{k+1} = \mathbf{c}_k + \mu \left[d_k - \mathbf{x}_k^T \mathbf{c}_k\right] \mathbf{x}_k. \tag{6.21}$$

The average coefficient error of the LMS algorithm follows the same recursion as the one for coefficient error of the gradient descent method:

$$E\left\{\mathbf{\Delta}_{k+1}\right\} = \bar{\mathbf{\Delta}}_{k+1} = \left[\mathbf{I} - \mu \mathbf{R}\right] \bar{\mathbf{\Delta}}_k. \tag{6.22}$$

The difference between (6.19) and (6.22) is that the LMS algorithm converges in the mean to the optimal solution, while the gradient descent method converges to the optimal solution for a given set of data. Also note that the LMS algorithm does not require a matrix-vector multiplication.

6.3.3 Least Squares Algorithms

There are numerous algorithms that solve for the deterministic counterpart of the Wiener solution. These algorithms are collectively called *least squares algorithms*. The estimate that minimizes the sum of the squared errors is called the least squares (LS) estimate and takes the form

$$\mathbf{c}_{LS} = \left[\mathbf{X}_k^T \mathbf{X}_k\right]^{-1} \mathbf{X}_k^T \mathbf{d}_k. \tag{6.23}$$

The LS estimate can be written as

$$\mathbf{c}_{LS} = \mathbf{\Phi}_k^{-1} \rho_k \tag{6.24}$$

by defining

$$\mathbf{\Phi}_k = \mathbf{X}_k^T \mathbf{X}_k, \tag{6.25}$$

$$\rho_k = \mathbf{X}_k^T \mathbf{d}_k, \tag{6.26}$$

where $\mathbf{\Phi}_k$ is known as the deterministic correlation matrix and ρ_k is the deterministic cross-correlation vector. The LS estimate has the geometrical interpretation of a projection of the desired data vector $\mathbf{d}_k$ onto the space spanned by the columns of the observation data matrix $\mathbf{X}_k$. This is the best least squares estimate of the desired output based upon the observed data.

The *recursive least squares* (RLS) algorithm iteratively updates the correlation matrix inverse $\mathbf{\Phi}_k^{-1}$. The standard RLS algorithm has the form

$$\mathbf{g}_k = \frac{\lambda^{-1}\mathbf{P}_{k-1}\mathbf{x}_k}{1 + \lambda^{-1}\mathbf{x}_k^T\mathbf{P}_{k-1}\mathbf{x}_k}, \tag{6.27a}$$

$$e_k = d_k - \mathbf{c}_{k-1}^T\mathbf{x}_k, \tag{6.27b}$$

$$\mathbf{c}_k = \mathbf{c}_{k-1} + e_k\mathbf{g}_k, \tag{6.27c}$$

$$\mathbf{P}_k = \lambda^{-1}[\mathbf{P}_{k-1} - \mathbf{g}_k\mathbf{x}_k^T\mathbf{P}_{k-1}], \tag{6.27d}$$

where $\mathbf{g}_k$ is the *gain vector* and $\mathbf{P}_k$ is the estimate of $\mathbf{\Phi}_k^{-1}$ using the matrix inversion lemma [27]. The parameter $0 < \lambda \leq 1$ allows the more recent samples to have greater influence on the output error, allowing for the tracking of non-stationary signals. The iteration (6.27a)–(6.27d) is initialized with $\mathbf{P}_0 = \delta^{-1}\mathbf{I}$ to guarantee invertibility of the correlation matrix, where $0 < \delta \ll 1$. The introduction of the forgetting factor λ and the initial condition, $\mathbf{P}_0 = \delta\mathbf{I}$, is equivalent to transforming the correlation matrix into

$$\mathbf{X}_k^T\mathbf{\Lambda}\mathbf{X}_k + \lambda^k\delta\mathbf{I}, \tag{6.28}$$

where $\mathbf{\Lambda} = \text{diag}[1, \lambda, \ldots, \lambda^{k-1}]$. The initial condition adds a small positive term, $\lambda^k\delta$, to each of the eigenvalues of the correlation matrix. This stabilizing effect eventually decays if the forgetting factor λ is less than unity, as is the case for time-varying system tracking.

A more general form of the least squares solution of (6.23) is

$$\mathbf{c}_{\text{PI}} = \mathbf{X}^\dagger\mathbf{d}, \tag{6.29}$$

where $\mathbf{X}^\dagger$ is the Moore–Penrose pseudo-inverse of the data matrix $\mathbf{X}$ [28]. The pseudo-inverse solution is unique and is the least squares solution of *minimum norm*. The pseudo-inverse solution is identical to the least squares solution only if the data matrix is full rank. Indeed, when the data matrix $\mathbf{X}$ is rank deficient, the correlation matrix inverse, $[\mathbf{X}^T\mathbf{X}]^{-1}$, does not exist. The pseudo-inverse, however, is well defined for any $m \times n$ matrix, regardless of its rank. The pseudo-inverse of the data matrix can be described by its *singular-value decomposition* (SVD). Let

$$\mathbf{X} = \mathbf{U}\mathbf{\Sigma}_x\mathbf{V}^T \tag{6.30}$$

be the SVD of the data matrix, where $\mathbf{U}$ is an $L \times L$ unitary matrix, $\mathbf{V}$ is an $N \times N$ unitary matrix, and $\mathbf{\Sigma}_x$ is an $L \times N$ matrix of *singular values*. The overdetermined case of (6.8b) produces an equivalent SVD decomposition where $\mathbf{U}$ is $L \times N$ and

$\mathbf{\Sigma}_x = \text{diag}(\sigma_{x1}, \sigma_{x2}, \dots, \sigma_{xN})$ is an $N \times N$ diagonal matrix, where $\sigma_{xi}^2 > 0$. The correlation matrix can also be diagonalized using the SVD decomposition in (6.30):

$$\mathbf{X}^T\mathbf{X} = \mathbf{V}\mathbf{\Phi}\mathbf{V}^T, \tag{6.31}$$

where

$$\mathbf{\Phi} = \mathbf{\Sigma}_x^2. \tag{6.32}$$

Equation (6.31) is the similarity transformation that produces the diagonal eigen-value matrix $\mathbf{\Phi}$. The *condition number* of the data matrix is defined as $\gamma = \sigma_{x\,max}/\sigma_{x\,min}$. A matrix is considered *ill conditioned* if the condition number is large relative to the numerical precision used. Inverting an ill-conditioned matrix usually causes numerical anomalies because of the limited dynamic range of the computing hardware. In practice, a condition number greater than ten imposes serious numerical problems on many methods of inversion. The singular values of the data matrix are the square roots of the eigenvalues of the correlation matrix (see (6.32)). Using (6.30), the pseudo-inverse of $\mathbf{X}$ is

$$\mathbf{X}^\dagger = \mathbf{V}\mathbf{\Sigma}_x^\dagger\mathbf{U}^T, \tag{6.33}$$

where $\mathbf{\Sigma}_x^\dagger = \text{diag}(\sigma_{x1}^\dagger, \sigma_{x2}^\dagger, \dots, \sigma_{xN}^\dagger)$, and $\sigma_{xi}^\dagger = \sigma_{xi}^{-1}$ only if $\sigma_{xi} \neq 0$ and $\sigma_{xi}^\dagger = 0$ otherwise. The number of nonzero singular values will denote the *rank* of the data matrix. Thus, if the data matrix has a null space, i.e., is rank deficient, then at least one of the singular values will equal zero. Components of the desired signal that reside in the null space of the data matrix are not observable, i.e., these components are not coupled to the observation vector. It is for this reason that the components of the solution $\mathbf{c}_{LS}$ residing in the null space of $\mathbf{X}$ will not contribute to the residual error $\mathbf{e}$ in (6.8b). However, it is still desirable to control the null-space behavior of $\mathbf{c}_{LS}$ in order to maintain certain characteristics of the estimate $\hat{\mathbf{d}} = \mathbf{X}\mathbf{c}_{LS}$. For example, in a nonstationary signal environment, the null space of the data matrix may abruptly change with time, with the result that previous null-space components of $\mathbf{c}_{LS}$ now operate on the data. Thus, the algorithm will produce a large residual error if the null-space components of the coefficient vector have previously become arbitrarily large. The solution to this problem is to ensure that there are no components of $\mathbf{c}_{LS}$ in the null space of $\mathbf{X}$. This is equivalent to requiring the least squares solution to be minimum norm, i.e., the pseudo-inverse solution. For this reason, the pseudo-inverse solution will be more robust than any other least squares solution.

The presence of additive noise in the system has the tendency to improve the condition number of the data matrix. That is, the noisy data matrix is of the form

$$X = X_{signal} + X_{noise}, \tag{6.34}$$

where X_{noise} represents a (probably) full-rank noise matrix of random values and X_{signal} is a rank-deficient signal matrix of rank r. The data matrix will now have singular values of the form $\Sigma_x = \text{diag}(\sigma_{s1} + \sigma_{n1}, \ldots, \sigma_{sr} + \sigma_{nr}, \sigma_{nr+1}, \ldots, \sigma_{nN})$. The inverse of the composite correlation matrix exists, since it is full rank; however, inversion of the singular values beyond σ_r essentially serves only to enhance the noise components in the data matrix. Thus, truncating the diagonal matrix Σ_x prior to inversion to eliminate noise-only components improves the solution [29]. This method requires ascertaining the rank of the correlation matrix from the data, usually a formidable task.

The predominant case for the data matrix is to have a full-rank signal matrix component and a full-rank noise matrix component. Assuming the noiseless case, the data matrix may still be *ill conditioned* even though it is not rank deficient, e.g., $\sigma_{s\,min} \ll 1$. As stated previously, this poses a problem in implementation and stability due to the numerical overflow in inverting very small numbers. The presence of noise will usually alleviate this problem by reducing the condition number of the data matrix. However, for those singular values, $\sigma_x = \sigma_s + \sigma_n$, satisfying $\sigma_s \ll \sigma_n$, direct inversion will result in noise enhancement. Since none of the signal-component singular values are zero, setting them to zero (as in the pseudo-inverse) is inappropriate. What is necessary is to apply a taper to the inversion of the singular values such that modes with low signal-to-noise ratios are attenuated. This process is called regularizing the spectrum, that is, smoothing the spectral components. Essentially, this amounts to minimizing the energy in the signal null-space components.

6.4 ILL-CONDITIONED DATA

Insight into what ill-conditioned data is is needed in order to proceed with the development of a technique to combat its effects. An easy way to do this is to examine the filtering problem from a frequency-domain viewpoint where the generation of ill-conditioned data will be interpreted as a frequency-nulling operation.

Consider a statistically white sequence d, of length n and variance σ_d^2, as the input to a transversal filter, $c = (c_0, c_1, \ldots, c_{l-1})^T$. On average, the input sequence has equal amplitudes at all its spectral components. The output sequence x, also of length n, is obtained as the convolution of the filter with the input sequence,

$$x = H_c d, \tag{6.35}$$

where $\mathbf{H}_c$ is the $n \times n$ convolution matrix. That is,

$$\mathbf{H}_c = \begin{bmatrix} h_0 & \cdots & h_{l-1} & 0 & \cdots & 0 \\ 0 & h_0 & \cdots & h_{l-1} & \cdots & 0 \\ \vdots & & \ddots & \ddots & & \vdots \\ 0 & 0 & \cdots & & 0 & h_0 \end{bmatrix}. \tag{6.36}$$

The convolution matrix is approximately circulant if $n \gg l$ [30]. Using (6.35), the correlation matrix of the filter output is

$$\mathbf{R}_{xx} = E\left\{\mathbf{x}\mathbf{x}^T\right\} = \mathbf{H}_c E\left\{\mathbf{d}\mathbf{d}^T\right\} \mathbf{H}_c^T = \sigma_d^2 \mathbf{H}_c \mathbf{H}_c^T. \tag{6.37}$$

Any circulant matrix can be diagonalized with a unitary matrix of the form [30]

$$\mathbf{F} = \frac{1}{\sqrt{N}} \begin{bmatrix} 1 & 1 & 1 & \cdots & 1 \\ 1 & f & f^2 & \cdots & f^{N-1} \\ 1 & f^2 & f^4 & \cdots & f^{2(N-1)} \\ \vdots & \vdots & \vdots & \ddots & \vdots \\ 1 & f^{N-1} & f^{2(N-1)} & \cdots & f^{(N-1)(N-1)} \end{bmatrix}, \tag{6.38a}$$

$$f = e^{2\pi i/N}. \tag{6.38b}$$

The form of (6.38) is identical to the discrete Fourier transform (DFT) matrix used to obtain a frequency estimate averaged over many snapshots of data [31]. The matrix $\mathbf{F}$ represents a forward DFT operation and the matrix $\mathbf{F}^T$ is the inverse DFT. The convolution matrix in (6.36) can be diagonalized using (6.38) as

$$\mathbf{H}_c = \mathbf{F}^T \mathbf{\Lambda} \mathbf{F}. \tag{6.39}$$

Substituting (6.39) into (6.37) yields

$$\mathbf{R}_{xx} = \sigma_d^2 \mathbf{F}^T \mathbf{\Lambda}^2 \mathbf{F}. \tag{6.40}$$

Thus, the eigenvalues of $\mathbf{R}_{xx}$ are interpreted as power spectral density (PSD) components of the filter autocorrelation. It should be noted that the DFT matrix cannot diagonalize a matrix that is not Toeplitz. The relationship between the eigenvalue spread (condition number) and the frequency characteristics of the input data is apparent from (6.40). The nulls in the spectrum of the input signal are directly analogous to small eigenvalues. The inversion process becomes unstable because the inverse requires the reciprocal of these small numbers.

When the small eigenvalues are dominated by noise components, the inversion will amplify the noise power.

The minimum eigenvalue of the data-correlation matrix for a transversal filter has been expressed as a function of the PSD of the input data [10]. Let Ω represent the continuous PSD of the data and let there be k nonvanishing derivatives at the minimum of Ω. Then the expression for the minimum eigenvalue of the correlation matrix, Φ, is [32]

$$\lambda_{\min} = \Omega_{\min} + \frac{\eta(k)\beta}{k!N^k} + O\left[\frac{1}{N^k}\right], \tag{6.41}$$

where β is the value of the kth derivative at $\Omega_{\min}$ and N is the length of the filter. The effect of (6.41) on the noise sensitivity of the least squares estimate can be seen by perturbing the input of the Wiener filter:

$$\left[\mathbf{R}_{xx} + \sigma_n^2\mathbf{I}\right]\mathbf{c}_p = \mathbf{r}_{xd}, \tag{6.42}$$

where $\sigma_n^2 \ll 1$ represents uncorrelated noise power. The sensitivity can now be expressed as [33]

$$\delta_c = \frac{\|\mathbf{c}_p - \mathbf{c}\|}{\|\mathbf{c}\|} \le \frac{\sigma_n^2}{\lambda_{\min}} + O(\sigma_n^4) \approx \frac{\sigma_n^2}{\lambda_{\min}}. \tag{6.43}$$

Combining (6.41) and (6.43) yields an approximate upper bound for the least squares sensitivity:

$$\delta_c \le \frac{\sigma_n^2}{\Omega_{\min} + \eta(k)\beta/k!N^k}. \tag{6.44}$$

The relationship (6.44) shows that the least squares filter coefficient estimate is very sensitive to low noise levels if the data has severe spectral nulls. The noise tends to occupy the frequency spectrum uniformly, leaking into the signal-deficient portions of the data matrix. The least squares algorithm cannot discern the noise components from the signal components; thus, the entire frequency range tends to get inverted, producing noise amplification. The approach of tapering, or weighting, the spectrum before inversion to avoid this problem is called *regularization*. In Section 6.5, we shall introduce a method that will provide a regularized solution by an adaptive procedure.

6.5   THE ROW-ACTION PROJECTION ALGORITHM

The conventional methods of adaptive filtering are steeped in analysis. The basic conclusions to be drawn from these analyses are that the LMS algorithm is effective where computational resources are at a premium and convergence speed is not critical. The least squares algorithms are useful in applications where convergence speed is critical. However, least squares techniques suffer from the accumulation of finite arithmetic errors and, more importantly, the instability associated with inverting an ill-conditioned matrix.

The solution of the Wiener–Hopf equations is often approached from the analysis of the statistical behavior of the update equations. However, the system of linear equations represented by (6.8b) also contains all the information necessary to arrive at the optimal solution c_{opt}. The techniques from matrix analysis and the inversion of linear systems of equations can be used equally well to solve for the optimal coefficients. Although these techniques are essentially equivalent to the statistical procedures, they offer geometrical insights based upon vector-space operations. These insights will allow us to modify and improve upon the standard LMS method.

The remainder of this chapter will focus on an algorithm that combines the good properties of both the LMS and LS algorithms while avoiding most of the pitfalls associated with these methods. In short, the advantages of fast convergence and numerical stability are combined into one algorithm—the *row-action projection algorithm* (RAP).

6.5.1 The Row-Action Projection Technique

The RAP algorithm approach for solving the system of equations in (6.8b) is an iterative improvement of the filter estimate based upon projections onto hyperplanes. Each equation in the system (6.8b) represents a hyperplane in the L-dimensional coefficient vector space. While the equations individually have many solutions, the combination of all the equations will have only one optimal solution. This is shown in Figure 6.2 for the simple case of a 3×2 system of equations. Starting from any arbitrary point in the coefficient space c_k, one can determine the filter coefficients that satisfy an equation by orthogonally projecting onto the hyperplane for that equation. This projection can be constructed by determining the unit normal vector to the hyperplane and the orthogonal distance of c_k to the hyperplane. The equation describing a single hyperplane is $c \cdot x = d$. From this equation, we see that the vector x is orthogonal to the hyperplane and the orthogonal distance from the origin to the hyperplane is d. The unit normal to the hyperplane is given as $x/\|x\|$. The distance between c_k and its orthogonal projection c_{k+1} is given by

$$\Delta = c_k \cdot \frac{x}{\|x\|} - c_{k+1} \cdot \frac{x}{\|x\|} = \frac{c_k \cdot x - d}{\|x\|}, \tag{6.45}$$

where $c_{k+1} \cdot x = d$, since c_{k+1} terminates on the hyperplane. The result in (6.45) can be used to form the final projection-update formula

$$c_{k+1} = c_k - \mu \Delta \frac{x}{\|x\|}, \tag{6.46}$$

SEQUENCE OF PROJECTIONS

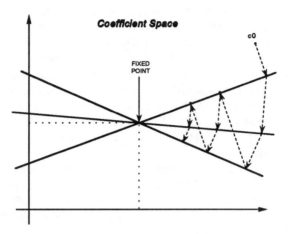

Figure 6.2 Each equation in the system describes a hyperplane in coefficient space. A sequence of orthogonal projections onto the set of hyperplanes solves the system of equations.

where μ is the step size, or relaxation parameter, that determines the projection distance. A step size of unity, $\mu = 1$, causes the projection-update vector to terminate on the hyperplane. This is illustrated in Figure 6.3. The goal of the RAP algorithm is to satisfy all the equations present in (6.8b). This global solution is approached by sequential multiple projections onto the hyperplanes. That is, given a system with $L + 1$ equations, the optimal solution vector is iteratively calculated using

$$\mathbf{c}_{k+1} = \mathbf{c}_k + \mu[d_j - \mathbf{c}_k^T \mathbf{x}_j]\frac{\mathbf{x}_j}{\|\mathbf{x}_j\|^2}, \qquad (6.47a)$$

$$\mathbf{c}_{k+1} = \left[\mathbf{I} - \mu\frac{\mathbf{x}_j \mathbf{x}_j^T}{\|\mathbf{x}_j\|^2}\right]\mathbf{c}_k + \mu d_j\frac{\mathbf{x}_j}{\|\mathbf{x}_j\|^2}, \qquad (6.47b)$$

where $\mathbf{c}_k$ is a previous estimate and $\mathbf{c}_{k+1}$ is the updated estimate that, if $\mu = 1$, exactly satisfies the jth equation (row) in (6.8b). The iteration index is related to the equation index by $j = i \bmod L + 1$. This cyclic update relationship is shown geometrically in Figure 6.2. The distance of the ith estimate to the jth hyperplane is

$$\epsilon_{k,j} = \frac{e_{k,j}}{\|\mathbf{x}_j\|}, \qquad (6.48)$$

where $e_{k,j} = [d_j - \mathbf{c}_k^T \mathbf{x}_k]$ and $\mathbf{x}_j/\|\mathbf{x}_j\|$ is the unit normal to the jth hyperplane.

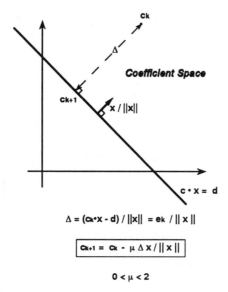

$$\Delta = (c_k \cdot x - d) / \|x\| = e_k / \|x\|$$

$$\boxed{c_{k+1} = c_k - \mu \Delta x / \|x\|}$$

$$0 < \mu < 2$$

Figure 6.3 Each coefficient vector update can correspond to a projection toward a hyperplane. The step size μ determines the projection distance.

6.5.2 RAP Operation

The basis of RAP is the repeated sequential application of (6.47a) on the rows (hyperplanes) of the data matrix. This iterative process converges to a pseudo-inverse solution for the unknown coefficients [34]. The steps comprising the RAP algorithm are as follows:

- Retain data vectors whose elements are the rows of the data matrix along with the desired output, i.e., $s_k = [x_k, d_k]$. For a transversal filter, this amounts to storing the input sample and the associated desired output sample at each sampling instant.
- Once the data buffer is full, make multiple updates of the adaptive coefficients by sequentially using the data vectors, s_k, and the update equation (6.47a).

The RAP algorithm is seen to be a block-by-block update procedure, that is, the indefinitely long data matrix is decomposed into blocks of size L. Each estimate is cyclically computed to form a local solution for the adaptive coefficients based upon that block of data. This allows fast tracking if the data is changing rapidly, that is, on the order of the block size. This approach, because it operates on each equation in the block individually, should not be confused with a block-projection approach, which uses matrix vector multiplication. The block-by-block action of the RAP algorithm is depicted in Figure 6.4.

The following optional steps may improve the performance of the algorithm:

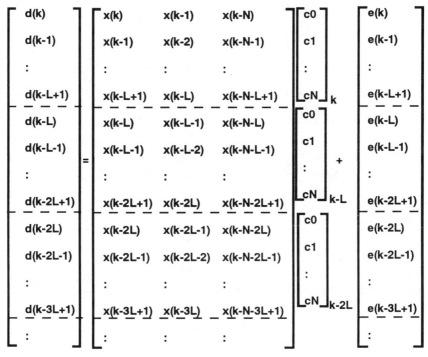

Figure 6.4 The data matrix is decomposed into smaller blocks by the RAP algorithm for processing. The solution for each block provides the starting point for the next block. In a parallel implementation, multiple blocks are processed simultaneously.

- Ascertain the hyperplane that produces the largest residual error, using (6.48), for each cycle through the data. Do not reuse this hyperplane to update the coefficients on the subsequent cycle. This step helps make the algorithm robust; that is, an inconsistent datum will consistently be disregarded by the coefficient-updating process (Figure 6.5).

- Start cycling through the data with a relatively large step size, i.e., $1 < \mu < 2$. Gradually diminish the step size so that it is small for the last cycle through the data, i.e., $0 < \mu < 1$. This is analogous to the *gear shifting* employed in the standard LMS algorithm and provides fast initial convergence with low excess error. It has been found in computer simulations that exponential step-size decay produces good results.

- Compute an accelerated update using three standard RAP updates. This step is equivalent to a pair-wise orthogonalization of the input data and will usually result in enhanced convergence performance (Figure 6.6).

- Compute an accelerated update that uses the difference of two successive block coefficient estimates. This essentially finds a vector in the direction of maximum change in the coefficient space (Figure 6.7).

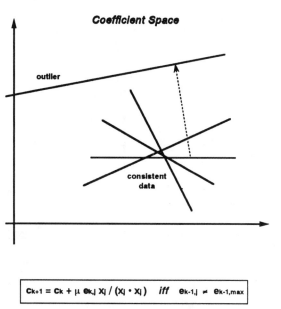

Figure 6.5 The cyclic reuse of the system matrix allows the RAP algorithm to detect outliers in the data. An outlier will have a projection distance that is always largest when the algorithm is near convergence.

6.5.3 Regularization Analysis

Regularization of ill-conditioned inversion problems was first proposed by Tikhonov and was later applied to signal restoration [35]. Regularization refers to the process by which sensitivity to ill-conditioned data is reduced in the inversion process by tapering the spectral components. However, the price paid for robustness is a reduction in the fidelity of the estimate. This can be shown as follows.

The trade-off between noise immunity and estimation accuracy can be shown by letting

$$c = X^{\dagger}d \tag{6.49}$$

be the solution to the system $d = Xc$. The regularized inversion is given by $G_n = \hat{X}^{\dagger}$ such that

$$\lim_{n \to \infty} G_n = X^{\dagger}. \tag{6.50}$$

The regularized estimate for a randomly perturbed observation is given by

$$c_n = G_n d', \tag{6.51}$$

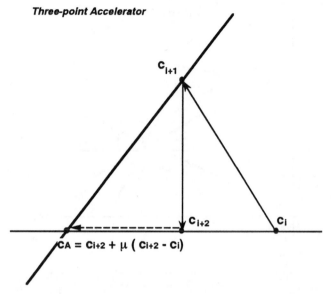

Figure 6.6 An acceleration technique that uses geometrical principles. Three coefficient updates are used to determine an improved estimate. This technique is equivalent to pairwise orthogonalization for a two-by-two system.

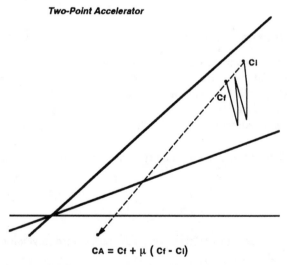

Figure 6.7 An acceleration technique that uses the difference of two successive block updates. This method updates the coefficients in the direction of maximum change.

where $\mathbf{d}' = \mathbf{d} + \mathbf{n}$. The error between these two estimates is

$$\|\mathbf{c}_n - \mathbf{c}\| = \|\mathbf{G}_n \mathbf{d}' - \mathbf{X}^\dagger \mathbf{d} + \mathbf{G}_n \mathbf{d} - \mathbf{G}_n \mathbf{d}\| \tag{6.52}$$

$$\|\mathbf{c}_n - \mathbf{c}\| \leq \|\mathbf{G}_n (\mathbf{d}' - \mathbf{d})\| + \|(\mathbf{G}_n - \mathbf{X}^\dagger) \mathbf{d}\| \leq \|\mathbf{G}_n\|\|\mathbf{n}\| + \|\mathbf{G}_n - \mathbf{X}^\dagger\|\|\mathbf{d}\| \tag{6.53}$$

The first term on the right-hand side of (6.53) represents the error due to the noise. This error is likely to increase without bound as $n \to \infty$ [36]. The second term on the right-hand side is the error due to the approximation of the pseudo-inverse, which goes to zero as $n \to \infty$.

The trade-off between noise attenuation and regularization error is apparent from (6.53). This relationship also highlights an interesting property of the regularization process: the estimation error first decreases, due to increased fidelity, and then it is likely to begin increasing, due to noise amplification. A reasonable question is, When should one stop the regularization process so that the two sources of error are in balance? The generally accepted guideline is known as the *discrepancy principle* [37]. According to the discrepancy principle, the value of the regularization parameter n is the one that satisfies

$$\|\mathbf{X}\mathbf{c}_n - \mathbf{d}'\| = \|\mathbf{n}\| = \sigma_n^2. \tag{6.54}$$

This is intuitively pleasing, since the fidelity-error power would equal the noise power if a perfect estimate were made. Continuing the regularization process beyond this point essentially forces the estimate $\mathbf{c}_n$ to reconstruct the noise, thus providing noise amplification.

The fact that the RAP algorithm is a regularized iterative update method can be seen by converting the row-action operator in (6.47a) into an equivalent matrix operator:

$$\mathbf{c}_l = \mathbf{H}\,\mathbf{c}_{l-1} + \tilde{\mathbf{d}}, \tag{6.55a}$$

$$\mathbf{H} = \mathbf{P}_k \mathbf{P}_{k-1} \cdots \mathbf{P}_{k-L+1}, \tag{6.55b}$$

$$\tilde{\mathbf{d}} = (\mathbf{P}_k \cdots \mathbf{P}_{k-L+2})\, \mu \frac{\mathbf{x}_{k-L+1}}{\|\mathbf{x}_{k-L+1}\|^2}\, d_{k-L+1} + \cdots + \mu \frac{\mathbf{x}_k}{\|\mathbf{x}_k\|^2}\, d_k, \tag{6.55c}$$

$$\mathbf{P}_k = \mathbf{I} - \mu \frac{\mathbf{x}_k \mathbf{x}_k^T}{\|\mathbf{x}_k\|^2}, \tag{6.55d}$$

where each increment of l represents an *entire* cycle through the data block of L equations, as in (6.8b). Further substitutions cast (6.55a)–(6.55d) into a recursive form for the pseudo-inverse of the data matrix $\mathbf{X}$ [38]:

$$\mathbf{c}_l = \mathbf{G}_l\, \mathbf{d}, \tag{6.56a}$$

$$\tilde{\mathbf{d}} = \mathbf{B}\,\mathbf{d}, \tag{6.56b}$$

$$\mathbf{B}\mathbf{X} = \mathbf{I} - \mathbf{H}, \tag{6.56c}$$

where G_l is an estimate of the pseudo-inverse of X after the lth cycle. The matrix B is obtained by substituting the vector

$$\eta_i = \begin{bmatrix} 0 \\ \vdots \\ 0 \\ 1 \\ 0 \\ \vdots \\ 0 \end{bmatrix} \tag{6.57}$$

for the scalar d_i in (6.55c) such that $\eta_i^T d = d_i$. Using this substitution, the matrix B can be written as

$$B = (P_k \cdots P_{k-L+2}) \, \mu \, \frac{x_{k-L+1}}{\|x_{k-L+1}\|^2} \, \eta_{k-L+1}^T + \cdots + \mu \, \frac{x_k}{\|x_k\|^2} \, \eta_k^T. \tag{6.58}$$

Substituting (6.56a)–(6.56c) into (6.55a), the pseudo-inverse recursion becomes

$$G_{l+1} = G_l + B[I - XG_l]. \tag{6.59}$$

Defining $E_l = I - G_l X$, with $B = G_0$, the error in the pseudo-inverse estimate can be stated in the equivalent forms

$$E_l = E_0[I - G_{l-1}X], \tag{6.60a}$$

$$E_l = E_0^{l+1}. \tag{6.60b}$$

The convergence of the inversion process (6.59) can be described by taking the singular-value decomposition of (6.60b), i.e.,

$$V[I - \Sigma_{gl}\Sigma_x]V^T = V[I - \Sigma_{g0}\Sigma_x]^{l+1}V^T. \tag{6.61}$$

Assuming that Σ_{gl} is diagonal, the inversion of the singular values of X follows:

$$\sigma_{gl} = \frac{1}{\sigma_x}\left[1 - (1 - \sigma_{g0}\sigma_x)^{l+1}\right]. \tag{6.62}$$

The general conclusion regarding (6.62) is that

$$\sigma_{gl} \rightarrow 1/\sigma_x \text{ if } 0 < \sigma_{g0}\sigma_x < 1. \tag{6.63}$$

If $|\mu| \ll 1$, then the first-order approximation of $G_0 = D$ becomes

$$G_0 \approx \mu X^T W, \tag{6.64}$$

where $\mathbf{W} = \text{diag}\left(1/\|\mathbf{x}_1\|^2, 1/\|\mathbf{x}_2\|^2, \ldots, 1/\|\mathbf{x}_L\|^2\right)$. The singular values of $\mathbf{G}_0$ in (6.64) are approximately given by

$$\sigma_{g0} \approx \frac{\mu}{N}\sigma_x, \tag{6.65}$$

where N is the number of adaptive coefficients. Substituting (6.65) into (6.62) yields

$$\sigma_{gl} \approx \frac{1}{\sigma_x}\left[1 - (1 - \frac{\mu}{N}\sigma_x^2)^{l+1}\right]. \tag{6.66}$$

The RAP algorithm is regularized via the iteration index, l, in (6.66). It provides a variable taper to the inversion of the singular values, allowing a trade-off between fidelity and noise attenuation. In general, this regularization gives more weight to inverting the stronger modes of the system because these modes are usually dominated by signal energy. The smaller singular values are inverted by the RAP algorithm only as the number of cycles through the block of equations increases. The RAP algorithm always produces zero as an estimate of a zero-valued singular value; that is, it goes toward the pseudo-inverse solution. The singular-value inversion characteristic of the exact least squares technique is such that all the singular values get inverted, even if some are zero. The initialization constant δ in (6.28) avoids the problem of division by zero; however, the mode corresponding to the null space is still greatly amplified, i.e., $\delta^{-1} \gg 1$. Since the RAP algorithm tends to attenuate small singular values, which can be dominated by noise energy, it will perform more effectively with regard to noise attenuation when the data matrix is ill conditioned. This point will be demonstrated by the numerical example given in the next section.

6.5.4 Regularized Inversion: Numerical Example

An example of regularization is given in Figure 6.8 for a rank-deficient data matrix with dimensions 50×11 and for $\mu = 0.2$. Note that the eigenvalues' inverses are plotted using $\phi_i = \sigma_i^2$. The singular-value matrix is given by $\Sigma = \text{diag}(1.0, 0.9, 0.8, \ldots, 0.1, 0.0)$. The smaller singular values are inverted by the RAP algorithm only as the number of cycles becomes large. The smallest singular value is identically zero, which corresponds to a rank-deficient matrix. The RAP algorithm always produces zero as an estimate of a zero-valued singular value, i.e., it tends toward the pseudo-inverse solution. The singular-value inversion characteristic of the RLS algorithm is given by the uppermost curves, indicating that all the singular values tend to get inverted, even if some are zero. The initialization constant δ in (6.28) avoids the problem of division by zero; however, the mode corresponding to the null space is still greatly amplified, i.e., $\delta^{-1} \gg 1$. Since the RAP algorithm tends to attenuate small singular values, which can be dominated by noise energy, it will perform more effectively with regard to noise attenuation when the data matrix is ill conditioned.

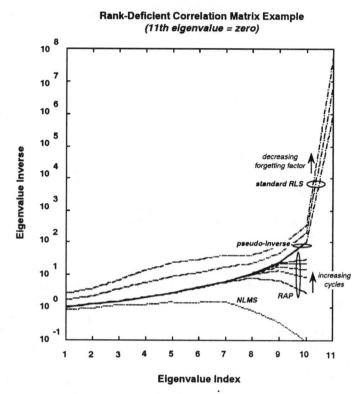

Figure 6.8 The regularizing action of the RAP algorithm is demonstrated. The RAP curves are plotted as 5, 10, 15, 20, and 25 cycles through the matrix. The RLS results represent forgetting factors of 1, 0.96, and 0.92.

6.5.5   Acceleration Techniques

In applying regularization to adaptive filtering there is concern about not only asymptotic convergence but also the rate of convergence. This section presents techniques that can be used to accelerate the rate of convergence of the RAP algorithm. Two simple acceleration techniques are presented that can be viewed as a modification of the regularization process given by (6.66).

A Two-Point Accelerator This accelerator uses the *difference* between two successive block-coefficient estimates to obtain a vector in the direction of maximum change (Figure 6.7). The accelerated update equations become

$$c_l^A = c_l + \alpha \left[c_l - c_{l-1}^A \right], \tag{6.67a}$$

$$c_l \leftarrow c_l^A, \tag{6.67b}$$

where the estimate c_l is obtained via (6.55a) and c^A is the accelerated estimate obtained after each cycle. The effect of (6.67) on the regularization process can be seen by rewriting it as

$$c_l^A = H_l c_{l-1}^A + B_l d_l + \alpha \left[H_l c_{l-1}^A + B_l - c_{l-1}^A \right], \tag{6.68a}$$

$$c_l^A = \left[(1 + \alpha)H_l - \alpha I \right] c_{l-1}^A + (1 + \alpha)B_l d_l. \tag{6.68b}$$

Making the substitution $B'_l = (1 + \alpha)B_l$ in (6.68b) yields

$$c_l^A = H'_l c_{l-1}^A + B' d_l, \tag{6.69}$$

where $H'_l = I - B'_l X$. The accelerated update is now in the form of the unaccelerated update (6.55a); thus, the regularization is the same except for the constant $1 + \alpha$:

$$\sigma_{gl}^A \approx \frac{1}{\sigma_x} \left[1 - \left(1 - \frac{\mu(1 + \alpha)}{N} \sigma_x^2 \right)^{l+1} \right]. \tag{6.70}$$

The bound for the acceleration parameter α is

$$0 < \alpha < \frac{2N}{\mu \sigma_{x,max}^2} - 1. \tag{6.71}$$

The action of α is to boost the lower-order singular values so that convergence for those modes is attained faster. The two-point accelerator will form one accelerated update, (6.69), for each standard update, (6.55a).

A Three-Point Accelerator The three-point accelerated update is given by

$$c_l^A = c_l + \alpha[c_l - c_{l-2}^A], \tag{6.72a}$$

$$c_l \leftarrow c_l^A, \tag{6.72b}$$

where c_l is given by (6.55a). This technique will form one accelerated update, (6.72a), for every three standard updates, (6.55a) (Figure 6.6). The motivation behind (6.72) is to obtain a better direction estimate by using older updates. This approach to acceleration can also be interpreted as a modification to the singular values in (6.66), although the derivation is not as straightforward as the one for the two-point accelerator.

The update in (6.72) can be converted to dependence on c_{l-2}^A, that is,

$$c_l^A = \left[\beta H^2 - (\beta - 1)I \right] c_{l-2}^A + \beta \left[I + H \right] d, \tag{6.73}$$

where $\beta = 1 + \alpha$. Using the transformations previously established, (6.73) can be written as a pseudo-inverse estimate recursion:

$$G_l^A = QG_{l-1}^A + PG_0^A, \tag{6.74a}$$

$$Q = \beta[I - G_0^A]^2 - (1 - \beta)I, \tag{6.74b}$$

$$P = \beta[2I + G_0^A X], \tag{6.74c}$$

$$G_0^A = B, \tag{6.74d}$$

$$G_1^A = B + B[I - XB]. \tag{6.74e}$$

The recursion (6.74a) can be written in terms of the initial pseudo-inverse estimate G_0^A as

$$G_l^A = \{Q^l[2I - X] + [Q^{l-1} + \cdots + Q + I]P\}G_0^A. \tag{6.75}$$

The terms in (6.75) can be diagonalized to yield

$$\sigma_{g\,l}^A \approx \frac{1}{\sigma_x}\left[1 - \left(1 - \frac{2(1 + \alpha)\mu}{N}\sigma_x^2\right)^{l+1}\right]. \tag{6.76}$$

The form of (6.76) is similar to that of (6.70). The major difference between the two is that the singular values are increased by a factor of two in (6.76). Thus, the expected rate of convergence is higher when using the three-point accelerator.

An interesting special case when using the three-point accelerator occurs for the case of a block size of two. The projections in (6.47a) will alternate between two hyperplanes in this case. The three-point accelerator will converge to the intersection of the two hyperplanes in one iteration if the acceleration parameter is

$$\alpha = \frac{\|c_l - c_{l-1}\|^2}{\|c_l - c_{l-2}\|^2}. \tag{6.77}$$

This procedure is equivalent to pairwise orthogonalization of the rows of the data matrix [39].

6.5.6   Augmented Regularization: Numerical Example

The acceleration techniques presented in Section 6.5.5 were applied to the example problem used in Section 6.5.4. The data matrix was slightly altered by the addition of a small positive term to the diagonal of the singular-value matrix— i.e., $\Sigma = \text{diag}(1.1, 1.0, 0.9, \ldots, 0.2, 0.1)$—and the step size was increased, $\mu = 0.5$. The additive term, which can represent an average noise level, makes the data matrix full rank, i.e., the condition number becomes $\gamma = 11$, or, equivalently, the eigenvalue spread becomes $\gamma^2 = 121$. Because the RAP algorithm will invert the smallest singular values in order from largest to smallest, the acceleration techniques are applied to reduce the convergence time associated with the smaller singular values. The application of the acceleration techniques is shown in Figure 6.9. The graph depicts the inversion of the smallest singular value for the standard RAP algorithm, the two-point accelerator, the three-point accelerator, and the combination of the two- and three-point accelerators. The combined accelerator uses the two-point technique after every block and the three-point technique after every third cyclic update. The true inverse of the smallest eigenvalue is $1/\phi_{\min} = 100$.

6.5.7 Error Analysis

There are two error metrics that are important when analyzing an adaptive algorithm. One is the estimation error and the other is the coefficient error. Often one error takes precedence over the other, depending on the application at hand. For example, in channel equalization minimizing the estimation error is important for reliable communications. The coefficient error is important in parameter-estimation problems, such as linear predictive coefficient coding of speech. Approximate expressions for the coefficient and estimation error for the RAP algorithm are given in the next two sections.

Coefficient Error

The coefficient error can be bounded by [40]

$$||\mathbf{c}_{k,\text{opt}} - \mathbf{c}_k|| = ||\epsilon_k|| \approx ||\mathbf{V}_k e^{-\frac{M\mu}{N}\Lambda_k}\Sigma_k^{-1}\mathbf{U}_k^T\mathbf{d}_k||, \qquad (6.78)$$

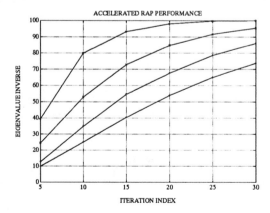

Figure 6.9 The application of the acceleration techniques for inversion of the smallest eigenvalue. (+): Standard RAP, (*): Two-point accelerator, (x): Three-point accelerator, (o): Two-point and three-point accelerator. The actual inverse is 100.

$$\|\epsilon\| \leq \|e^{-\frac{M\mu}{N}\Lambda_k}\Sigma_k^{-1}\|\|\mathbf{d}_k\|,\tag{6.79}$$

where the unitary matrices satisfy

$$\|\mathbf{V}_k\| = \|\mathbf{U}_k\| = 1.\tag{6.80}$$

The parameter N is the length of the adaptive coefficient vector and M is the number of cycles over each block. The slowest time constant in (6.79) is approximately M times the slowest time constant of the normalized LMS algorithm.

Estimation Error

The estimation error can be bounded in much the same way as the coefficient estimate error [40]:

$$\|\mathbf{e}_k\| \approx \|\mathbf{U}_k e^{-\frac{M\mu}{N}}\mathbf{U}_k\mathbf{d}_k\|,\tag{6.81}$$

$$\|\mathbf{e}_k\| \leq \|e^{-\frac{M\mu}{N}}\|\|\mathbf{d}_k\|.\tag{6.82}$$

The estimation error is less sensitive to the eigenvalue spread of the correlation matrix than is the coefficient error. This is readily seen by comparing (6.79) to (6.82). This is due to the faster inversion of the signal-dominated eigenmodes inherent in the regularization process.

6.5.8 Robustness

One major drawback of exact least squares techniques is that the estimates they produce tend to be sensitive to statistical outliers in the data. This is due to the underlying assumption that the noise distribution is Gaussian and has tails that decay as e^{-x^2}. An example of this sensitivity can be found in linear regression, where the goal is to fit the observed data to the model

$$y_i(x_i) = ax_i + b + e_i, \qquad i = 1, 2, 3, \ldots, N,\tag{6.83}$$

where the unknowns a and b are the slope and intercept, respectively. The exact least squares performance index is stated as

$$\mathcal{J}(a, b) = \sum_{\substack{i=1 \\ i \neq j}}^{N} e_i^2 + e_j^2,\tag{6.84}$$

where

$$e_j^2 \gg e_i^2 \qquad \forall \quad i \neq j.\tag{6.85}$$

Thus, the statistical outlier y_j in the data will tend to dominate the performance index and bias the solution. Removing the outlier from the data set and then performing a regression fit will improve the solution relative to the underlying linear model.

The interpretation of an outlier, when cast into the language of the RAP algorithm, is that of a fixed point shared by all but one hyperplane. This is shown in Figure 6.5. In terms of the matrix description, an outlier transforms the problem into

$$\mathbf{d} + \check{\mathbf{d}}_j = \mathbf{Xc}, \tag{6.86a}$$

$$\check{\mathbf{d}}_j = \begin{bmatrix} 0 \\ \vdots \\ 0 \\ \check{d} \\ 0 \\ \vdots \\ 0 \end{bmatrix}, \tag{6.86b}$$

where $\check{d}_j$ is the lone outlier located in the jth position. The iterative nature of the RAP algorithm allows it to monitor large projection distances and disregard those hyperplanes which consistently produce large residual errors. This is accomplished via the following procedure:

Determine the hyperplane that produces the largest residual for the nth cycle over the block of equations. At the $(n+1)$th cycle, do not update the coefficients using the hyperplane that yielded the largest error on the previous cycle. However, use the entire block to determine the hyperplane that corresponds to the largest residual error for the $(n+1)$th cycle. Continue in this manner until all the cycles are completed.

The above process will eventually isolate the hyperplane that is inconsistent with the rest of the block and remove it from the update process. This is easy to see if the algorithm is at the fixed point of the consistent set of hyperplanes; that is, the projection distance to all the consistent hyperplanes will be identically zero and the projection distance to the inconsistent hyperplane will be large and nonzero. The modified update equation produced by the above procedure is

$$\mathbf{c}_{k+1} = \mathbf{c}_k + \frac{\mu \epsilon_{k,j}}{\|\mathbf{x}_j\|^2} \mathbf{x}_j \qquad \text{iff} \quad \epsilon_{k,j} \neq \max \epsilon_{k-1}, \tag{6.87}$$

where $\epsilon_{k,j}$ is defined in (6.48).

6.5.9   Computational Complexity and Stability

The objective of this section is to show that computational complexity of the RAP algorithm is comparable to the computational complexity of the FRLS algorithm.

The computational complexity of the RAP algorithm can be derived in a manner similar to the computational complexity of the NLMS algorithm. Each coefficient vector update using (6.47a) requires $O(2N)$ multiply/adds and one division. Thus, the computational requirement for each input sample is $O(2NK)$ multiply/adds and one division, where K is the total number of cycles for each block of equations. This estimate assumes that the normalization, $1/\|\mathbf{x}\|$, can be computed recursively. If this assumption is not true, then the complexity becomes $O(2NK + N)$ multiply/adds per input sample, with still only one division. The RAP algorithm is stable, i.e., convergent, if the step size satisfies $0 < \mu < 2$. The stability is independent of the eigenvalue spread of the correlation matrix.

The FRLS algorithms have computational complexity in the range from $O(7N)$ to $O(40N)$ multiply/adds. However, the faster versions of FRLS are prone to be unstable and are usually monitored to detect and eliminate the onset of instability. Some procedures include utilizing a standard LMS algorithm during stabilization of the FRLS algorithm [12]. The slower versions of the FRLS algorithm use normalization techniques to alleviate numerical round-off, thus introducing many more divisions and square root operations not found in the lowest-complexity FRLS implementations [11].

The computational complexity of various adaptive algorithms, in multiply/adds per input sample, is shown in Table 6.1. The following comments pertain to that table:

- The $O(N)$ complexity of the RAP algorithm is *not* dependent upon the data structure. This contrasts with the FRLS algorithm, which relies on a tapped delay-line model to obtain $O(N)$ complexity. This is important in applications where the data matrix cannot be modeled as a Toeplitz system, as in, for example, adaptive array processing.

- As the multiply/add complexity of the RAP algorithm increases, the number of required divisions remains unity and the number of square roots remains zero. The FRLS, in comparison, requires an increasing number of divisions and square roots as the multiply/add complexity increases. Also, the faster

TABLE 6.1 Computational complexity comparison of adaptive filtering algorithms. The numbers represent complex arithmetic operations per input sample.

Algorithm	Multiplications	Divisions	Square Roots
RAP	$2KN + K$	1 (real)	no
FRLS	$20N + 5$	3	yes (for stability)
LMS	$2N + 1$	0	no
RLS	$2.5N^2 + 4.5N$	2	yes (for stability)

versions of FRLS usually incorporate recovery techniques to circumvent stability problems, incurring overhead costs in the algorithm implementation. The number of cycles, K, in the RAP algorithm usually does not exceed 20. This assumption is justified by experience with numerical simulations. Thus, the total computational complexity of the RAP algorithm compares favorably to the total computational complexity of the FRLS algorithms.

- The lower bound on the complexity of the RAP algorithm, $2N/1/0$, is smaller than that for the FRLS algorithm, $7N/2/0$. That is, the RAP algorithm can still be used when the computational resources are insufficient for an FRLS implementation. This has been shown to be advantageous in applying the RAP algorithm to perform frequency-subband processing with dynamic allocation of computational resources [41].

- Although the chart shows only complexity-per-sample, a better performance metric is (complexity-divide-convergence speed). The RAP algorithm and the FRLS algorithm still provide good performance when using this metric. However, the standard RLS algorithm, using the matrix-inversion lemma, has a high complexity-per-sample cost, while the LMS algorithm will usually incur a high samples-for-convergence cost.

6.5.10 Parallel Implementation

The task of parallel implementation of a signal-processing algorithm is greatly simplified if the algorithm can be decomposed into a sequence of multiplies and additions [42]. The RAP algorithm is decomposable in two respects: the rows of the data matrix can be operated on in any order that preserves the local stationarity, and each coefficient update consists of a vector inner product, a scalar multiply/add, and a vector addition.

A block diagram of a coarse-grain parallel implementation of the RAP algorithm is shown in Figure 6.10. The processing units consist of local memory, an I/O processor, and an arithmetic processor. Assume that there are P separate processing units, each accepting an input sample every Pth sampling time. Thus, each processing unit stores L rows of the data matrix spaced P samples apart, cycles through the L equations, and produces an estimate independently of the other processing units. The separate estimates are combined to produce an aggregate coefficient estimate that is passed to the adaptive filter.

6.6 SUMMARY AND CONCLUSION

Summary

A novel adaptive technique called the row-action projection (RAP) algorithm was presented. The RAP algorithm has its basis in the theory of matrix inversion and is, in fact, an iterative technique for solving a system of linear equations. The RAP algorithm has wide applicability, since many of the techniques and

PARALLEL IMPLEMENTATION

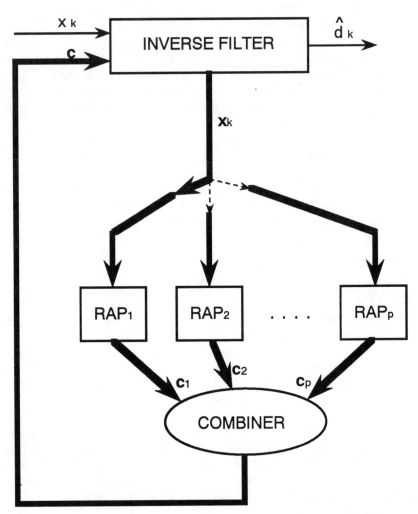

Figure 6.10 The system of equations can be parsed to multiple processing units. Each processor produces an independent coefficient-vector estimate that is eventually combined with the other estimates.

objectives in estimation theory and adaptive filtering can be linked to the problem of inverting a system of linear equations.

The general inversion process is known to be potentially unstable when the system is ill conditioned. The least squares inversion technique is not robust for ill-conditioned systems, due to the vanishing eigenvalue spectrum of the correlation matrix of the data. In addition, the stable stochastic-gradient techniques exhibit slow convergence when confronted with a large eigenvalue spread. The RAP algorithm achieves a compromise between these two techniques by using a performance metric fundamentally different from the least squares techniques. The RAP algorithm converges to a regularized inverse, rather than the direct inverse, of the data-correlation matrix. This property of the algorithm provides stability due to the noninversion of the zeros in the eigenvalue spectrum. The RAP algorithm also makes multiple uses of the data to achieve better convergence than the stochastic gradient techniques.

Noise-corrupted data tends to improve the condition of the system matrix. However, the direct inversion of all the eigenmodes of the system leads to noise amplification at the output of the adaptive filter. Direct inversion is a property of the least squares techniques and contributes to their unstable behavior. The solution for the adaptive filter must be modified to avoid noise amplification. The process of *regularization* is used to trade off estimation fidelity and noise amplification. The RAP algorithm is shown to be a regularized solution controlled by the iteration index. Adaptive filtering via *controlled* regularization is a unique property of the RAP algorithm.

The implementation of the RAP algorithm has some desirable features. It is *low cost*, in the sense that the number of arithmetic operations is linear in the length of the filter. The RAP algorithm is *fast*, in the sense that it converges quickly. Most importantly, the RAP algorithm is *stable* for all conditions of the data matrix. The combination of these properties makes the RAP algorithm unique.

Conclusion

Based upon the material presented in this chapter, the following conclusions can be made about the RAP algorithm:

- **It is stable for ill-conditioned data**. It permits wide applicability for situations where frequency nulled data is common.
- **It is a regularized inversion technique**. It allows a controlled trade-off between estimation accuracy and noise attenuation.
- **It has linear computational complexity, independent of data structure**. This permits its use in situations where the measurement matrix is non-Toeplitz; for example, array processing.

- **It has a simple implementation.** Stable operation obviates the use of rescue devices and auxiliary processors. It is also highly decomposable for parallel implementation.
- **It has fast convergence speed.** It requires many fewer samples than the LMS algorithm to converge and is comparable to the exact least squares techniques.
- **It has fast tracking speed.** Its finite block memory allows rapid adaptation to abrupt changes in the operating environment.

ACKNOWLEDGMENTS

The author acknowledges the support of the Rutgers University Center for Computer Aids for Industrial Productivity (CAIP) during portions of the preparation of this manuscript. The author was also supported through a Harpole–Pentair Faculty Fellowship while at Iowa State University.

REFERENCES

1. C. F. Gauss. *Theoria Motus Corporum Coelestium in Sectionibus Conicus Solem Ambietum (1809)*. (Trans.) New York: Dover, 1963.
2. S. T. Alexander. Fast adaptive filters: A geometrical approach. *IEEE Acoust. Speech Signal Process.*, 3(4):18–28, 1986.
3. J. R. Treichler, C. R. Johnson, Jr., and M. G. Larimore. *Theory and Design of Adaptive Filters*. New York: Wiley, 1987.
4. B. Widrow et al. Adaptive noise canceling: Principles and applications. *Proc. IEEE*, December 1975.
5. R. D. Gitlin and F. R. Magee, Jr. Self-orthogonalization adaptive equalization algorithms. *IEEE Trans. Commun.*, COM-25(7):666–672, 1977.
6. R. W. Chang. A new equalizer structure for fast start-up digital communication. *Bell Syst. Techn. J.*, 50:1969–2013, 1971.
7. G. Ungerboeck. Theory on the speed of convergence in adaptive equalizers for digital communication. *IBM J. Res. Dev.*, November 1972.
8. E. A. Lee and D. G. Messerschmitt. *Digital Communication*. Kluwer Academic, 1988.
9. S. Haykin. *Adaptive Filter Theory*. Englewood Cliffs, N.J.: Prentice-Hall, 1986.
10. G. A. Ybarra and S. T. Alexander. Effects of ill-conditioned data on least squares adaptive filters. In *IEEE International Conference on Acoustics, Speech, and Signal Processing*, pp. 1387–1390, 1988.
11. J. M. Cioffi and T. Kailath. Fast recursive least squares transversal filters for adaptive filtering. *IEEE Trans. Acoust. Speech Signal Process.*, ASSP-32:304–337, 1984.
12. E. Eleftheriou and D. D. Falconer. Restart methods for stabilizing FRLS adaptive equalizers in digital HF transmissions. In *IEEE Global Communications Conference*, pp. 48.4.1–48.4.5, December 1984.

13. G. E. Bottomley and S. T. Alexander. A theoretical basis for the divergence of conventional recursive least squares filters. In *IEEE International Conference on Acoustics, Speech, and Signal Processing*, 1989.

14. D. T. M. Slock and T. Kailath. Numerically stable fast recursive least-squares transversal filters. In *IEEE International Conference on Acoustics, Speech, and Signal Processing*, pp. 1365–1368, 1988.

15. J. M. Cioffi. Limited-precision effects in adaptive filtering. *IEEE Trans. Circ. Syst.*, 34:821–833, July 1987.

16. D. Kim and W. E. Alexander. Stability analysis of the fast RLS adaptation algorithm. In *IEEE International Conference on Acoustics, Speech, and Signal Processing*, pp. 1361–1364, 1988.

17. A. Benallal and A. Gilloire. A new method to stabilize fast RLS algorithms based on a first-order model of the propagation of numerical errors. In *IEEE International Conference on Acoustics, Speech, and Signal Processing*, pp. 1373–1376, 1988.

18. R. D. Gitlin and S. B. Weinstein. Fractionally-spaced equalization: An improved digital transversal equalizer. *Bell Syst. Techn. J.*, 60:275–296, 1981.

19. F. Ling and J. G. Proakis. Adaptive lattice decision feedback equalizers—their performance and application to time variant multipath channels. *IEEE Trans. Commun.*, COM-33(4):348–356, 1985.

20. L. B. Milstein. Interference suppression to aid acquisition in direct-sequence spread-spectrum communication. *IEEE Trans. Commun.*, 36(11):1200–1207, 1988.

21. S. Kay. Recursive maximum likelihood estimation of autoregressive processes. *IEEE Trans. Acoust. Speech Signal Process.*, 31:56–65, 1983.

22. J. Makhoul. Linear prediction: A tutorial review. *Proc. IEEE*, 63(4):561–580, 1975.

23. J. L. Flanagan et al. Speech coding. *IEEE Trans. Commun.*, COM-27:710–736, 1979.

24. S. J. Orfanidis. *Optimum Signal Processing*. New York: McGraw-Hill, 1988.

25. H. Robbins and S. Monro. A stochastic approximation method. *Ann. Math. Statist.*, 22:400–407, 1951.

26. B. Widrow and Jr. M. Hoff. Adaptive switching circuits. In *IRE WESCON Conv. Rec.*, pp. 96–104, 1960.

27. S. T. Alexander. *Adaptive Signal Processing: Theory and Applications*. New York: Springer-Verlag, 1986.

28. A. Ben-Israel and T. Greville. *Generalized Inverses: Theory and Applications*. New York: Wiley, 1974.

29. G. Long, F. Ling, and J. G. Proakis. Fractionally-spaced equalizers based on singular value decomposition. In *IEEE International Conference on Acoustics, Speech, and Signal Processing*, pp. 1514–1517, 1988.

30. J. M. Ortega. *Matrix Theory: A Second Course*. New York: Plenum Press, 1987.

31. L. R. Rabiner and B. Gold. *Theory and Application of Digital Signal Processing*. Englewood Cliffs, N.J.: Prentice-Hall, 1975.

32. S. V. Parter. On the extreme eigenvalues of truncated toeplitz matrices. *Bull. Am. Math. Soc.*, 67:191–196, 1961.

33. G. H. Golub and C. F. Van Loan. *Matrix Computations*. Baltimore, Md.: Johns Hopkins Press, 1983.

34. D. Youla. Generalized image reconstruction by the method of alternating projections. *IEEE Trans. Circ. Syst.*, 25, 1978.

35. A. N. Tikhonov and V. Y. Arsenin. *Solutions to Ill-Posed Problems.* Washington, D.C.: V. H. Winston and Sons, 1977.

36. C. K. Rushforth. Signal restoration, functional analysis, and Fredholm integral equations of the first kind. In H. Stark, ed., *Image Recovery: Theory and Application,* chap. 1. Orlando, Fla.: Academic Press, 1987.

37. C. W. Groetsch. *The Theory of Tikhonov Regularization for Fredholm Equations of the First Kind.* London: Pitman, 1984.

38. K. Tanabe. Projection method for solving a singular system of linear equations and its applications. *Numer. Math.,* **17**:203–214, 1971.

39. G. C. Goodwin and K. S. Sin. *Adaptive Filtering Prediction and Control.* Englewood Cliffs, N.J.: Prentice-Hall, 1984.

40. J. F. Doherty. *Regularized Adaptive Processing of Ill-Conditioned Data.* Ph.D. thesis, Rutgers University, 1990.

41. S. L. Gay and R. J. Mammone. Acoustic echo cancellation using RAP on the DSP16. In *IEEE International Conference on Acoustics, Speech, and Signal Processing,* April 1990.

42. L. S. Haynes et al. A survey of highly parallel computing. *IEEE Comput. Mag.,* **15**(1):9–24, 1982.

7

ADAPTIVE IMAGE RECOVERY

Shyh-shiaw Kuo[1]

AT&T Bell Laboratories
600 Mountain Avenue
Murray Hill, New Jersey 07947
email: skuo@research.att.com

7.1 INTRODUCTION

The image-recovery algorithm introduced in Chapter 4 was shown to be a very realistic approach to the problem of image recovery [1, 2]. In this chapter, we demonstrate that the algorithm can be applied to enhance the resolution of computerized tomographic (CT) images.

The resolution of CT images is limited by the bandlimited reconstruction process and the use of smoothing windows, which eliminate high spatial frequencies from the image. The problem can be transformed into one of image restoration [3–5] and solved by the algorithm introduced in Chapter 4. The algorithm restores those missing spectral components and thus increases the spatial resolution of the image. The method also provides a *zoom-in* capability that yields a high-resolution estimate of a specified region of the image. The *zoom-in* feature could be of great utility in medical and commercial applications of tomographic image reconstruction. Computer simulations demonstrate the new method to be very effective in increasing the resolution of designated regions of the reconstructed image.

Performance of the filtered back-projection (FBP) method of image reconstruction [6, Chap. 3] approaches optimality in the mean squared sense for the parallel-beam reconstruction algorithm [7, 8]. It is desirable to maintain this level

---

[1]This research was done when Dr. Kuo was a Ph.D. candidate at Rutgers University.

of performance when the effective processing bandwidth W_p is less than the image bandwidth W. This situation may arise when the reconstruction filter of the FBP method is deliberately truncated to control the signal-to-noise ratio (SNR) of the reconstructed images, or when the width of the detector is relatively large ($W_p < W$).

The tradeoff between the root mean squared error (rms) and spatial resolution in the reconstructed image is given by [9],

$$\epsilon \propto \left[\frac{W_p}{W}\right]^{3/2} \sigma, \tag{7.1}$$

where ϵ is the rms error of the reconstructed image and σ is the standard deviation of the measurement noise. It is obvious from (7.1) that the rms error can be reduced by decreasing the processing bandwidth W_p. However, this will decrease the spatial resolution due to the elimination of the high-frequency information. The reconstruction error is also dependent on the intensity of the beam, as shown by Brooks and Dichiro [7, eq. (10)]. Thus, another way to reduce the reconstruction error is to increase the intensity of the beam. Unfortunately, a decrease in reconstruction error by a factor of two requires an increase in dose by a factor of eight, i.e., the relationship is third order [7]. Hence, this is generally prohibitive in practice. In addition to the physics underlying (7.1), the resolution of the reconstructed image is also limited by the process of windowing the sampled data. A smoothing window, such as a hamming window [10], is generally applied during the reconstruction process in order to reduce the ringing artifact in the reconstructed image. The use of a smoothing window results in an additional loss of resolution of the reconstructed image.

At this point, we would like to clarify the ambiguity of the use of the word *projection*. In tomographic imaging, the word *projection* usually denotes a collection of line integrals in which the integration paths are straight lines passing through the object. In projection onto convex sets (POCS), the word *projection* is used to indicate a mathematical operation which yields the point in some set which is closest to a given point. In order to avoid confusion of the term *projection*, we shall use the term *projection* as it applies to POCS and use the term *raysum* [11] to denote the physical projection of the tomographic imaging process.

The problem of enhancing the resolution of the FBP-reconstructed images is formulated in Section 7.2. The approach of applying the RAP algorithm to the problem is presented in Section 7.3. Numerical results are given in Section 7.4.

7.2   PROBLEM FORMULATION

To introduce the notation which will be used in this chapter and to give the details of the method of implementation, a brief review of the FBP method of image reconstruction will be given. It is well known [6, 12] that the back-projection image $f_b(x, y)$ can be written in terms of a convolution of the original image $f(x, y)$ with the two-dimensional symmetric kernel $1/r$, i.e.,

$$f_b(x, y) = f(x, y) ** \frac{1}{r}, \tag{7.2}$$

where $r = \sqrt{x^2 + y^2}$ and $**$ is the two-dimensional convolution. The frequenc-domain representation of (7.2) in polar coordinates is

$$\rho \cdot F_b(\rho, \theta) = F(\rho, \theta), \tag{7.3}$$

where ρ is the two-dimensional Fourier transform of r.

Hence, the original image $f(x, y)$ can be reconstructed from the back-projection image by applying a 2-D filter whose Fourier-domain representation in polar coordinates is given by

$$\mathbf{H}(\rho) = \rho. \tag{7.4}$$

The two-dimensional transform can be implemented by calculating N one-dimensional transforms, where N is the number of raysums. The one-dimensional transforms form linear slices of the two-dimensional transform, as stated by the Fourier slice theorem [6]. The inverse 2-D transform can then be taken to reconstruct the image. The FBP method circumvents the need for the 2-D transform. The FBP algorithm consists of the following steps:

1. Measure the raysum $P_\theta(R)$, which is a line integral of $f(x, y)$ at a distance R from the origin where the perpendicular to the line is at an angle θ [12, 13], i.e.,

$$P_\theta(R) = \int \int f(x, y)\delta(x \cos \theta + y \sin \theta - R) \, dx \, dy \tag{7.5}$$

If the sampling interval is τ, then the raysum data can be represented as

$$P_\theta(m\tau), \qquad m = \frac{-N}{2}, \ldots, 0, \ldots, \frac{N}{2} - 1. \tag{7.6}$$

2. Apply the filtering operation

$$Q_\theta(R) = \int_{-W}^{W} S_\theta(\omega)|\omega||H_w(w)|e^{j2\pi\omega R} \, d\omega, \tag{7.7}$$

where $S_\theta(\omega)$ is the Fourier transform of the raysum $P_\theta(R)$ and $H_w(w)$ is a 1-D rectangular window function that characterizes the processing bandwidth

W_p. If the processing bandwidth is equal to the bandwidth of the image, i.e., $W_p = W$, then $H_w(w)$ is unity.

3. The FBP-reconstructed image $f_{\text{FBP}}(x, y)$ can be obtained by summing the filtered raysums, $Q_\theta(R)$, over the image plane (this is the back-projection process):

$$f_{\text{FBP}}(x, y) = \int_0^\pi Q_\theta(x \cos \theta + y \sin \theta) \, d\theta, \qquad (7.8)$$

or, in its discrete version,

$$f_{\text{FBP}}(x, y) = \frac{\pi}{K} \sum_{i=1}^K Q_{\theta_i}(x \cos \theta_i + y \sin \theta_i), \qquad (7.9)$$

where the K angles θ_i are those for which the raysums $P_\theta(R)$ are known.

In order to avoid the *dishing* and *dc shift* artifacts [6, Chap. 3], the filtered raysums $Q_\theta(n\tau)$ have been calculated as follows,

$$Q_\theta(n\tau) = \text{IFFT}\left\{[\text{FFT}\, P_\theta(n\tau)_{\text{ZP}}] \times [\text{FFT}\, h_f(n\tau)_{\text{ZP}}] \times H_w(w) \times H_{\text{sw}}(w)\right\}, \qquad (7.10)$$

where the subscript ZP denotes zero padding, FFT is the fast Fourier transform, $H_{\text{sw}}(w)$ is the smoothing window and $h_f(n\tau)$ is the impulse response of the filter function, i.e., the inverse Fourier transform of $|w|$.

Consider the situation where the effective processing bandwidth W_p is less than the bandwidth W of the true image. In this case, the 1-D window function in (7.7) is given by

$$H_w(w) = \text{rect}\left(\frac{w}{W_p}\right) = \begin{cases} 1, & \text{for } |w| < W_p/2, \\ 0, & \text{for } |w| > W_p/2. \end{cases} \qquad (7.11)$$

This corresponds to a modification of (7.3) by a 2-D window function given by

$$\text{circ}\left(\frac{\rho}{2\rho_0}\right) = \begin{cases} 1, & \text{for } |\rho| < \rho_0, \\ 0, & \text{for } |\rho| > \rho_0. \end{cases} \qquad (7.12)$$

If we denote the 2-D Fourier transform of the FBP-reconstructed image as $G(\rho, \theta)$, then we have

$$G(\rho, \theta) = F(\rho, \theta) \cdot \text{circ}\left(\frac{\rho}{2\rho_0}\right) \cdot H_{\text{sw}}(\rho), \qquad (7.13)$$

where $\mathbf{H}_{sw}(\rho)$ is a two-dimensional smoothing window. The space-domain representation of (7.13) is

$$g(x,y) = f(x,y) ** h(x,y), \tag{7.14}$$

where $g(x,y)$ is the bandlimited reconstructed image via the FBP algorithm and $h(x,y)$ is the inverse Fourier transform of the multiplication of the two window functions in (7.13), i.e., $h(x,y) = \mathcal{F}^{-1}[\text{circ}(\rho/2\rho_0) \cdot \mathbf{H}_{sw}(\rho)]$.

The problem of finding $f(x,y)$ when $g(x,y)$ is given, as specified by (7.14), will be called a spatial resolution enhancement of the FBP-reconstructed image. This problem is now one of the image-restoration problems that can be solved numerically by first transforming (7.14) into a discrete form:

$$g(i,j) = \sum_m \sum_n h(i-m, j-n)f(m,n). \tag{7.15}$$

It is also convenient to write the above equation as a set of linear equations,

$$\mathbf{g} = \mathbf{Hf}, \tag{7.16}$$

where $\mathbf{g}$ and $\mathbf{f}$ are the lexicographic row-ordered vectors of the discretized versions of g and f, respectively, and $\mathbf{H}$ is composed of the impulse response of the reconstruction system, $h(i,j)$. The matrix $\mathbf{H}$ is generally rank-deficient or ill conditioned, due to the presence of spectral zeros in the continuous operator. In addition, the set of linear equations given by (7.16) will be inconsistent due to noise and modeling inaccuracies. Thus, the direct inverse is not applicable for solving the above set of linear equations. However, it has been shown in Chapter 4 that this type of problem can be successfully solved by using the adaptive RAP algorithm, along with the use of constraints on $\hat{\mathbf{f}}$. The goal of enforcing these constraints is to reduce the possibility of large deviations in the estimate $\hat{\mathbf{f}}$ from what is known *a priori* about the actual image $\mathbf{f}$.

We would like to distinguish RAP from algebraic reconstruction techniques (ART), which are very similar methods but are used for different purposes. ART, as the name implies, is a technique for image reconstruction and works directly on the raysums. RAP is a technique for image restoration and works on the mathematically derived model given by (7.14). The linear equations used in ART and RAP are set up according to the physical imaging system and the mathematical model, respectively. The conventional update equation in ART [11, eq. (31)] is based on the estimated raysums and weights, which depend on the length of the ith ray through the pixel. The update equation generally has to be modified according to the physical imaging system [6, Chap. 7]. However, the update equation (4.19) in RAP only depends on the distance from the previous estimate to the next equation [14]. In addition, unlike ART, the image-restoration problem solved by RAP is decomposable. This will be shown in the next section.

7.3 IMPLEMENTATION

In solving the proposed problem, the RAP algorithm can also be efficiently implemented by operating on the 2-D image and the 2-D point-spread function (PSF) directly, as given by (4.25). In addition, (4.25) can be simplified for this shift-invariant problem:

$$
\hat{\mathbf{f}}^{(k+1)} =
\begin{cases}
\hat{f}^{(k)}(m,n) + \lambda \dfrac{\epsilon(i,j)}{\|\mathbf{h}\|^2} h(i-m, j-n), & \text{if } \hat{f}^{(k)}(m,n) \in S_h, \\[2ex]
\hat{f}^{(k)}(m,n), & \text{otherwise,}
\end{cases}
\tag{7.17}
$$

where S_h is the support of $\mathbf{h}$ and

$$
\epsilon(i,j) = g(i,j) - \sum_{m,n \in S_h} h(i-m, j-n)\hat{f}^{(k)}(m,n),
$$

$$
\|\mathbf{h}\|^2 = \sum_{m,n \in S_h} h(m,n)^2.
$$

The problem of spatial resolution-enhancement of FBP-reconstructed images can be solved according to the following steps:

1. **System identification.** The impulse response $h(x,y)$ of the reconstruction system can be calculated by using the inverse Fourier transform of the transfer function between the FBP-reconstructed image and a known test image:

$$
h(x,y) = \text{IFFT}\left[\frac{F_{\text{FBP}}(u,v)}{F_{\text{test}}(u,v)}\right],
\tag{7.18}
$$

where $F_{\text{FBP}}(u,v)$ and $F_{\text{test}}(u,v)$ represent the 2-D Fourier transforms of the FBP-reconstructed image and the test image, respectively. If the noise level is significant, then the impulse response of the system can be found by averaging the various estimated PSFs, which are obtained from (7.18) by using different test images. Alternatively, the impulse response $h(x,y)$ can be found via the FBP algorithm itself, i.e., (7.9) and (7.10). Since the projection of an impulse is still an impulse, the $[\text{FFT} P_\theta(n\tau)_{\text{ZP}}]$ term is unity in (7.10), i.e.,

$$
Q_\theta(n\tau) = \text{IFFT}\left\{[\text{FFT}\, h_f(n\tau)_{\text{ZP}}] \times \text{rect}\left(\frac{w}{W_p}\right) \times H_{\text{sw}}(w)\right\}
\tag{7.19}
$$

Thus, the impulse response $h(x, y)$ can be expressed as

$$h(x, y) = \frac{\pi}{K} \sum_{i=1}^{K} Q_\theta (x \cos \theta_i + y \sin \theta_i).$$

(7.20)

The magnitudes of the sidelobes of $h(x, y)$ are generally much lower than that of the mainlobe. Keeping them increases the size of the PSF, which correspondingly increases the number of required computations at each iteration of RAP. Thus, the truncated version of $h(x, y)$ is used during the restoration process in order to reduce the computational complexity.

2. **Working region selection.** By taking advantage of the local projection operator of the RAP algorithm, we can specify a region of the FBP-reconstructed image for increased resolution. This zoom-in feature not only reduces the computational complexity by reducing the number of pixels to be processed, but also provides a practical utility, because resolution enhancement is generally desired for some specific region in the reconstructed image. The process of applying the RAP method to the designated region can be expressed as a cycling strategy. The projection operator $P_{C_{zoom}}$ is defined by

$$P_{C_{zoom}} \mathbf{f} = \begin{cases} \mathbf{f} + \dfrac{g(i, j) - \langle \mathbf{h}, \mathbf{f} \rangle}{\|\mathbf{h}\|^2} \mathbf{h}, & \text{if } g(i, j) \in Z_R, \\ \mathbf{f}, & \text{otherwise,} \end{cases}$$

(7.21)

where $\mathbf{h}$ denotes the normal vector of the hyperplane that represents the (i, j)th pixel of $\mathbf{g}$, and Z_R is the designated zoom-in region.

3. **The local minimum and maximum bounds estimation.** The minimum and maximum bound constraints can be expressed as a POCS projection operator as

$$P_{C_b} f = \begin{cases} f_{\max}, & \text{if } f(i, j) > f_{\max}, \\ f_{\min}, & \text{if } f(i, j) < f_{\min}, \\ f(i, j), & \text{otherwise.} \end{cases}$$

(7.22)

Since the dynamic range of a designated region of the image is generally less than that of the entire image, the local minimum and maximum bounds have to be estimated for the designated region. Based on the fact that the rate of convergence is improved as the accuracy of the estimated bounds increases, the minimum and maximum bounds for Z_R can be estimated by the following steps:

Step 1: Z_R is segmented to separate the high-intensity region from the background grey levels.

Step 2: k is set equal to 1.

Step 3: The RAP method is applied to Z_R for k iterations. The smoothness and the global minimum and maximum bound constraints are imposed. The initial estimate $\mathbf{f}^{(0)}$ is set to the reconstructed image $\mathbf{g}$.

Step 4: According to the segmentation in Step 1, the average value of the background in the estimated image obtained from the previous step is computed and is used as the tentative minimum bound $\hat{f}_{\min}$. The largest value of region Z_R in the estimated image is found and is used as the tentative maximum bound $\hat{f}_{\max}$ if it is smaller than the global maximum bound.

Step 5: The RAP method is again applied to Z_R with *tentative* minimum and maximum bounds, $\hat{f}_{\min}$ and $\hat{f}_{\max}$, imposed. The initial estimate $\mathbf{f}^{(0)}$ is again set to $\mathbf{g}$. This is followed by M iterations of RAP, where M is a small fixed number. The sum of the seminorm is calculated:

$$L_k = \sum_{m=1}^{M} S_m, \qquad (7.23)$$

where S_m denotes the seminorm after the mth iteration. The seminorm is defined as the sum of the residuals

$$S_m = \sum_{i,j \in Z_R} |g(i,j) - \langle \mathbf{h}, \mathbf{f}^{(m)} \rangle|, \qquad (7.24)$$

where $\mathbf{f}^{(m)}$ denotes the estimated image after the mth iteration.

Step 6: If $L_k - L_{k-1} > 0.1$ then STOP; i.e., the local minimum and maximum bounds estimation stage is terminated when the sum of the seminorm increases. Otherwise, set $k = k + 1$ and go to Step 3.

The resulting tentative minimum and maximum bounds will be used as the $f_{\max}$ and $f_{\min}$ in (7.22) for the designated region Z_R.

4. **Image Restoration.** The true image can be restored by the following iterative process:

$$\mathbf{f}^{(k+1)} = M_s P_{C_b} P_{C_e} P_{C_b} P_{C_{\text{zoom}}} \mathbf{f}^{(k)}, \qquad k = 0, 1, \ldots, \qquad (7.25)$$

where M_s denotes the nonexpansive mapping of smoothness constraint described in (4.28), P_{C_e} is the projection operator defined in (4.34) and $\mathbf{f}^{(0)} = \mathbf{g}$. All the RAP projections are implemented in 2-D, as described in (7.17) and Section 4.4.

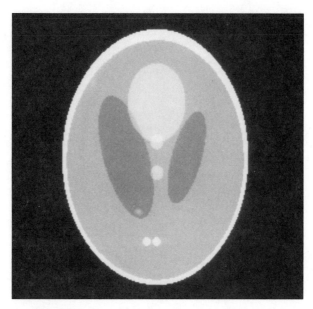

Figure 7.1 The 256 × 256 head-phantom target image.

7.4   NUMERICAL RESULTS

The true image is the 256×256 computer-simulated head-phantom image [6] shown in Figure 7.1(a). The rectangular window shown at the bottom of the target image is one of the designated zoom-in regions used in the following experiments. Since the highest sidelobe of the impulse response of the reconstruction system is 40 dB down from its mainlobe, the impulse response $h(i,j)$ is truncated to 31×31 pixels [5]. The relative error ϵ is used as a quantitative error measure:

$$\epsilon = 100 \times \frac{||\mathbf{f} - \hat{\mathbf{f}}||}{||\mathbf{f}||},$$

where $\hat{\mathbf{f}}$ is the estimate of $\mathbf{f}$. As is standard in the FBP-reconstruction algorithm, a hamming window is used to reduce the ringing artifact in the reconstructed image. The performance of the new algorithm is demonstrated by the following experiments:

Experiment I

This experiment demonstrates that the new algorithm can be used to increase the resolution of FBP-reconstructed images when the number of raysums and/or samples are insufficient for the desired resolution. Figure 7.2 is the FBP-reconstructed image via 100 raysums with 99 parallel beams in each raysum. The

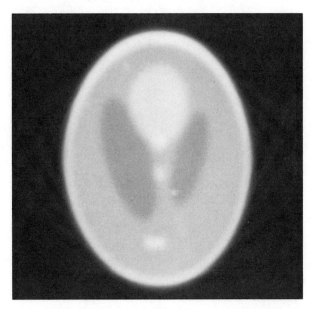

Figure 7.2 The standard FBP-reconstructed image.

blurring and streaks in the reconstructed image are due to the aliasing of the spectrum [6, Chap. 5]. However, the new algorithm can be used to enhance the resolution of the blurred reconstructed image. For example, the two spots at the bottom can be successfully restored by using the new algorithm, as shown in Figure 7.3. The edge, which is windowed by the box in Figure 7.3 is also enhanced. The relative errors of the zoom-in regions in Figures 7.2 and 7.3 are listed in Table 7.1.

Experiment II

The second experiment is designed to demonstrate the new algorithm's ability to reduce the reconstruction noise and increase the resolution of the reconstructed image at the same time. Figure 7.4 is the standard FBP reconstructed image with 30 dB white Gaussian noise added to each raysum. It is obvious from Figure 7.4 that the reconstructed image is quite noisy. However, the rms error of the reconstructed image can be reduced by decreasing the processing bandwidth W_p [9]. Figure 7.5 is the bandlimited FBP-reconstructed image from the same noisy raysum data used in Figure 7.4. Each raysum is bandlimited to $W_p = 0.4/\tau$, where τ is the sampling interval in each raysum. It is clear that the noise in the reconstructed image has been reduced, but that the resolution of the image is also reduced. For example, the two spots at the bottom are not discernible in Figure 7.5. But the two spots can be restored successfully by using the new algorithm. The edge windowed by a box is also enhanced. The resulting image is shown in Figure 7.6. Thus, the new algorithm combined with the bandlimited

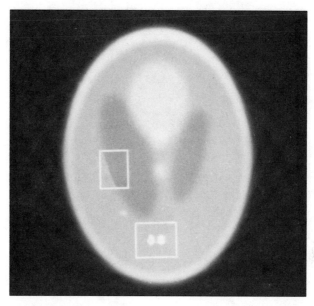

Figure 7.3 The resolution of the image inside the boxes has been enhanced using the new algorithm.

FBP algorithm can reduce the rms error and simultaneously increase the resolution of the reconstructed image. The relative errors of the zoom-in regions in Figures 7.4, 7.5, and 7.6 are summarized in Table 7.2.

7.5 SUMMARY AND CONCLUSION

A new projection method for enhancing the resolution of reconstructed tomographic images is presented. The proposed problem is transformed into one of image restoration and solved by regularized deconvolution. The new algorithm does not require the storage of the entire lexicographically stacked degradation matrix **H** because of the 2-D implementation of the RAP algorithm. This reduces not only the memory space requirements but also the computational complexity. Hence, it is possible to process large images adaptively or in parallel. Since the RAP algorithm is a subset of the POCS method and nonexpansive mappings [1, 15, 16], it allows the addition of convex constraints and nonexpansive mapping

TABLE 7.1 The relative errors of the designated regions in Figure 7.2.

Images	FBP Algorithm	New Algorithm
edge	3.9%	2.2%
two spots	12.2%	6.6%

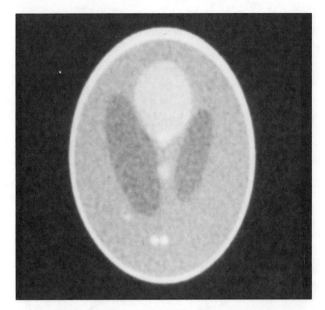

Figure 7.4 The standard FBP-reconstructed image.

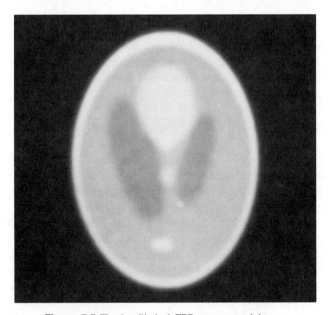

Figure 7.5 The bandlimited FBP-reconstructed image.

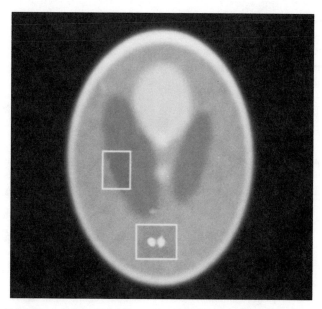

Figure 7.6 The resolution of the image has been enhanced using the new algorithm.

strategies to reduce ambiguity in the solution and accelerate the speed of convergence. In addition, the new method does not require any global operations, such as a 2-D FFT, to be performed on the entire image as in the Gerchberg–Papoulis [17, 18] method and the conventional implementation of POCS. Even though the theoretical implications of the new algorithm have yet to be explored, computer-simulated results indicate that the proposed method can be used to enhance the resolution of conventional tomographic images without increasing the dose or number of data points collected. The method could be used in conjunction with the standard FBP method to provide an interactive zoom-in capability. This zoom-in feature could be of great utility in medical and commercial applications of tomographic image reconstruction.

TABLE 7.2   The relative errors of the designated regions in Figure 7.3.

Images	Standard FBP $(W_p = 1/\tau)$	Bandlimited FBP $(W_p = 0.4/\tau)$	New Algorithm
edge	5.4%	4.1%	3.2%
two spots	7.9%	11.4%	6.0%

ACKNOWLEDGMENTS

This research reported here was made possible through the support of the New Jersey Commission on Science and Technology and the Center for Computer Aids for Industrial Productivity at Rutgers University.

REFERENCES

1. S. Kuo and R. J. Mammone. Image restoration by convex projections using adaptive constraints and the l_1 norm. *IEEE Trans. Signal Process.*, 1992, in press.

2. S. Kuo and R. Mammone. Refinement of EM restored images using adaptive POCS. In *Proceedings of the IEEE International Conference on Acoustics, Speech, and Signal Processing*, 1991.

3. S. Kuo, R. Mammone, J. Doherty, and C. Podilchuk. Resolution enhancement of reconstructed images by using image restoration techniques. In *Proceedings of the IEEE International Conference on EMBS*, November 1990.

4. S. Kuo and R. Mammone. Resolution enhancement of ct images using the adaptive row action projection method. In *Proceedings of the SPIE/SPSE International Conference on Biomedical Image Processing*, February 1991.

5. S. Kuo and R. J. Mammone. Resolution enhancement of tomographic images using the row action projection method. *IEEE Trans. Med. Imaging*, December 1991.

6. A. C. Kak and M. Slaney. *Principles of Computerized Tomographic Imaging*. New York: IEEE Press, 1988.

7. R. A. Brooks and G. Dichiro. Statistical limitations in x-ray reconstruction tomography. *Med. Phys.*, 3:237–240, 1976.

8. O. J. Tretiak. Noise limitation in x-ray computed tomography. *J. Comput. Assist. Tomog.*, 2:477–480, 1978.

9. G. Kowalski. Reconstruction of objects from their projections; the influence of measurement error on the reconstruction. *IEEE Trans. Nucl. Sci.*, 24:850–864, 1977.

10. F. J. Harris. On the use of windows for harmonic analysis with the discrete fourier transform. *Proc. IEEE*, 66(1):51–83, 1978.

11. P. Oskoui-Fard and H. Stark. Tomographic image reconstruction using the theory of convex projections. *IEEE Trans. Med. Imaging*, 7(1):45–58, 1988.

12. G. T. Herman. *Image Reconstruction from Projections*. New York: Academic Press, 1980.

13. P. R. Edholm and G. T. Herman. Linograms in image reconstruction from projections. *IEEE Trans. Med. Imaging*, 6:301–307, 1987.

14. S. Kaczmarz. Angenaherta auflosung von systemen linearer gleichungen. *Bull. Acad. Polon. Sci. Lett.*, A35:355–357, 1937.

15. J. F. Doherty and R. J. Mammone. A row-action projection algorithm for adaptive filtering. In *Proceedings of the IEEE International Conference on Acoustics, Speech, and Signal Processing*, April 1990.

16. C. I. Podilchuk and R. J. Mammone. Row and block action projection techniques for image restoration. In *Signal Recovery and Synthesis III*, Cape Cod, Mass.: Optical Society of America, June 1989.

17. A. Papoulis. A new algorithm in spectral analysis and band-limited extrapolation. *IEEE Trans. Circuits Syst.*, **CAS-22**(9):735–741, 1975.

18. R. W. Gerchberg. Super-resolution through error reduction. *Opt. Acta*, **21**:709–720, 1974.

8

ADAPTIVE TIME-VARYING SPECTRAL ESTIMATION

Adam B. Fineberg

CAIP Center
Rutgers University
Piscataway, New Jersey 08855–1390
email: fineberg@caip.rutgers.edu

8.1 INTRODUCTION

Traditionally, methods of spectral estimation of discrete signals have been based on the Z-Transform, the discrete Fourier transform (DFT), or time-series modeling, such as autoregressive (AR) and autoregressive moving average (ARMA) models [1]. These methods are based on an implicit assumption that the signal of interest is from a wide-sense stationary (WSS) process. There are many applications in which it is desired to form a spectral estimate of a signal from a quasi-stationary process. Many techniques have been studied that attempt to determine the signal distribution in both frequency and time [2]. These methods include short-time stationary methods, Wigner distributions, wavelet transforms, and ambiguity functions. For a time-varying signal, these methods determine the average frequency content over some relatively small time interval. In the limit, they attempt to determine the instantaneous frequency content in the interval from t to $t + \Delta t$ as $\Delta t \to 0$. The analysis of time-varying signals—such as those found in speech, frequency-hopped communications, and computer visualization of moving objects—has typically been treated using spectrograms [3], averaged periodograms [4], and weighted covariance estimators [5]. These methods are based on the assumption that the underlying process is stationary for some determinable Δt.

The optimal selection of the length of the short-time stationary interval and the analysis overlap rate is a difficult and oft-times intractable problem. The length of the interval determines the frequency resolution of the spectral estimate and the analysis overlap rate determines the tracking ability (i.e., how well the spectrum represents the frequency content of a time-varying signal at any given instant in time). Thus, there is an inherent trade-off between frequency resolution and the tracking rate.

The spectrogram and windowed DFT analyses are probably the most widely used methods for short-time spectral analysis [1]. It is well known that a long observation time is required for good frequency resolution; however, this produces poor time localization [6, 7]. Time localization can be improved by using a window that is compact in the time domain; however, this will broaden the spectral peaks in the frequency domain and thus decrease the frequency localization. The uncertainty principle [6] states that it is not possible to find a waveform that is compact in both the time and frequency domains simultaneously. This limits the resolution attainable by linear windowing techniques. However, frequency resolution can be improved when constrained optimization techniques that use some *a priori* information about the signal are applied [8].

A method of frequency tracking that obtains the time versus frequency distribution of one or more nonstationary narrowband signals is based upon the discrete spectral decomposition technique described in Chapter 2. To form a basis for understanding and comparison, some conventionally used techniques will be discussed.

8.2 CONVENTIONAL FREQUENCY TRACKING

As already stated, many techniques have been developed over the years to track the instantaneous frequency of a signal. The term "instantaneous frequency" is used to describe the frequency at a given instant in time. Although this concept is not well defined mathematically, we know from the study of many signals that it is a desirable quantity to determine (i.e., FM demodulations and frequency-hopping transmitters). Instantaneous frequency is typically defined as the derivative of the phase of the analytic signal. If we obtain a true distribution of time versus frequency, then a more intuitive definition might be that instantaneous frequency is the expected value of the frequency at each instant in time. For this discussion of frequency tracking, however, the analytic-signal formulation will be used.

8.2.1 Analytic Signal Method

The analytic signal is the representation of a real-valued signal in terms of a complex-valued signal with a one-sided spectrum. The analytic signal representation of a real-valued discrete signal $y(k)$ is given by

$$z(k) = y(k) + j\hat{y}(k), \tag{8.1}$$

where $\hat{y}(k)$ is the Hilbert transform of $y(k)$ [9], whose frequency components are in phase quadrature with those of $y(k)$. This allows us to write $z(k)$ in a compact exponential form:

$$z(k) = a(k)e^{j\phi(k)}, \tag{8.2}$$

where $a(t)$ is the time-varying amplitude and $\phi(k)$ is the phase function.

The amplitude function of $z(k)$ is given by

$$a(t) = \sqrt{\left(\Re(z(t)^2) + \Im(z(t)^2)\right)} \tag{8.3}$$

and the phase function of $z(k)$ is obtained by

$$\phi(k) = \tan^{-1}\left[\frac{\Im(z(k))}{\Re(z(k))}\right] = \tan^{-1}\left[\frac{\hat{y}(k)}{y(k)}\right]. \tag{8.4}$$

The instantaneous frequency $\beta(k)$ is obtained from the phase function $\phi(k)$ by taking the derivative:

$$\beta(k) = \frac{d\phi(k)}{dk} = \frac{1}{2\pi}\frac{[\hat{y}'(k)y(k) - y'(k)\hat{y}(k)]}{[y^2(k) + \hat{y}^2(k)]}. \tag{8.5}$$

Since there is a single estimate of the instantaneous frequency, it is implicit in this technique that the signal has only a single component. A multicomponent signal is one in which the time-frequency distribution shows two or more nonoverlapping concentrations of energy. These concentrations then each have a local center frequency and local bandwidth that correspond to the expected value and standard deviation for each component. Many signals of interest are multicomponent (i.e., speech, FDMA communications, and music) and a tracking technique should therefore be able to determine the instantaneous frequency for each component.

The analytic signal method performs very well when only one component is present. This can be clearly seen in Figure 8.1. When two or more components are present, it only produces an "average" frequency estimate, as shown in Figure 8.2. More complicated schemes with adaptive frequency-selective filtering can be used to overcome this problem. This is accomplished by adaptively filtering each component from the composite signal. Each filtered subband is processed as if it were from a single-component signal. Even if the adaptive filters can be designed to perform perfectly, this approach is computationally infeasible for many applications.

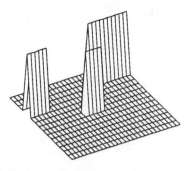

Figure 8.1 Single-tone frequency tracking of an analytic signal.

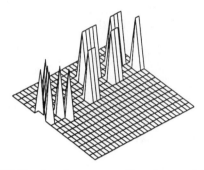

Figure 8.2 Two-tone frequency tracking of an analytic signal.

Another shortcoming of this approach is that the nonlinear calculations and differentiation in (8.4) and (8.5) are very sensitive to additive noise. This sensitivity is seen in the simulation results shown in Figure 8.3. The performance of the instantaneous frequency estimate from the analytic signal in the presence on noise has been well studied [10].

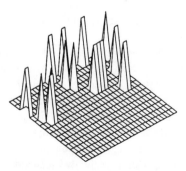

Figure 8.3 Single-tone frequency tracking of analytic signal with 10 dB SNR.

8.2.2 The Wigner Distribution

The Wigner distribution is one of the oldest methods applied to time-frequency representations of finite energy harmonic signals [2]:

$$\mathbf{W}(t, \omega) = \frac{1}{2\pi} \int y^* \left(t - \frac{\tau}{2}\right) e^{-j\tau\omega} y \left(t + \frac{\tau}{2}\right) d\tau. \tag{8.6}$$

The Wigner distribution can be used for frequency tracking by finding the peaks in the time-frequency distribution. It is known [2] that these peaks follow the derivative of the phase of the analytic signal for a pure frequency-modulated signal. The Wigner distribution is therefore a good estimator of the instantaneous frequency.

The Wigner distribution is closely related to local autocorrelation methods and, in fact, can be represented by a scaled version of the Fourier transform of the symmetric autocorrelation. The Wigner distribution was the first practical method to obtain a time-frequency distribution. However, since the autocorrelation and, thus, the Wigner distribution are bilinear and real, this representation suffers from oscillatory cross-terms generated by negative frequency components that corrupt the distribution [11]. These cross-terms are evidenced when a multicomponent signal is analyzed. The cross-terms will appear between every two components; and therefore, when more than two components are present, it is not easily determined (except for the two outermost components) which are real and which are artifacts.

Mitigation of these artifacts has been shown to be possible by modifying the kernel of the Wigner distribution, as with the Choi–Williams method [2]. This relieves the distribution of nearly all cross terms for carefully constructed kernels. However, many signals would require *a priori* knowledge of the signal structure to design such a kernel.

Another closely related distribution is the ambiguity function, which is the characteristic function of the Wigner distribution. Since it, too, is of the same form (bilinear and real) it suffers from the same spurious peaks. Other distributions studied, such as the Page and the Rihaczek, also suffer from cross-term corruption, since all require nonlinear operations. Thus these methods offer limited tracking capability when a multicomponent signal is present.

8.3 SIGNAL DECOMPOSITION

This method provides a fast sample-by-sample Fourier decomposition of the input signal and allows for enhanced resolution and accuracy when it is known *a priori* that the signal can be constrained to be within a subset of some given finite set of discrete sinusoids. The method is shown to provide greater resolution than that generally expected by the stability time-bandwidth product [1]. In addition, this technique provides a real-time method of tracking the instantaneous frequency of a signal. The tracking capability of a frequency-modulated (FM) signal is compared to that given by the analytic signal and the Wigner distribution methods. This technique is useful for tracking multiple narrowband signals and has applications in demodulation of digital FM [12], multiple target tracking, formant tracking in speech signals, tracking of frequency-hopping signals, dynamic subband analysis, and analysis of nonlinear dynamic systems [13].

The technique to be presented differs from other commonly used methods in that it has the capability of providing enhanced resolution and, because it is linear, it can track multiple tones with no intermodulation distortion. Most signal-subspace partitioning techniques (i.e., autocorrelation and eigenvector methods) require exact *a priori* knowledge (or at least very good knowledge) of the dimension of the signal subspace. The new technique only requires an estimate of the number of subspace regions which contain some signal energy. Since this technique will provide a sample-by-sample update, a dynamic block size can be heuristically determined. For each time sample, only the number of neighboring samples which are required to reduce the estimation error are used. Thus, in rapidly varying signals only time samples in a relatively small interval are used; therefore, a short effective block size will be selected. Whereas, for slowly varying portions of a signal, many samples will help reduce the error and, therefore, a longer effective block size will be implemented. This overcomes the tracking versus resolution trade-off discussed previously.

8.3.1 Least Squares Signal Decomposition

The problem of determining the discrete decomposition of a signal at a given time can be formulated as a weighted least squares problem. The decomposition then provides a constrained quadratic optimization:

minimize $\quad \epsilon^T A \epsilon$

subject to
$$
\begin{pmatrix} y_1 \\ y_2 \\ \vdots \\ y_m \end{pmatrix} = \begin{pmatrix} w_{11} & w_{12} & \cdots & w_{1n} \\ w_{21} & w_{22} & \cdots & w_{2n} \\ \vdots & \vdots & \ddots & \vdots \\ w_{m1} & w_{m2} & \cdots & w_{mn} \end{pmatrix} \cdot \begin{pmatrix} x_1 \\ x_2 \\ \vdots \\ x_n \end{pmatrix} + \begin{pmatrix} \epsilon_1 \\ \epsilon_2 \\ \vdots \\ \epsilon_m \end{pmatrix}, \quad (8.7)
$$

where the matrix A is a positive semi-definite symmetric matrix, y is a vector containing samples of the observed time-domain signal on an interval T_{obs}, the columns of the W matrix are the kernel vectors which characterize the decomposition, and the x vector contains the decomposition coefficients [14]. The solution vector x yields the optimum decomposition of the given sequence y onto the kernel vectors.

The kernel vectors corresponding to a discrete Fourier decomposition (those which constitute the DFT matrix) are composed of sampled sinusoids. In the case of harmonic Fourier decomposition, the kernel vectors are orthogonal and therefore constitute a basis for the decomposition space. Many other kernel vectors may be chosen that would reflect the character of the requested decomposition space. If the selected kernel vectors are not orthogonal, then they may constitute a frame rather than a basis. A frame is a discrete set of vectors $\{w_i\}$ in a Hilbert space for which the inequality

$$B_1\|\mathbf{x}\|^2 \le \sum_i |\langle w_i, \mathbf{x}\rangle|^2 \le B_2\|\mathbf{x}\|^2 \tag{8.8}$$

on the norm of $\mathbf{x}$ holds (i.e., $B_1 > 0$, $B_2 < \infty$) [15]. If $B_1 = B_2 = 1$, then a frame forms an orthonormal basis; otherwise, the vectors will be linearly dependent. Kernel vectors that are Gaussian-modulated sinusoids, those which constitute the Gabor transform [16], determine a frame rather than a basis, since they are no longer orthogonal. Many other kernel vectors have been studied. Those of particular interest in recent years are those which are made up of either translates and modulates of a given function, or those which are translates and dilates of a function. These are commonly referred to as wavelet frames [17, 18].

If the kernel vectors are chosen such that they do not span the space of $\mathbf{y}$ (i.e., the column space of W), then the system of equations given by (8.7) will be underdetermined and the matrix W will be rank deficient.

This approach focuses on the use of kernel vectors that relate to a harmonic or nonharmonic Fourier decomposition. The kernel vectors may therefore be nonorthogonal and linearly dependent. Since it has been shown [15] that a system of complex exponentials, either harmonic or nonharmonic, constitutes a frame for every sampled time sequence $y(k)$ of a real signal, the kernel vectors w_i will constitute a frame and possibly a basis.

If $A = I$, $\mathbf{y}$ is sampled at exactly the Nyquist rate, and $\mathbf{x}$ is sampled at the resolution limit of $1/T_{\text{obs}}$, a unique least squares solution exists for the harmonic decomposition of (8.7), since the kernel vectors are a basis. This solution is given by

$$\mathbf{x} = [W^T W]^{-1} W^T \mathbf{y}. \tag{8.9}$$

Since this decomposition has a basis, the matrix inverse is guaranteed to exist. In this case, $\mathbf{x}$ is the DFT of the vector $\mathbf{y}$. If A is a diagonal matrix with diagonal elements given by the values of a sampled window function, then the least squares solution yields the windowed DFT of the vector $\mathbf{y}$. It should be noted that the DFT provides the least squares solution to the discrete signal-decomposition problem. Although this is not immediately obvious from the traditional derivation, it is important to realize that—even if the harmonic spacing requirements can not be met, due to sampling limitations—the DFT will provide a reasonable (and, in the least squares sense, optimum) result.

8.3.2 High-Resolution Decomposition

Oversampling of the frequency domain creates nonharmonic kernel vectors that are no longer a basis (but still constitute a frame). This would provide enhanced frequency resolution if a unique decomposition were available. This is often difficult because the matrix W will become rank deficient. In this case, there is no unique least squares solution, since the inverse of the Grammian matrix $[W^T W]^{-1}$ [19] does not exist; that is, $W^T W$ is also rank deficient.

When the inverse does not exist, the pseudo-inverse, as presented in Chapter 2, is applicable [19]. Therefore, a unique solution is guaranteed only if the dimension of the null space of W is zero. This occurs only when $W^T W$ is nonsingular. For the case where the Grammian matrix is rank deficient, a family of least squares solutions corresponding to $A = I$ will exist and is given by

$$\mathbf{x} = W^\dagger \mathbf{y} + [I - W^\dagger W]\mathbf{z} = P\mathbf{y} + Q\mathbf{z}. \tag{8.10}$$

The discrete Fourier nonharmonic analysis of a signal onto an oversampled frequency space can be written in the form of (8.7) as a least squares decomposition where

$$W = \begin{pmatrix} e^{j\omega_1 t_1} & e^{j\omega_2 t_1} & \cdots & e^{j\omega_n t_1} \\ e^{j\omega_1 t_2} & e^{j\omega_2 t_2} & \cdots & e^{j\omega_n t_2} \\ \vdots & \vdots & \ddots & \vdots \\ e^{j\omega_1 t_m} & e^{j\omega_2 t_m} & \cdots & e^{j\omega_n t_m} \end{pmatrix} \tag{8.11}$$

The solution to (8.7) using the oversampled frequency matrix W given by (8.11) provides enhanced resolution; however, since the W matrix is rank deficient, the solution is not unique. Therefore, it is difficult to determine which of the family of solutions described by (8.10) is the correct decomposition and thus provides the solution which contains the enhanced resolution. To find a unique solution, the dimensionality of the solution space must be constrained, thereby reducing the ambiguity in the null space. A weighted least squares formulation (i.e., $A \neq I$) is used to perform the reduction in dimensionality.

8.3.3 Constrained Subspace Partitioning

This formulation can be converted to a least squares formulation (i.e., $A = I$) and solved using the tools already described in previous chapters. This is accomplished by making the substitution

$$\eta = A^{1/2}\epsilon, \tag{8.12}$$

where $A^{1/2}$ is the square root matrix of the matrix A. Since A is positive semidefinite symmetric, $A^{1/2}$ is positive definite Hermitian [19]. Since A can be chosen to be diagonal, $A^{1/2}$ will also be diagonal and, thus, symmetric. Hence, its inverse, $A^{-1/2}$, is also known to exist. The formulation of the problem given by (8.7) can then be rewritten as

$$\text{minimize} \quad \eta^T \eta$$

$$\text{subject to} \qquad \mathbf{y} = W\mathbf{x} + A^{-1/2}\boldsymbol{\eta}. \tag{8.13}$$

The constraint equations can then be premultiplied by $A^{1/2}$ to obtain

$$A^{1/2}\mathbf{y} = A^{1/2}W\mathbf{x} + \boldsymbol{\eta}. \tag{8.14}$$

The following substitution of variables can then be made:

$$\mathbf{y'} = A^{1/2}\mathbf{y} \qquad \text{and} \qquad W' = A^{1/2}W, \tag{8.15}$$

and the constrained quadratic optimization problem of (8.7) is given by

$$\text{minimize} \qquad \boldsymbol{\eta}^T\boldsymbol{\eta}$$
$$\text{subject to} \qquad \mathbf{y'} = W'\mathbf{x} + \boldsymbol{\eta} \tag{8.16}$$

A least squares solution, as given by (8.9) and (8.10), where the substitution of the primed quantities given by (8.15) is made, can then be found. When it is known *a priori* that the decomposition has only a few contributing components, the weighting matrix $A^{1/2}$ may be chosen such that it is a diagonal matrix with diagonal elements given by

$$a_i = \begin{cases} 1, & \text{for } i \in \mathcal{A}, \\ 0, & \text{otherwise,} \end{cases} \tag{8.17}$$

where $\mathcal{A}$ is the subset of contributing components. If a Fourier kernel is used, this subset refers to the frequency regions of interest and would correspond to one or more narrowband signals. To ensure that a unique solution is available, $\mathcal{A}$ needs to be a subset of the set $\mathcal{S}$, where

$$\mathcal{S} = \{\text{space spanned by } W\}. \tag{8.18}$$

If $\mathcal{A}$ is known *a priori*, then (8.16) can be solved over $\mathcal{A}$ rather than over $\mathcal{S}$. Therefore, the null space is constrained and a unique least squares solution for

$$\mathbf{y}_{\mathcal{A}} = W_{\mathcal{A}}\mathbf{x}_{\mathcal{A}} \tag{8.19}$$

is given by (8.9) to be

$$\mathbf{x}_{\mathcal{A}} = \left[W_{\mathcal{A}}^T W_{\mathcal{A}} \right]^{-1} W_{\mathcal{A}}^T \mathbf{y}_{\mathcal{A}}. \tag{8.20}$$

Since $\mathcal{A}$ is strictly a subset of $\mathcal{S}$ (i.e., $\mathcal{S} = \mathcal{A} + \mathcal{N}$, where $\mathcal{N} = \bar{\mathcal{A}}$) the signal is of finite support. Therefore solving (8.16) over $\mathcal{N}$ will not provide any additional information. This subspace reduction technique has been found to be effective in enhancing the resolution of a spectral estimate [20].

The block least squares solution given by (8.20) optimizes the fit to a few of the equations at the expense of the others. The values chosen for the weights are therefore trade-off parameters similar to a conventional window design. This block method optimizes the frequency localization of a few frequency bands but suffers from increased side-lobes outside the selected frequency bands. Thus, if the weights are properly chosen, a nonuniform distribution of frequency resolution is attainable, where certain bands have enhanced frequency resolution and others—which are of no interes—have poor resolution.

8.4   PROJECTION METHOD OF DECOMPOSITION

A computationally efficient technique for obtaining a solution to a possibly rank-deficient system of linear equations is the row-action projection (RAP) method described in Chapter 2. This method has advantages in implementation because of the following properties [21]:

1. No changes are made to the original system matrix (W).
2. No operations are performed on the system matrix as a whole.
3. Each iterative step requires access to only one row of the matrix.
4. In a single iterative step, the only iterate needed is the immediate predecessor.

In the cases of interest, the matrix W of (8.7) may be rank deficient and ill conditioned, so the system of equations is typically inconsistent due to measurement noise. Therefore, the inverses of W and of the Grammian matrix $W^T W$ do not exist.

The RAP method forms a solution to the system of linear equations by iteratively performing an orthogonal projection from a solution of one equation onto the hyperplane defined by the next equation. Since each equation in (8.7) represents a hyperplane in an n-dimensional Hilbert space, the RAP method iteratively projects the solution toward the region of the intersection of all hyperplanes. Due to the uncertainty introduced by measurement noise, the hyperplanes may not meet at a common point but may form a simplex region. This region represents an area of solutions to the system of equations. The use of constraints reduces the null-space ambiguity and shrinks the size of the simplex region, which reduces the ambiguity of the solution.

The RAP update equation is given by

$$\mathbf{x}^{i+1} = \mathbf{x}^i + \lambda_k \frac{\mathbf{y}_k - \langle W_k, \mathbf{x}^i \rangle}{|W_k|^2} W_k, \tag{8.21}$$

where $\mathbf{x}^i$ is the value of the solution after the ith iteration, λ_k is a relaxation parameter, $\mathbf{y}_k$ is the kth sample of the time signal, and W_k is the kth row of the W matrix. For each iteration i, the solution is updated by projecting onto the kth equation. This projection equation is known as the Kaczmarz row-action projection [22] algorithm and is equivalent to the normalized least mean squares (NLMS) method [23] in adaptive filtering and the algebraic reconstruction technique (ART) [24] in computer tomography.

Each hyperplane represents a linear surface that is also convex. Thus, RAP consists of alternating orthogonal projections onto convex surfaces and is, therefore, a projection onto convex sets (POCS) method [25]. Since RAP is a form of POCS, any constraint that forms a convex set can be incorporated into the method. It has been shown [26] that many useful constraints form closed convex sets and, therefore, can be used to improve the solution. Some common convex constraints are bandlimiting, time-limiting, bounded energy, nonnegativity, and upper and lower limits. These constraints have been found to be very powerful tools in improving the trade-off between resolution and frequency tracking [8] by reducing the null-space ambiguity.

The addition of convex constraints to the row-action projection algorithm requires that the solution be alternately projected onto a hyperplane and then projected onto a constraint set, i.e.,

$$\hat{\mathbf{x}}^i = P_C \{\mathbf{x}^i\}, \tag{8.22}$$

where P_C is the concatenation of the convex projection operators given by

$$P_C = P_{C_1} P_{C_2} \cdots P_{C_r}, \tag{8.23}$$

where P_{C_j} is the jth constraint, $\mathbf{x}^i$ is the outcome of the row-action projection, and $\hat{\mathbf{x}}^i$ is the constrained solution used for the next iteration of the row-action projection. Therefore, the constrained row-action projection algorithm is given by

$$\mathbf{x}^{i+1} = P_C \{\mathbf{x}^i\} + \lambda_k \frac{y_k - \langle W_k, P_C \{\mathbf{x}^i\} \rangle}{|W_k|^2} W_k. \tag{8.24}$$

This constrained row-action projection algorithm can be used with a simple constraint, such as positivity. Although not very powerful for spectral estimation, this constraint still improves the convergence of the algorithm. This can be seen from the following example, which tries to identify two closely spaced cosine tones. The columns of the system matrix are discrete samples of cosines at frequencies $2\pi/6$, $3\pi/6$, $4\pi/6$, and $5\pi/6$. It can easily be seen from inspection of (8.25a) that the two tones are located at the central two frequencies:

$$\begin{pmatrix} -0.5 \\ -1.5 \\ -1.0 \\ 0.5 \end{pmatrix} = \begin{pmatrix} 0.5 & 0.0 & -0.5 & -0.866 \\ -0.5 & -1.0 & -0.5 & 0.5 \\ -1.0 & 0.0 & -1.0 & 0.0 \\ -0.5 & 1.0 & -0.5 & -0.5 \end{pmatrix} \begin{pmatrix} x_1 \\ x_2 \\ x_3 \\ x_4 \end{pmatrix} \tag{8.25a}$$

The typical solution to this system of equations is found by computing the inverse of the system matrix. In this example, the matrix is singular and, therefore, the inverse does not exist. The Grammian matrix, given by $[W^T W]$, is very ill conditioned; therefore, the inverse is unstable and the least squares solution would most likely not produce a good result. A solution can be found using the constrained row-action projection algorithm by starting with a zero-solution vector, x^0, and iterating until the error is small. For this problem, the error will be considered small when the magnitude of the projection distance given by

$$\bar{d} = \frac{\text{abs}(y - \langle W, x \rangle)}{|W|^2} \tag{8.25b}$$

is less than 0.05. This error threshold must be chosen to match the problem of interest. The iterations are given by (8.24) and proceed as follows:

$$x^1 = \begin{pmatrix} 0.0 \\ 0.0 \\ 0.0 \\ 0.0 \end{pmatrix} + -\frac{0.5}{1.25} \begin{pmatrix} 0.5 \\ 0.0 \\ -0.5 \\ -0.866 \end{pmatrix} = \begin{pmatrix} -0.2 \\ 0.0 \\ 0.2 \\ 0.34 \end{pmatrix}. \tag{8.25c}$$

The positivity constraint can now be applied by setting all negative elements of the solution vector to zero:

$$x^2 = \begin{pmatrix} 0.0 \\ 0.0 \\ 0.2 \\ 0.34 \end{pmatrix} + -\frac{1.57}{1.75} \begin{pmatrix} -0.5 \\ -1.0 \\ -0.5 \\ 0.5 \end{pmatrix} = \begin{pmatrix} 0.45 \\ 0.89 \\ 0.65 \\ -0.11 \end{pmatrix}, \tag{8.25d}$$

$$x^3 = \begin{pmatrix} 0.45 \\ 0.89 \\ 0.65 \\ 0.0 \end{pmatrix} + \frac{0.1}{2.0} \begin{pmatrix} 1.0 \\ 0.0 \\ -1.0 \\ 0.0 \end{pmatrix} = \begin{pmatrix} 0.4 \\ 0.89 \\ 0.6 \\ 0. \end{pmatrix}. \tag{8.25e}$$

At this point, the error is equal to, but not less than, the error threshold; therefore, the iterations will continue:

$$x^4 = \begin{pmatrix} 0.4 \\ 0.89 \\ 0.6 \\ 0.0 \end{pmatrix} + \frac{0.11}{1.75} \begin{pmatrix} -0.5 \\ 1.0 \\ -0.5 \\ -0.5 \end{pmatrix} = \begin{pmatrix} 0.37 \\ 0.95 \\ 0.57 \\ -0.03 \end{pmatrix}. \tag{8.25f}$$

Since all equations have been used and the error is still too large, the algorithm continues by iterating over the equations a second time:

$$x^5 = \begin{pmatrix} 0.37 \\ 0.95 \\ 0.57 \\ 0.0 \end{pmatrix} + -\frac{0.4}{1.25} \begin{pmatrix} 0.5 \\ 0.0 \\ -0.5 \\ -0.866 \end{pmatrix} = \begin{pmatrix} 0.21 \\ 0.95 \\ 0.73 \\ 0.28 \end{pmatrix}, \tag{8.25g}$$

$$x^6 = \begin{pmatrix} 0.21 \\ 0.95 \\ 0.73 \\ 0.28 \end{pmatrix} + -\frac{0.22}{1.75} \begin{pmatrix} -0.5 \\ -1.0 \\ -0.5 \\ 0.5 \end{pmatrix} = \begin{pmatrix} 0.27 \\ 1.08 \\ 0.8 \\ 0.21 \end{pmatrix}, \tag{8.25h}$$

$$x^7 = \begin{pmatrix} 0.27 \\ 1.08 \\ 0.80 \\ 0.21 \end{pmatrix} + \frac{0.07}{2.0} \begin{pmatrix} 1.0 \\ 0.0 \\ -1.0 \\ 0.0 \end{pmatrix} = \begin{pmatrix} 0.24 \\ 1.08 \\ 0.77 \\ 0.21 \end{pmatrix}. \tag{8.25i}$$

At this point, a simple threshold can easily detect the presence of two tones in the second and third positions. If the iterations are continued, the solution may improve; however, a suitable solution has been found. Additional constraints can be found; they may be more powerful and, therefore, produce a more ideal solution. The error threshold and the constraints both need to be matched to the problem and the data available.

An example of a commonly used constraint—one which allows for signal subspace partitioning—is the finite bandwidth constraint, which is enforced by the projection operator given by

$$P_{C_i}\{x^i\} = \begin{cases} 0, & \text{if } |\omega| > \omega_{max}, \\ 0, & \text{if } |\omega| < \omega_{min}, \\ x^i, & \text{otherwise.} \end{cases} \tag{8.26}$$

If several bandlimiting constraints are concatenated, then the adaptive signal subspace partitioning can be performed and any arbitrary regions formed.

The adaptive signal subspace partitioning of the frequency domain allows for improved resolution of the estimate without sacrificing tracking rate. The adaptive strategy can take advantage of the quasi-stationary nature of the signal and use it to reduce the computational cost. Since the decomposition is formed sample by sample, changes in the frequency subspace are tracked within a few time samples. When applicable, additional convex constraints can be appended to the decomposition to further accelerate the convergence of the algorithm and increase the tracking rate.

8.4.1 Subspace Determination

Several methods have been investigated to determine the extent of the subspace A. In the ideal case, the signal-component regions are known *a priori* and the subspace A can be chosen to produce optimal results. When these regions are not known *a priori*, a heuristically developed adaptive frequency space partitioning method can be used.

To determine the signal subspace of interest, an initial estimate equivalent to a low-resolution DFT is calculated. This estimate is determined from only the

```
BEGIN:
        Do while i < mᵢ (some small number of time samples)
                Perform Row-Action Projection (x, W, X, i)
        End Do
        Let W' = columns of W corresponding to Threshold(X) = 1
        BEGIN Row-Action Projection:
                Store A_peak, BW_est, E_tot, Bin_peak
                Do until no new samples
                        Perform Row-Action Projection (x, W', X, i)
                        Find A_peak', BW_est', E_tot', Bin_peak'
                        If A_peak' < threshold * A_peak  Goto BEGIN
                        If E_tot' < threshold * E_tot  Goto BEGIN
                        If Bin_peak' ≠ Bin_peak then
                                Shift W' by Bin_peak - Bin_peak'
                                Goto BEGIN Row-Action Projection
                        End if
                        If BW_est' ≠ BW_est then
                                Dilate W' by BW_est'/BW_est
                                Goto BEGIN Row-Action Projection
                        End if
                End Do
        End BEGIN Row-Action Projection
End BEGIN
```

Figure 8.4 Pseudocode for adaptive partitioning.

first few samples of data and produces a low resolution decomposition onto a large frequency space. A threshold is applied to the result to determine those regions which contain significant energy and are of sufficient bandwidth. These regions define the desired signal subspace and can then be used as the initial estimate of the signal subspace $\mathcal{A}$.

After initialization, the representative characteristics of the subspace are tracked with each subsequent decomposition. The characteristics are the peak amplitude, estimated bandwidth, and total energy within each region. When these features vary significantly from their initial estimates, the direction of movement is determined by matching the movement of the peaks and the estimated bandwidths. The subspace regions are shifted and dilated according to the shift in center frequency and the change in bandwidth about the center frequency. If there is a significant change in the signal energy or the peak amplitudes, it is assumed that a component has been either added or removed and a new initial positioning is determined. An example of pseudocode for implementing such an adaptive partitioning strategy is given in Figure 8.4.

8.5 COMPUTATIONAL COMPLEXITY

The computational complexity is discussed in terms of the number of multiplications required for the application of a particular method to the signal of interest. The computational complexity of the general form of the analytic-signal method is

$$2MN + 6N \qquad (8.27)$$

multiplications, where M is the number of time samples and N is the number of frequency samples. Using the fact that the Hilbert transform matrix is square yields $M = N$. In addition, since the Hilbert transform can be implemented by calculating the inverse Fourier transform of the one-sided spectrum, the more efficient fast Fourier transform (FFT) can be used, thus reducing the computational cost to

$$C = 2N \log_2(N) + 6N \qquad (8.28)$$

multiplications. Typically, when studying the instantaneous spectrum of multiple narrowband tones, a large number of time and frequency samples are desired to provide good resolution in both time and frequency domains. Therefore, if N is large, the computational complexity of the analytic-signal method can be expressed as being on the order of $N \log_2(N)$, or $O(N \log_2(N))$.

Since the Wigner distribution (in the discrete case) is given by the sum over all τ, where $|\tau| \le M/2$, for each M and N, the computational complexity of the general form of the Wigner distribution is

$$C = M^2 N \qquad (8.29)$$

multiplications, where typically $M = N$. Since it is obvious from (8.6) that the Wigner distribution can be implemented in part by an FFT, the computational complexity can be reduced to $O(N^2 \log_2(N))$. This method is therefore more computationally expensive than the analytic-signal method.

The use of adaptive frequency space partitioning and the dynamic block size strategy requires only modest computational complexity, but provides great savings on the computational complexity of the projection algorithm. Each projection onto the full column space of W requires N multiplications. By partitioning the frequency space, the column space of W is reduced to be on the order of the number of tones P actually present in the composite signal **y**. Typically, at least M iterations are required to reduce the estimation error to an acceptable level. By dynamically adapting the size of the analysis blocks, the row iterations are reduced to the minimum required to model the nonstationarity of the signal **y**. For all simulations, this was found to be on the order of

$$C = 2PM/N \qquad (8.30)$$

multiplications, where typically $M \approx N$. Therefore, for all simulations in Section 8.5, the total number of multiplications for any observation interval was found to be $O(P^2)$, where $P \ll N$. Since in a spectrum composed of a few narrowband signals typically $P \le \log_2(N)$, the computational complexity of the RAP method is found to be significantly less than that of either the analytic-signal or Wigner distribution methods.

8.6   NUMERICAL RESULTS

Several figures of merit are used to illustrate the superresolving capabilities and tracking performance of the new method. First, we shall consider the superresolving capabilities in terms of the detection probability for tones spaced closer together than the uncertainty distance. Receiver operating characteristic (ROC) curves showing the probability of detection and probability of false alarm versus signal-to-noise ratio (SNR) are given at various resolutions. Then we shall present the frequency-tracking performance of the new method. To show tracking ability, comparisons of FM tracking with the analytic-signal and the Wigner distribution methods are shown for single- and multitone signals.

The probability-of-detection curves show the likelihood of detecting a tone in the correct frequency bin even though the uncertainty principle imposes an ambiguity of several bins. The probability-of-false-alarm curves show the immunity of the algorithm to the production of spurious and ambiguous peaks in the estimate, which falsely increase the probability of detection. Therefore, if an algorithm has a high probability of detection and a low probability of false alarm when the bin spacing is closer than the uncertainty distance, it can be said to be superresolving. The adaptive RAP method is shown to be superresolving to one-eighth (1/8) the uncertainty distance even in the presence of strong noise.

The probability of detection shown in Figures 8.5 and 8.6 represent the optimal detection using the block method outlined in (8.16), since the active frequency bands were known *a priori*. The performance of the adaptive RAP method without any *a priori* knowledge of active bands, as given by (8.24) and Figure 8.4

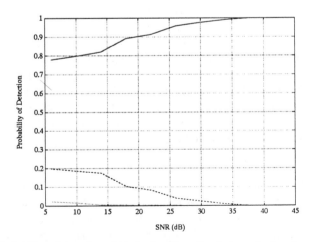

Figure 8.5 Probability of detection of block algorithm. Resolution = 4 × time-bandwidth product.

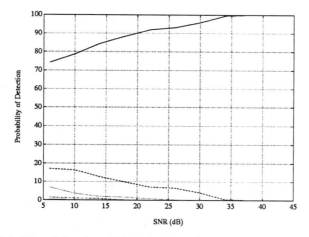

Figure 8.6 Probability of detection of block algorithm. Resolution = 8 × time-bandwidth product.

for the same resolutions is shown in Figures 8.7 and 8.8. Each of these simulations represents the performance of the algorithm averaged over variable relative phase and noise conditions. Unlike many discrete spectral-estimation techniques that are sensitive to relative phase variations due to destructive interference, this technique is robust to varying phase and noise conditions.

The performance is shown for the superresolving cases of 4 and 8 times the time-bandwidth product. For these cases, the effective resolution exceeds the stability-time-bandwidth product since the *a priori* information reduces the variance in the spectral estimate by eliminating the nonactive frequency bands. The detection process is a square-law detection process that produces a chi-square

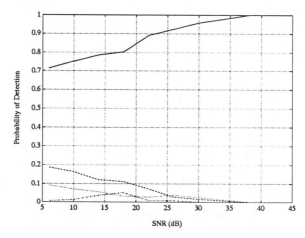

Figure 8.7 Probability of detection of row-action projection algorithm. Resolution = 4 × time-bandwidth product.

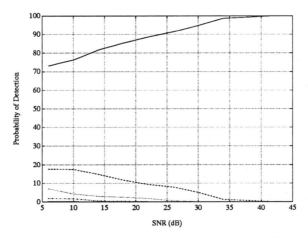

Figure 8.8 Probability of detection of row-action projection algorithm. Resolution = 8 × time-bandwidth product.

statistic. Since the nonactive frequency bands have been zeroed, the degrees of freedom of the statistic have been reduced, thereby increasing the probability of detection for a given probability of false alarm. The analysis of detection probabilities with a square-law detector is well known [27].

The adaptive-partitioning strategy allows the row-action projection method to perform almost as well as the block method down to 20 dB SNR, where the noise begins to affect the adaptability of the algorithm. When the noise power is relatively large, it is difficult to reduce the frequency space, since the signal and the noise become indistinguishable in the coarse estimation phase of the partitioning algorithm. It should also be mentioned that, for high SNR (> 30 dB), the frequency space partitioning is very close to the actual bandwidths (within 2 or 3 frequency bins per single tone) and thus performs comparably. Development of adaptive algorithms for bandwidth estimation is a current research area.

The probability-of-false-alarm rate shown in Figures 8.9 and 8.10 represent the best case for false-alarm rate, since the block method was used with perfect knowledge of the active frequency bands. This yields a best case for the probability of false alarm, since most possible false alarm candidate bins have already been eliminated. Since it was seen above that the adaptive partitioning strategy and, therefore, the frequency-space partitioning works very well for reasonable SNR, it is expected that the probability of false alarm is very close to the best case. In fact, the probability of false alarm for signals with SNR above 35 dB are almost identical to the best-case examples. This can be seen in Figures 8.11 and 8.12. These examples are also shown for the superresolving cases of 4 and 8 times the time-bandwidth product. It should be noticed that, even for very poor SNR (~ 10 dB), the probability of false alarm may be acceptable for many applications.

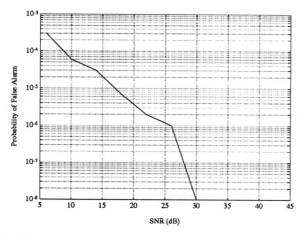

Figure 8.9 Probability of false alarm of block algorithm. Resolution = 4 × time-bandwidth product.

The tracking analysis performance of the proposed algorithm is illustrated in computer simulations by estimating the instantaneous frequency content of a nonstationary signal. The tracking performance for abrupt transitions and blanking is of particular interest. Tracking one narrowband signal with a time-varying instantaneous frequency is studied and it is found that the analytic-signal method produced the best result, as can be seen by comparing Figures 8.13–8.15. The analytic-signal method performs a little better than the RAP method because it allows for continuous phase and, hence, continuous frequency estimation. Since the RAP method implements a decomposition into discrete sinusoids, tracking the instantaneous frequency across discontinuities will always take a finite number

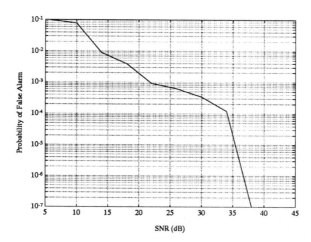

Figure 8.10 Probability of false alarm of block algorithm. Resolution = 8 × time-bandwidth product.

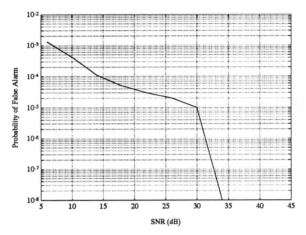

Figure 8.11 Probability of false alarm of RAP algorithm. Resolution = 4×time-bandwidth product.

of samples to converge. The Wigner distribution, as shown in Figure 8.14, suffers from energy present in the distribution where there should be none. Modifying the Wigner distribution with the Choi–Williams method [2] helps to reduce some of this spurious energy. It is shown in Figure 8.15 that the adaptive-frequency space-partitioning algorithm only requires a few samples to track from one frequency to the next. This very quick tracking could be used very effectively to track the fast frequency variations used in fast frequency-hopped communications [28]. Since subtle changes are tracked quickly, effective tracking of speech formants is also possible.

The multitone simulations were performed for two and five tones. The two-tone performance of the analytic signal is shown in Figure 8.16. Since it is

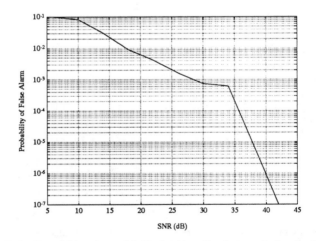

Figure 8.12 Probability of false alarm of RAP algorithm. Resolution = 8×time-bandwidth product.

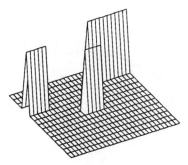

Figure 8.13 Single-tone frequency tracking of analytic signal.

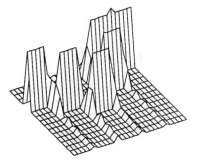

Figure 8.14 Single-tone frequency tracking of Wigner distribution.

tacitly assumed in the analytic-signal method that only a single tone is present, the results are not very useful for this case. The two-tone performance of the Wigner distribution is shown in Figure 8.17 and is corrupted by cross-terms; however, some data can be extracted by thresholding and, possibly, by using *a priori* information. This postprocessing increases the already high computational cost. The performance of the ro-action projection method is shown in Figure 8.18 and clearly shows the positions in time and frequency for both tones.

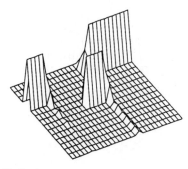

Figure 8.15 Single-tone frequency tracking of RAP algorithm.

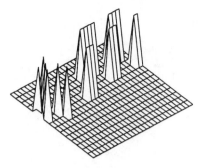

Figure 8.16 Two-tone frequency tracking of analytic signal.

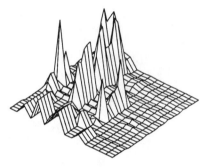

Figure 8.17 Two-tone frequency tracking of Wigner distribution.

The fast tracking ability of the row-action projection method is clearly demon-strated. The performance of the Wigner distribution with five tones is shown in Figure 8.19 and indicates that the cross-term corruption becomes too severe to extract any useful spectral or temporal information about the signal. The five-tone performance of the row-action projection method is shown in Figure 8.20 and, since it is linear, suffers little degradation due to the multitone nature of the signal.

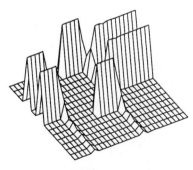

Figure 8.18 Two-tone frequency tracking of RAP algorithm.

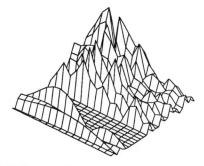

Figure 8.19 Five-tone frequency tracking of Wigner distribution.

8.7 SUMMARY AND CONCLUSIONS

A new approach to fast tracking of narrowband signals has been presented. This new method uses a row-action projection technique with adaptive-frequency space partitioning. A sample-by-sample update strategy provides a dynamic block size and allows for superresolution in the frequency domain with little loss of time localization. The algorithm is shown to be a subset of the method of projection onto convex sets (POCS) and therefore allows the addition of convex constraints to reduce the null-space ambiguity and increase the rate of convergence.

Numerical results demonstrate the superresolving capability and fast tracking rate, as well as stability in the presence of additive noise. Comparisons with the analytic-signal and Wigner distribution techniques indicate that the new method is highly competitive, particularly when several narrowband signals are present simultaneously and when noise is present. The method also offers computational advantages over the other two methods particularly for the sparse-spectrum case of interest here. Potential applications of this new method include formant tracking in speech signals, tracking of frequency-hopping communications, and FM demodulation.

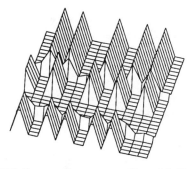

Figure 8.20 Five-tone frequency tracking of RAP algorithm.

ACKNOWLEDGMENTS

The research reported here was made possible, in part, through the support of the New Jersey Commission on Science and Technology and the Computer Aids for Industrial Productivity (CAIP) Center at Rutgers University.

REFERENCES

1. S. L. Marple. *Digital Spectral Analysis with Applications*. Englewood Cliffs, N.J.: Prentice-Hall, 1987.

2. L. Cohen. Time-frequency distributions—A review. *Proc. IEEE*, **77**(7), 1989.

3. J. L. Flanagan. *Speech Analysis, Synthesis, and Perception*, 2nd ed. New York: Springer-Verlag, 1972.

4. P. D. Welch. The use of fast Fourier transform for the estimation of power spectra: A method based on time averaging over short, modified periodograms. *IEEE Trans. Audio Electroacoust.*, **15**, June 1967.

5. R. B. Blackman and J. W. Tukey. *The Measurement of Power Spectra from the Point of View of Communications Engineering*. New York: Dover, 1958.

6. A. Papoulis. *Signal Analysis*. New York: McGraw Hill, 1977.

7. D. Slepian and H. O. Pollak. Prolate spheroidal wave functions, Fourier analysis and uncertainty—I. *Bell Syst. Tech. J.*, **40**:43–63, 1961.

8. R. J. Mammone. Spectral extrapolation of constrained signals. *J. Opt. Soc. Am.*, **73**(11), 1983.

9. R. J. Mammone, R. J. Rothacker, C. I. Podilchuk, S. Davidovici, and D. L. Schilling. Estimation of carrier frequency, modulation type and bit rate of an unknown modulated signal. In *Proceedings of the IEEE International Conference ICC-87*, Seattle, Wash., June 1987.

10. J. E. Mazo and J. Salz. Theory of error rates for digital fm. *Bell Syst. Tech. J.*, 1511–1535, November 1966.

11. N. M. Marinovic and G. Eichmann. An expansion of Wigner distribution and its applications. In *Proceedings of the IEEE International Conference on Acoustics, Speech, and Signal Processing 85*, Tampa, Fla., March 1985.

12. H. Taub and D. Schilling. *Principles of Communication Systems*. New York: McGraw Hill, 1986.

13. A. Libchaber and J. Maurer. A Rayleigh-Benard experiment: Helium in a small box. In T. Riste, ed., *Non-linear Phenomena at Phase Transitions and Instabilities*. New York: Plenum, 1982.

14. A. B. Fineberg and R. J. Mammone. Adaptive time-varying spectral analysis for multiple narrowband signals. In *Proceedings of the Fifth ASSP Workshop on Spectrum Estimation and Modeling*, Rochester, NY, October 1990.

15. R. M. Young. *An Introduction to Nonharmonic Fourier Series*. New York: Academic Press, 1980.

16. D. Gabor. Theory of communication. *J. IEE (London)*, **93**, 1946.

17. I. Daubechies. The wavelet transform, time-frequency localization and signal analysis. *IEEE Trans. Inf. Theory*, 1989.

18. G. Strang. Wavelets and dilation equations: A brief introduction, September 1989.

19. J. M. Ortega. *Matrix Theory—A Second Course*. New York: Plenum Press, 1987.

20. A. B. Fineberg and R. J. Mammone. An adaptive technique for high resolution time-varying spectral estimation. In *Proceedings of the IEEE International Conference on Acoustics, Speech, and Signal Processing*, Toronto, Ontario, May 1991.

21. Y. Censor. Row-action techniques for huge and sparse systems and their applications. *SIAM Rev.*, **23**(4), 1981.

22. P. P. B. Eggermont, G. T. Herman, and A. Lent. Iterative algorithms for large partitioned linear systems with applications to image reconstruction. *Linear Alg. Appl.*, **40**, 1981.

23. B. Widrow and S. D. Stearns. *Adaptive Signal Processing*. Englewood Cliffs, N.J.: Prentice-Hall, 1985.

24. G. T. Herman. *Image Reconstruction from Projections: The Fundamentals of Computerized Tomography*. New York: Academic Press, 1980.

25. L. G. Gubin B. T. Polyak and E. V. Raik. The method of projections for finding the common point of convex sets. *U.S.S.R. Comput. Math. Math. Phys.*, **7**:1–24, 1967.

26. D. C. Youla. Generalized image restoration by the method of alternating orthogonal projections. *IEEE Trans. Circ. Syst.*, **25**(9):694–702, 1978.

27. C. W. Helstrom. Distribution of the filtered output of a quadratic rectifier computed by numerical contour integration. *IEEE Trans. Inf. Theor.*, **32**(4), 1986.

28. L. B. Milstein and R. A. Iltis. Signal processing for interference rejection in spread spectrum communications. *IEEE Acoust. Speech Signal Process. Mag.*, **3**(2), 1986.

9

COMMUNICATION CHANNEL EQUALIZATION FOR DIGITAL DATA

John F. Doherty[1]

Electrical and Computer Engineering
Iowa State University
Ames, Iowa 50011
email: doherty@isuee1.ee.iastate.edu

9.1 DIGITAL EQUALIZATION

The deleterious effect of frequency selectivity on a communications system's performance is well known [1]. The term "frequency selectivity" can be used to denote the general condition that the data contains spectral nulls. The usual approach to reducing frequency selectivity is to use a channel equalizer in the receiver, which acts as an inverse filter. The frequency-selectivity phenomenon can arise naturally due to bandlimited channel dispersion or multipath propagation. Spectral nulling can also be induced at the receiver by using a fractionally spaced equalizer, which uses a sampling rate higher than the Nyquist rate. Spread-spectrum jammer excision and signal bandlimiting for frequency efficiency are other cases of intentionally induced spectral nulling [2, 3].

The simplest of the equalizer structures is the linear transversal equalizer, where the equalizer consists of a single finite impulse response filter operating on the sampled received data. Unfortunately, the linear transversal equalizer lacks the ability to undo severe frequency annihilation of the transmitted signal without an adverse enhancement of the contaminating noise. One might expect better frequency response performance from a linear feedback equalizer, where a feedback filter is added to process the unquantized equalizer output. The use of

---

[1]This research was done when Dr. Doherty was a Ph.D. candidate at Rutgers University.

the equalizer output in this manner affords little performance advantage, especially when the added complexity is considered [4]. An approach that provides significant performance gains is the decision-feedback equalizer (DFE), where the equalizer output is passed through a nonlinear decision device before it is fed back. The decision device simply selects the nearest discrete level that is known to exist for the transmitted symbol sequence. Thus, the use of *a priori* information about the transmitted symbols is used to recover loss due to spectral nulls.

The optimum equalizer filter coefficients are usually computed adaptively with the incoming data, since the channel characteristics are not known beforehand and are usually time varying. Thus, the adaptive estimation algorithm used must be able to track the variations in the channel response. The efficacy of conventional adaptive algorithms is hampered by data whose correlation matrix exhibits large eigenvalue spread. The least-mean squares (LMS) algorithm suffers from slow convergence when the eigenvalue spread is large [5]. The recursive least squares (RLS) methods exhibit numerical instability due, in part, to the inversion of an ill-conditioned correlation matrix [6]. The eigenvalue spread has a direct dependence on the spectral content of the sampled data. The presence of additive noise can reduce the eigenvalue spread, but it is also renders the equalizer prone to increased noise at the input of the decision device [7].

This chapter deals with the stable operation of an adaptive equalizer operating with data containing spectral nulls. The technique of iterative row-action projection (RAP) discussed in Chapter 6 will be shown to have performance advantages over conventional techniques because of its inherent regularization property. Regularization is a technique in system inversion by which the sensitivity to nulls in the signal spectrum is removed.

9.1.1 Intersymbol Interference

Intersymbol interference is a phenomenon observed by the adaptive equalizer caused by frequency distortion of the transmitted signal (Figure 9.1). This distortion is usually caused by the frequency-selective characteristics of the transmission medium. However, it can also be due to deliberate time dispersion of the transmitted pulse to affect realizable implementations of the transmit filter. The purpose of the equalizer is to remove deleterious effects of the ISI on symbol detection. The ISI generation mechanism is described next, with a description of equalization techniques to follow.

It has been shown that a bandpass-transmitted pulse train has an equivalent low-pass representation [4],

$$s(t) = \sum_{n=0}^{\infty} A_n p(t - nT) \tag{9.1}$$

where $\{A_n\}$ is the information-bearing symbol set, $p(t)$ is the equivalent low-pass transmit pulse waveform, and T is the symbol rate. The signal at the input of

INTERSYMBOL INTERFERENCE (ISI)
Band Limiting Transmission

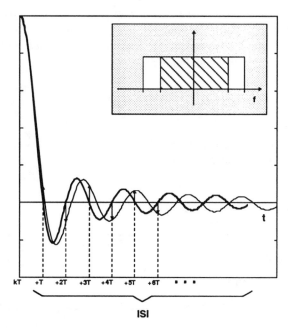

Figure 9.1 Frequency nulling disrupts the symbol-spaced zero crossings of the transmitted pulse. The frequency band of the channel is shown as a shaded region and the frequency extent of the pulse is the unshaded region. The low-pass nature of the channel nulls the transmitted pulse.

the receiver filter is

$$r(t) = \sum_{n=0}^{\infty} A_n \int_{-\infty}^{+\infty} p(t - nT)g(t - nT - \tau)\,d\tau + n(t), \qquad (9.2)$$

where $g(t)$ is the equivalent low-pass bandlimited impulse response of the channel and $n(t)$ is the additive noise. The optimum receiver filter, $w(t)$, is the *matched filter*, which is designed to give maximum correlation with the received pulse [8]. The output of the receiver filter—that is, the signal seen by the sampler—can be written as

$$x(t) = \sum_{n=0}^{\infty} A_n h(t - nT) + \nu(t), \qquad (9.3a)$$

$$h(t) = \int_{-\infty}^{+\infty} r(t)w(t - \tau)\,d\tau, \qquad (9.3b)$$

$$\nu(t) = \int_{-\infty}^{+\infty} n(t)w(t - \tau)\,d\tau, \tag{9.3c}$$

where $h(t)$ is the response of the receiver filter to the received pulse, representing the overall impulse response between the transmitter and the sampler, and $\nu(t)$ is a filtered version of the channel noise. The input to the equalizer is a sampled version of (9.3a); that is, sampling at times $t = kT$ produces

$$x(kT) = \sum_{n=0}^{\infty} A_n h(kt - nT) + \nu(kT) \tag{9.4}$$

as the input to the discrete time equalizer. By normalizing with respect to the sampling interval and rearranging terms, (9.4) becomes

$$x_k = \underbrace{h_0 A_k}_{\text{desired symbol}} + \underbrace{\sum_{\substack{n=0 \\ n \neq k}}^{\infty} A_n h_{k-n}}_{\text{intersymbol interference}} + \nu_k \tag{9.5}$$

The Nyquist Criterion

Assume that the actual transmission channel has a low-pass description,

$$G(f) = 1, \qquad |f| \leq B, \tag{9.6}$$

where B is the bandwidth of the channel. The output of the matched filter then becomes

$$H(f) = |P(f)|^2. \tag{9.7}$$

Since the channel impulse response is bandlimited, it can be expanded into the series [9]

$$h(kT - nT) = \sum_{m=-\infty}^{\infty} h(mT) \frac{\sin(\pi(k - n - m))}{\pi(k - n - m)}, \tag{9.8}$$

where $T = 1/2B$ is the *Nyquist rate*. Obviously, in order to eliminate the ISI terms in (9.5), all the samples of the channel impulse response must be zero for $m \neq 0$. Thus, any received pulse shape that has periodic zero crossings at a rate of $1/T$ produces no ISI at the sampler output. This condition for no ISI reception is called the *Nyquist criterion*. The simplest pulse that satisfies the Nyquist criterion is

$$h(t) = \frac{\sin(\pi t/T)}{\pi t/T}, \tag{9.9a}$$

$$H(f) = \begin{cases} T, & |f| \le B, \\ \\ 0, & |f| > B, \end{cases} \tag{9.9b}$$

where $h(t)$ in (9.9a) is referred to as a *sinc pulse*. The sinc pulse is convenient for analysis; however, its infinite duration makes it unrealizable. Realizable pulse shapes with controlled ISI are called *partial response signals*. One of the most widely used partial response signals is the raised cosine pulse:

$$h(t) = \frac{\sin(\pi t/T)}{\pi t/T} \frac{\cos(\beta\pi t/T)}{1 - (2\beta t/T)^2} \tag{9.10a}$$

$$H(f) = \begin{cases} T, & 0 \le |f| \le (1-\beta)B, \\ \\ \dfrac{T}{2}\left[1 - \dfrac{\sin(\pi T(f-B))}{\beta}\right], & (1-\beta)B \le |f| \le (1+\beta)B, \end{cases} \tag{9.10b}$$

where β is called the *roll-off parameter* and controls the amount of spectral leakage outside the band $|f| \le B$.

Signal Ill Conditioning by Oversampling

This is a convenient point at which to reexamine the signal ill conditioning brought about by sampling the received pulse higher than the Nyquist rate, as in fractionally spaced equalization. The spectrum of the sampled pulse in (9.3b) is the *folded spectrum* given by

$$H_s(f) = \frac{1}{T} \sum_{m=\infty}^{\infty} H(f + mB), \tag{9.11}$$

where $H_s(f)$ is the spectrum of the sampled received pulse. The sampler replicates the spectrum of the continuous received pulse at a spacing of B Hz. Thus, if the continuous spectrum is zero beyond the folding frequency—that is, $H(f) = 0$, $|f| > B/2$—then the continuous received pulse can be completely recovered from the sampled received pulse.[2] Assume that the bandwidth of the channel is less than twice the nominal folding frequency. If the sampling rate is doubled, then the spectrum replication rate is halved so that

$$H_s(f) = 0, \qquad B_{\text{channel}} \le |f| \le B, \tag{9.12}$$

where B_{channel} is the bandwidth of the channel and B is the folding frequency at the doubled sampling rate. For this example, doubling the sampling rate causes a frequency-nulled signal to appear at the input to the equalizer, as given by (9.12).

---

[2]This is a statement of Nyquist's sampling theorem.

9.1.2 Equivalent Discrete-Time Channel Model

The information transmitted by a digital communication system is comprised of a set of discrete symbols. Likewise, the ultimate form of the received information is cast into a discrete form. However, the intermediate components of a digital communications system operate with continuous waveforms that carry the information. The major portions of the communications link are the transmitter pulse-shaping filter, the modulator, the channel, the demodulator, and the receiver filter. It will be advantageous to transform the continuous part of the communication system into an equivalent discrete-time channel description for simulation purposes. The discrete formulation should be transparent to both the information source and the equalizer when evaluating performance. The equivalent discrete-time channel model is attained by combining the transmit filter, the channel filter, and the receiver filter into a single continuous filter, that is,

$$H(f) = W(f)G(f)P(f). \tag{9.13}$$

The action of the sampler is to discretize the aggregate filter. Use of the equivalent discrete-time channel as a means to simulate the performance of digital communications systems was popularized by Proakis [4] and has found subsequent use throughout the communications literature [10, 11].

9.2 EQUALIZER DESIGN CRITERIA

The criterion the equalizer bases its estimate on determines the noise immunity of the its output. The most obvious choice of performance criterion is the minimization of the average probability of error. However, the nonlinearity of the probability of error with respect to the adaptive equalizer coefficients makes it impractical. Thus, there are two other prevalent criteria for equalizer design, the *zero-forcing* criterion and the *minimum mean square error* criterion.

9.2.1 Zero-Forcing Equalizer

The only goal of the zero-forcing equalizer is to cancel the ISI terms in (9.5). The rationale behind this is that ISI is the limiting factor in many communications systems where the signal-to-noise ratio is high. Significant performance gains would be expected in such situations if the distortion of ISI were eliminated. The zero-forcing equalizer takes the form

$$C(f) = \frac{1}{H(f)}, \tag{9.14}$$

where $C(f)$ is the transfer function of the equalizer. It is apparent that the equalizer, in cascade with the channel, produces an all-pass equivalent filter; that is,

$$H(f)C(f) = 1. \tag{9.15}$$

The zero-forcing all-zero equalizer can only approximate the all-pole channel inverse. The output mean square error for an infinite-length zero-forcing equalizer is [4]

$$E\left\{[A_n - \hat{A}_n]^2\right\} = \sigma_e^2 = \frac{1}{2B} \int_{-B}^{B} \frac{\sigma_n^2}{H_s(f)}\, df, \tag{9.16}$$

where $\hat{A}_n$ is the estimate of the nth symbol, σ_n^2 is the additive noise power of the channel, and $H_s(f)$ is the spectrum of the discrete-time equalizer input signal. The main point about (9.16) is that the integrand becomes very large whenever the channel response contains spectral nulls, that is, whenever the input signal is ill conditioned. Thus, the zero-forcing equalizer produces large noise amplification when the channel is poorly conditioned. The mean square error performance criterion is often used because it directly accounts for the noise in the input signal.

9.2.2   Minimum Mean Square Error Equalizer

The goal of the minimum mean square error (MMSE) equalizer is to minimize the distortion introduced by the ISI *and* the additive noise. The performance criterion of the MMSE is

$$\min E\left\{[A_n - \hat{A}_n]^2\right\} = \min J(n) = J_{\min} \tag{9.17}$$

A lower bound for $J_{\min}$ can be obtained if the infinite-length equalizer is again considered. The MMSE for this case is [12]

$$J_{\min} = \frac{1}{2B} \int_{-B}^{B} \frac{\sigma_n^2}{H_s(f) + \sigma_n^2}\, df. \tag{9.18}$$

The form of (9.18) would be identical to (9.16), if not for the presence of the noise term in the denominator. This difference provides a performance gain for the MMSE equalizer when operating with a highly noisy channel.

9.3   LINEAR EQUALIZATION

Perhaps the most straightforward discrete time equalizer structure is that of a tapped delay line transversal filter, also called a linear equalizer. The linear equalizer forms a weighted sum of the sampler output to estimate the desired signal. The form of the linear equalizer is identical to that of the adaptive filter in Chapter 6. The formulation of the problem is repeated here:

$$
\begin{bmatrix} \bar{d}_k \\ \bar{d}_{k-1} \\ \vdots \\ \bar{d}_{k-L+1} \end{bmatrix} = \begin{bmatrix} \mathbf{x}_k^T \\ \mathbf{x}_{k-1}^T \\ \vdots \\ \mathbf{x}_{k-L+1}^T \end{bmatrix} \begin{bmatrix} c_0 \\ \vdots \\ c_{N-1} \end{bmatrix} + \begin{bmatrix} e_k \\ e_{k-1} \\ \vdots \\ e_{k-L+1} \end{bmatrix}, \tag{9.19}
$$

where $\mathbf{x}_k^T = [r_k r_{k-1} \cdots r_{k-N}]$ is the state vector of the equalizer at time k and r_k is the kth output of the channel. The solution for the adaptive coefficients in (9.19) must proceed with some special attention to determining the desired response $\bar{d}_k$. In fact, the equalizer can only approximate the desired response by using *a priori* information about the transmitted symbols.[3] This a priori information is usually implemented in the form of a decision device or level slicer. The output of the equalizer is a continuous-valued[4] signal; however, the transmitted symbols are discrete valued. This can be represented as

$$
\hat{d}_k = A_k + n_k, \tag{9.20}
$$

where A_k is the desired symbol and n_k is the continuous valued error. If the equalizer is performing well, then

$$
\bar{d}_k = Q\left[\hat{d}_k\right] = A_k, \tag{9.21}
$$

where $Q[\cdot]$ is the decision device, which chooses the nearest symbol so that $|\hat{d}_k - A_k|$ is minimized. Thus, if the noise satisfies

$$
n_k < |A_{k,i} - A_{k,j}| \qquad \forall\ i \neq j, \tag{9.22}
$$

then

$$
\bar{d}_k = A_k \qquad \forall\ k, \tag{9.23}
$$

where the indices $\{i,j\}$ range over the symbol alphabet. The pitfall with (9.23) is that the equalizer must be near its optimum setting in order to produce a residual error that satisfies (9.22); however, it cannot converge to its optimum setting if the decision device is producing poor estimates. The way to circumvent this dilemma is to train the equalizer by transmitting a known preamble. This *training sequence* allows time for the equalizer to gain a good estimate of the coefficients.[5] The time variation of the channel usually necessitates periodic retraining of the equalizer.

---

[3] Perfect knowledge of the transmitted symbols would obviate the need for an equalizer.
[4] To within the arithmetic precision used.
[5] This is also referred to as the *open-eye* condition [13].

Another detail of the equalizer design is accounting for the group delay of the channel. The group delay allows the maximum channel response to be delayed so that there appears to be ISI *before* and after the detected symbol. The equalizer must account for the group delay in order to cancel all the ISI. Intuitively, the equalizer should have its impulse response centered about its center tap such that it contains all the ISI components due to a transmitted symbol. This is accomplished by imposing a delay in the desired signal path that causes a delay in symbol detection. Since this delay is the same for each symbol, it is usually negligible.

The operation of the RAP algorithm can now be stated for the case of the linear equalizer,

$$\mathbf{c}_{k,i+1} = \mathbf{c}_{k,i} + \mu \epsilon_i \frac{\mathbf{x}_j}{\|\mathbf{x}_j\|^2}, \tag{9.24a}$$

$$\epsilon_i = \bar{d}_j - \mathbf{c}_{k,i}^T \mathbf{x}_j, \tag{9.24b}$$

$$\bar{d}_j = \begin{cases} d_{j-m} \left(= A_{j-m}\right), & \text{if training}, \\ \\ Q\left[\hat{d}_{j-m}\right], & \text{otherwise}, \end{cases} \tag{9.24c}$$

$$i = 1, 2, \ldots, L, \ldots, KL \quad \text{(iteration index)},$$

$$j = k - (i \bmod L) \quad \text{(equation index)}, \tag{9.24d}$$

$$k = L, 2L, 3L, \ldots \quad \text{(time index)},$$

where $j = i \bmod L$ and m is the delay needed to offset the group delay of the channel.

9.4   DECISION FEEDBACK EQUALIZATION

The decision-feedback equalizer utilizes the previously detected symbols in an attempt to predict the ISI caused by these symbols. The DFE is considered a nonlinear technique due to the nonlinear decision device in the feedback path. The minimum MSE for an infinite-length DFE can be expressed as [14]

$$J_{\min} = \frac{1}{2B} \exp \left\{ \int_{-B}^{B} \ln \left[\frac{\sigma_n^2}{H_s(f) + \sigma_n^2} \right] df \right\}. \tag{9.25}$$

Comparing (9.25) to (9.18) shows that the DFE MMSE will be lower than the linear equalizer case due to the exponentiation of the argument.

9.4.1   DFE Matrix Formulation

The structure of a decision-feedback equalizer, including the equivalent discrete time communication channel, is shown in Figure 9.2. The task of finding the

DISCRETE TIME CHANNEL MODEL AND EQUALIZER

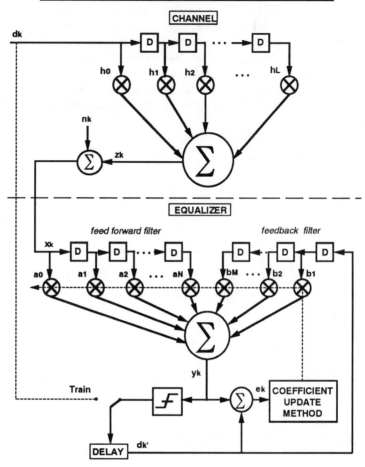

Figure 9.2 Block diagram of equivalent discrete-time communication system showing the channel model and decision-feedback equalizer. All quantities are complex and the channel coefficients, h_n, are time varying.

optimum DFE coefficients can be described by casting the problem into a system of linear equations:

$$
\begin{bmatrix} \tilde{d}_k^T \\ \tilde{d}_{k-1}^T \\ \vdots \\ \tilde{d}_{k-L+1}^T \end{bmatrix} = \begin{bmatrix} \mathbf{r}_k^T & \tilde{\mathbf{d}}_{k-1}^T \\ \mathbf{r}_{k-1}^T & \tilde{\mathbf{d}}_{k-2}^T \\ \vdots & \vdots \\ \mathbf{r}_{k-L+1}^T & \tilde{\mathbf{d}}_{k-L}^T \end{bmatrix} \begin{bmatrix} \mathbf{a}_k \\ \mathbf{b}_k \end{bmatrix} + \begin{bmatrix} e_k \\ e_{k-1} \\ \vdots \\ e_{k-L+1} \end{bmatrix} \tag{9.26}
$$

and

$$\mathbf{r}_k^T = [r_{k-N+1}, r_{k-N+2}, \ldots, r_{k-1}, r_k],\tag{9.27a}$$

$$\bar{\mathbf{d}}_{k-1} = [\bar{d}_{k-1}, \bar{d}_{k-2}, \ldots, \bar{d}_{k-M+1}, \bar{d}_{k-M}],\tag{9.27b}$$

where $\bar{d}_k$ is determined using (9.24c). The $N \times 1$ column vector $\mathbf{a}_k$ and the $M \times 1$ column vector $\mathbf{b}_k$ represent the feedforward and feedback coefficients, respectively. The received sample at time k is r_k, which consists of the channel output corrupted by additive noise. The vector $\hat{\mathbf{d}}_{k-1}$ represents the past M feedback values at time k. The DFE is said to be in decision-directed mode when $\hat{d}_k$ is taken as the output of the nonlinear decision device. The DFE is in training, or reference-directed, mode when $\hat{d}_k$ is explicitly made identical to the transmitted sequence d_k. The vectors $\mathbf{a}_k$ and $\mathbf{b}_k$ are the coefficients of the feedforward and feedback filters, respectively, at time k. The notation in (9.19) and (9.26) can be written in the compact matrix form,

$$\bar{\mathbf{d}}_{k,L} = \mathbf{X}_{k,L}\mathbf{c}_k + \mathbf{e}_{k,L},\tag{9.28}$$

by making the obvious associations. This means that the RAP algorithm can be applied to both equalizers by making the appropriate choices for the matrix and vector quantities. Note that the parameter L determines the number of rows of the time-varying matrix $\mathbf{X}_{k,L}$.

The assumption will be made that the correct symbols are presented to the feedback portion of the DFE. This condition validates the use of the system of linear equations in (9.28). If this assumption is not made, the DFE is only approximated by the linear equations [15]. In practice, the correct symbol assumption is valid most of the time due to high signal-to-noise operating conditions and the use of training sequences. The effect of incorrect feedback by the nonlinear decision device is usually short, contiguous bursts of errors that do not significantly degrade performance [16].

The RAP processor will accumulate the most recent L rows of the data matrix and then calculate KL updates of the DFE coefficients based on (9.24). The DFE will be provided with one set of updated coefficients for every L input samples. The method of choosing the number of iterations, KL, will be presented in Section 9.6.

9.4.2 Signal Fading and Error Propagation

Two phenomena that have deleterious effects on adaptive DFE performance are signal fading and error propagation. The signal-fading problem occurs when the channel output power becomes negligibly small. This may occur when the receiver experiences multipath destructive interference or direct path obstruction. The result is that the eigenvalue spread of the correlation matrix becomes excessively large and noise dependent. The RAP algorithm, when presented with a low-power input signal, can converge to a steady-state coefficient vector whose feedforward portion, a_k, is negligibly small. The reason for this is that $c_k = [\mathbf{0} \mid b_k]^T$ is a reasonable minimum-error solution for the DFE coefficient vector in the decision-directed mode. For example, $c_k = [\mathbf{0} \mid 0, 0, \ldots, 0, 1]^T$ produces errorless limit-cycle behavior because the output of the decision device and the state of the feedback tapped delay line are constant. The decision-directed DFE can never escape from this fixed point. An effective solution to this problem is to keep the energy of the feedforward coefficients above a predetermined threshold, T. A decision-directed DFE cannot produce any meaningful output when operating with a signal severely faded relative to the background noise level. Thus, the threshold value can be chosen to keep the power at the output of the feedforward filter at, or above, the input signal power without adversely affecting the DFE performance. This requirement translates into a constraint on the feedforward coefficients:

$$a_k^T a_k \geq T \geq 1. \tag{9.29}$$

Performing the extra step,

$$a_{k,KL}{}' = \frac{1}{\sqrt{T}} a_{k,KL} \qquad \text{iff} \quad a_{k,KL}^T a_{k,KL} < T \tag{9.30}$$

in (9.24) keeps the feedforward section of the DFE active in case the signal returns from the fade. This added operation is a projection onto a convex set (POCS) and is widely used in restoration problems to constrain the solution to a set of acceptable solutions [17].

The problem of error propagation occurs when the decision device passes incorrect decisions to the feedback section of the DFE. Subsequent updates of the coefficient vector are based upon these erroneous data, causing further degradation of the DFE performance. The effects of error propagation are usually manifested in short error bursts at the DFE output [11]. A simple and effective solution to this problem is to insert periodic training sequences into the data.[6] As a practical matter, the training sequences should be as short as possible. Typically 15% training symbols is considered acceptable overhead [19]. The RAP algorithm has a built-in mechanism to perform periodic retraining of the DFE coefficients efficiently; namely, the finite span of data over which it operates. The algorithm essentially forgets the erroneous state of the DFE in favor of the correct state inherent in the training information. After P iterations, the transient associated with the initial condition c_0 is given by [20]

$$\mathbf{H}^P c_0, \tag{9.31}$$

---

[6]Another error-propagation solution for the RAP algorithm, based on projection techniques, is presented in [18].

where the eigenvalues of H are all less than unity. Thus, $\lim_{P \to \infty} \mathbf{H}^P = \mathbf{0}$. The decay of the initial value $\hat{c}_0$ can be increased by increasing μ during retraining. The retraining process can be further streamlined by equating the number of retraining symbols to the length of the data matrix, L. This decouples the reference-directed sequence from the decision-directed sequence in the update process.

9.5   ECHO CANCELLATION

A problem closely related to channel equalization is echo cancellation. The echo phenomenon arises in both electrical and acoustical energy transmission. The electrical echo phenomenon is due primarily to impedance mismatches at the hybrid transformers of transmission lines. The acoustical echo problem arises when the transmitted signal arrives over multiple propagation paths. This typically occurs in a teleconference telephone connection.

The acoustical and electrical echo cancellation problems have the following significant differences:

1. The acoustical echo response is much longer than the electrical echo response, typically 125 and 4 msec, respectively.

2. The acoustical echo response changes much more rapidly than the electrical echo response.

3. The background noise is stronger in the acoustical environment.

4. Speaker-microphone amplifier gains may work in conjunction with the acoustical path losses to produce amplifier saturation, also known as "singing."

The unique problems associated with acoustical echo cancellation place it beyond the scope of this chapter. The interested reader may find the application of the RAP algorithm to subband acoustic echo cancellation in the scientific literature [21].

9.6   ALGORITHM NOISE SENSITIVITY

The total number of iterations, KL, in (9.24) is chosen so that K satisfies

$$K = \min \begin{cases} m \ni \bar{\epsilon}_{m+1}^2 \ge a\epsilon_m^2, \\ K_{\max}, \end{cases} \tag{9.32}$$

where

$$\bar{\epsilon}_m^2 = \sum_{i=(m-1)L+1}^{mL} \epsilon_i^2 \tag{9.33}$$

is the average squared error for the mth pass through the data. The first choice for the parameter K in (9.32) simply stops the adaptive process at the first instance of estimate degradation. This stopping criterion is founded in the theory of solving ill-posed problems [22]. The other constraint in (9.32), $K \leq K_{\max}$, is enforced so that the RAP algorithm does not exceed the maximum allowable processing load for a particular application. The cause of estimation degradation in adaptive equalization can be linked to noise sensitivity when frequency nulling of the transmitted signal has occurred.

The presence of spectral nulls in the channel frequency response increases the eigenvalue spread of the equalizer data correlation matrix. In fact, for spectrally nulled data,

$$\|\langle \mathbf{X}_{k,L}^T \mathbf{X}_{k,L} \rangle^{-1}\| \approx \sigma_\nu^{-2}, \tag{9.34}$$

where σ_ν^2 represents the average noise power and $\langle \cdot \rangle$ is the expectation operator. Note that the absence of noise in the data will make the RLS estimate highly unstable, i.e., $\|\langle \mathbf{X}_{k,L}^T \mathbf{X}_{k,L} \rangle^{-1}\| \to \infty$. Also, the lack of signal energy makes the least squares coefficient estimate sensitive to noise-dominated components of the data.

The RAP algorithm circumvents this problem by computing a regularized estimate of the adaptive coefficient vector [20]. In terms of the eigenvalues, ρ_i, of the data correlation matrix, $\Phi_{k,L} = \mathbf{X}_{k,L}^T \mathbf{X}_{k,L}$, an iterative regularized inversion leads to the mapping

$$\hat{\rho}_{i,l}^{-1} = \Omega_l(\rho_i)\rho_i^{-1}, \tag{9.35a}$$

$$\Omega_l(0) = 0, \qquad l < \infty, \tag{9.35b}$$

where $\Omega_l(\rho)$ is the *regularization function* which decreases with decreasing ρ. The RAP algorithm has the regularization function [23, 24]

$$\Omega_l(\rho_i) = \left[1 - \left(1 - \frac{\mu}{N+M}\rho_i\right)^{l+1}\right]^2, \qquad l \geq 1, \tag{9.36}$$

where each increment of l represents L updates to the coefficient vector and $0 < \mu < 1/\rho_{\max}$. There is no danger of division by zero in (9.35) since l is finite and

$$\Omega_l(\rho_i) \sim \rho_i^2 \qquad \text{as} \quad \rho_i \to 0 \tag{9.37}$$

Nonregularized inversion of a spectrally nulled correlation matrix produces estimates that are sensitive to random noise in the signal [25]. This can be seen by looking at the error vector, which is defined as $\varepsilon_l = \hat{\mathbf{d}}_{\text{optimum}} - \hat{\mathbf{d}}_l$:

$$\|\varepsilon_l\|_{\text{max}} \sim \|\varepsilon_l\|_{\text{fidelity}} + \|\varepsilon_l\|_{\text{noise}}, \tag{9.38}$$

where

$$\begin{aligned}
\|\varepsilon_l\|_{\text{fidelity}} &\ll 1 \qquad \text{for} \quad l \gg 1, \\
\|\varepsilon_l\|_{\text{noise}} &\gg 1 \qquad \text{for} \quad l \gg 1.
\end{aligned} \tag{9.39}$$

The regularization process improves the fidelity of the estimate as the iteration index, l, increases. However, degradation due to noise also increases with the iteration index. The RAP algorithm uses this property to stop iterating when its local estimate begins to degrade. This is the basis for the determination of K in (9.32). The RAP algorithm thus provides a solution that is insensitive to the noise-dominated components of the data. A similar approach has been suggested that requires the computation of the singular-value decomposition [7]. However, this technique requires an estimate of the number of significant eigenvalues. The RAP algorithm improves upon this result by avoiding the costly matrix computations and by using a simpler iterative technique that allows easy control of the inversion process. The criterion given by (9.32) effectively determines the correct number of significant eigenvalues used in the RAP algorithm.

9.7 SIMULATION RESULTS

This section provides computer simulation results that demonstrate the utility of the RAP algorithm for both linear and decision-feedback equalization.

9.7.1 Linear Equalization Results

This section presents simulation results obtained for the RAP algorithm in a time-varying channel-tracking application using an adaptive transversal equalizer. The RAP algorithm will be contrasted to the RLS algorithm to compare tracking ability and robustness.

The simulations presented in this section were obtained using a transversal filter structure, as shown in the generic setup of Figure 9.3. The transversal equalizer is obtained by removing the feedback section. The adaptive filter length was 21 coefficients. This number represents a trade-off between reducing the approximation error and the speed and cost of adaptation. The unknown system is modeled as an HF (time-varying) communications channel. This is accomplished by modeling the channel as an FIR filter with coefficients obtained by low-pass filtering white Gaussian noise. The 3-dB point of the low-pass filter used was $f_s/4000$, where f_s is the sampling rate of the communication system.

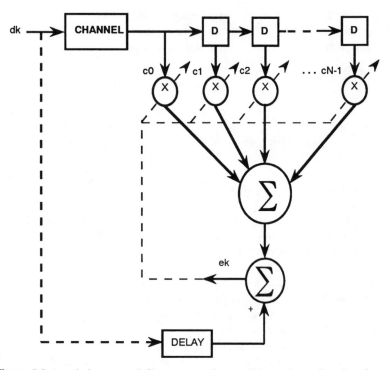

Figure 9.3 A typical transversal filter structure for equalizing a linear distorting channel.

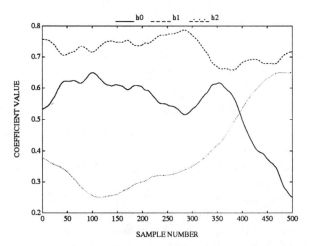

Figure 9.4 The time variation of the channel filter taps for the linear equalizer simulations.

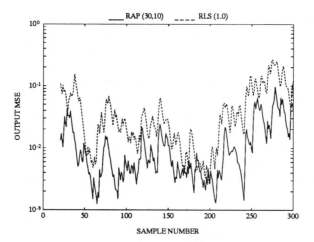

Figure 9.5 A comparison of RAP and RLS for $\lambda = 1$. The output error of the RAP algorithm is 5.4 dB better, on average.

This model has been used previously to describe a HF link [26, 11]. The variation of the channel is depicted in Figure 9.4. The symbol sequence used was a binary antipodal sequence randomly chosen from the set $\{-1, 1\}$.

The tracking performance of the RAP algorithm is compared to that of the RLS in Figures 9.5 and 9.6. In Figure 9.5, the *forgetting factor* equals unity, i.e., $\lambda = 1$. The expected poor tracking of the RLS algorithm is borne out in Figure 9.5. The forgetting factor is reduced in Figure 9.6 to $\lambda = 0.9$ to provide a performance improvement. The RAP algorithm operated with a block size of 30 equations with 10 cycles and a step size $\mu = 1$. This configuration is denoted as RAP(30,10). The block size was chosen to approximately match the duration

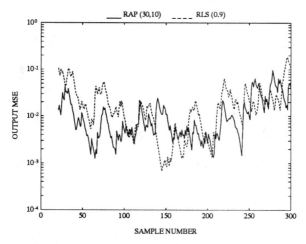

Figure 9.6 A comparison of RAP and RLS for $\lambda = 0.9$. The output error of the RAP algorithm is 2.0 dB better, on average.

of the stationarity in the channel. The number of cycles was chosen to provide a computational cost comparable to other fast adaptive algorithms.

The performance of both algorithms for a rank-deficient data matrix is shown in Figure 9.7. The rank deficiency was artificially induced by setting the second column of the data matrix equal to the first column. This only affects the tails of the impulse response of the equalizer, which are small compared to the center tap. The tails of the impulse response occur at the beginning and end of the coefficient vector, since the reference signal is delayed to center the impulse response of the equalizer. Physically, the presence of nulls in the frequency response of the channel induces ill conditioning in the data matrix [26]. The average signal-to-noise ratio (SNR) at the input to the equalizer was measured at 25.8 dB. The noise dominates the lowest-order eigenvalue of the correlation matrix and is amplified when the lowest eigenvalue gets inverted, as in the RLS algorithm. In contrast, the RAP algorithm attenuates the noise-only modes of the data matrix. This noise-attenuation property of the RAP algorithm is evident in Figure 9.7 by its significant performance advantage over the RLS algorithm.

The robustness of the RAP algorithm was demonstrated by a simulation in which large noise spikes, i.e., shot noise, were injected into the additive noise of the channel. The shot noise consisted of noise samples with amplitudes of 100 standard deviations spaced at intervals of 50 samples. The first spike occurs at sample 40. The signal and noise are otherwise identical to the previous simulation. Figure 9.8 shows the quick recovery time and marginally robust performance of the standard RAP algorithm. The simple procedure of residual monitoring described in Chapter 6 produces significant performance improvement, as shown in Figure 9.9. For comparison, the performance of the RLS algorithm for the same conditions is shown in Figure 9.10. The performance degradation is much more pronounced than the RAP algorithm degradation. Much of this

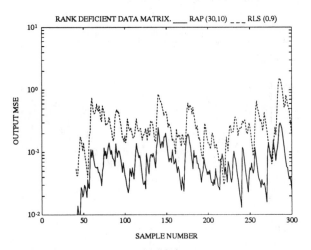

Figure 9.7 The performance of RAP and RLS for a rank-deficient data matrix. The output error of the RAP algorithm is 6.0 dB better, on average The average SNR is 25.8 dB.

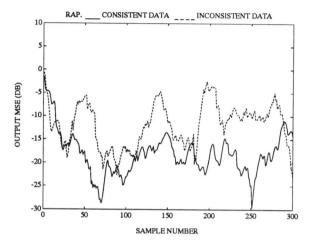

Figure 9.8 The performance of RAP in the presence of shot noise. The shot noise causes a degradation of 6.3 dB. The average SNR (without shot noise) is 25.8 dB.

difference can be attributed to the effective memory of the two algorithms. The effective memory of the RAP algorithm is essentially the block size, L. Thus, the effect of past data is retained only through the evolution of the filter coefficients. In contrast, the RLS algorithm has its effective memory given by the time constant of the exponential decay of the data, $\tau = 1/(1-\lambda)$. Since all the weighted data is used to form coefficient updates, the effect of a statistical outlier will generally last longer on the RLS algorithm than on the RAP algorithm.

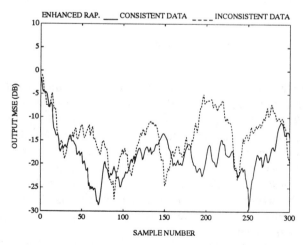

Figure 9.9 The performance of RAP in the presence of shot noise disregarding large residuals. The shot noise causes a degradation of 4.3 dB, an average improvement of 2.0 dB from the standard RAP algorithm.

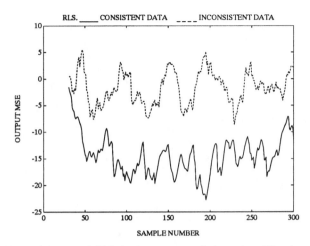

Figure 9.10 The performance of RLS in the presence of shot noise. The shot noise causes a degradation of 12.8 dB. The RAP algorithm shows a 4.6 dB average improvement over the RLS algorithm.

9.7.2 Decision-Feedback Equalization Results

In this section, the RAP algorithm is compared to the RLS and LMS algorithms for updating a complex-baseband fractionally spaced decision-feedback equalizer. The standard RLS algorithm is used in order to isolate instability due to spectral nulls in the data. The fast RLS algorithms are not used in order to avoid the instability issues they cause.

The baseband received signal at the DFE is modeled as a finite sum of $L' + 1$ reflected plane waves:

$$r(t) = \sum_{n=0}^{L'} h_n(t)s(t - \tau_n) + \nu(t), \qquad (9.40)$$

where the complex path attenuations, $h_n(t)$, have Rayleigh distributed amplitudes, $s(t)$ is the transmitted signal, and $\nu(t)$ is zero-mean white Gaussian noise with variance σ_ν^2. The model in (9.40) has been used for describing propagation in a mobile communication environment. The model chosen for simulation is an urban mobile environment, which is characterized by an attenuated line-of-sight path. There is evidence to indicate that the time-delay profile of this type of channel is only a few symbols in duration. A reasonable model gives $L' = 2$ in (9.40) with path delays of $\tau_n = nT$ and with relative powers of –3 dB, 0 dB, and –6 dB, respectively. The mobile speed is set at 50 mph, a nominal maximum for an urban environment.

The transmitted signal is QAM, with symbol alphabet $d_k = \{\pm1, \pm1\}$ and raised cosine pulse shaping with 100% excess bandwidth. Training symbols are used for the first 600 symbols to allow algorithm settling. Thereafter, 30 symbols of every 300 transmitted are used for training, incurring a 10% training overhead. The symbol spacing is set at $T = 7.4 \times 10^{-6}$ sec. The feedforward section of the DFE has six fractionally spaced $T/2$ coefficients and the feedback section has two symbol-spaced coefficients. The fractionally spaced equalizer is used to avoid the problems of sampling phase offset. The adaptive algorithms are configured as follows:

LMS	step size, $\mu = 0.008$,	
RLS	forgetting factor, $\lambda = 0.98$,	
RAP	samples/block, $L = 30$	iterations/block, $K_{max} = 10$,

These values were empirically determined to be optimum for this simulation.

The performance metrics measured are the average output mean square error (MSE) and the probability-of-symbol error, P_E, as they vary with signal-to-noise ratio (SNR). The numerical results are obtained using an average of 10 ensembles of 10,000 symbols, where the symbol and noise sequences change and the channel variation remains the same. The time-varying SNR, γ_k, and the mean SNR, $\bar{\gamma}$, are defined as

$$\gamma_k = \sum_{n=0}^{L'} \frac{h_{n,k}^* h_{n,k}}{\sigma_\nu^2}, \tag{9.41a}$$

$$\bar{\gamma} = \frac{1}{10,000} \sum_{k=1}^{10,000} \gamma_k. \tag{9.41b}$$

A plot of (9.41a) with $\sigma_\nu^2 = 1$ is shown in Figure 9.11

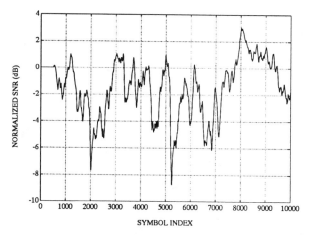

Figure 9.11 The normalized time-varying SNR used in the simulations. This plot corresponds to a noise power of $\sigma_\nu^2 = 0$ dB and a mean input SNR of $\bar{\gamma} = -1.5$ dB.

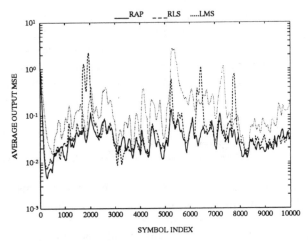

Figure 9.12 A comparison of the output MSE for the RAP, RLS, and LMS algorithms for an average input SNR of $\bar{\gamma} = 28.5$ dB.

A high-noise ($\bar{\gamma} = 8.5$ dB) and a low-noise ($\bar{\gamma} = 28.5$ dB) simulation were performed; the average MSE results are presented in Figures 9.12 and 9.13. The RLS-output MSE in the high-noise simulation reflects the noise sensitivity inherent in that algorithm for spectrally nulled data. This can be interpreted as large amounts of noise power passing through the frequency nulled portions of the signal spectrum to the output of the DFE. The unusually large output MSE is also caused by virtually continuous error propagation in the RLS DFE. This is demonstrated by the comparison to an RLS DFE with 100% training symbols in Figure 9.14. The LMS algorithm exhibits less noise sensitivity; however, it

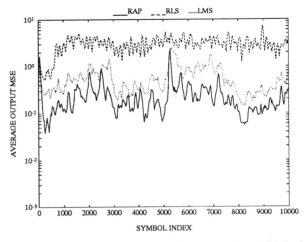

Figure 9.13 A comparison of the output MSE for the RAP, RLS, and LMS algorithms for an average input SNR of $\bar{\gamma} = 8.5$ dB.

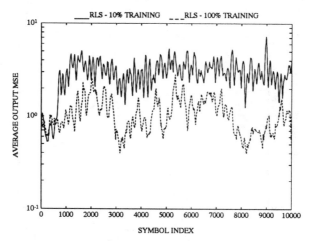

Figure 9.14 A comparison of the output MSE for the RLS algorithm with 10 and 100% training symbols. The average input SNR of $\bar{\gamma} = 8.5$ dB.

shows poor tracking capability. On average, the RAP algorithm output MSE is 11.7 dB lower than the RLS algorithm output MSE and 4.2 dB lower than the LMS algorithm output MSE. Note the rapid recovery of the RAP algorithm from the fade after symbol 5000. The average for the parameter K in this simulation is 9.1 iterations/block.

The output MSE results for the low-noise simulation show that RLS performance is hampered by large deviations induced by frequency nulls in the data. The RAP algorithm exhibits no instability and provides an MSE that is 1.4 dB lower than the RLS algorithm, including the deviations. The gain over the LMS algorithm is 5.6 dB. The average for K in this simulation is 5.8 iterations/block.

The probability of error, P_E, calculations are performed using mean SNR values ranging from 5 dB to 35 dB. The results are shown in Figure 9.15. Note that the performance of the LMS algorithm levels out due to its poor tracking capability.

Table 9.1. shows the computational complexity of the RAP, RLS, LMS, and fast RLS (FRLS) [27] algorithms. A plot of the functions in Table 9.1 is given in Figure 9.16 with $K = 10$, a typical value. Each divide is given the same complexity as a multiplication; in practice, however, implementing a division is several times more complex than a multiplication.

9.8 CONCLUSION

The row-action projection (RAP) algorithm was presented for adaptive equalization of spectrally nulled data. This frequency nulling occurs in a wide variety of communication systems, such as fractionally spaced equalization, reduced-bandwidth signaling, and spread-spectrum jammer excision. The RLS algorithm

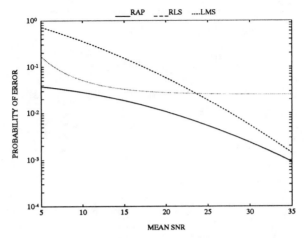

Figure 9.15 A comparison of the probability-of-symbol error for the RAP, RLS, and LMS algorithms.

TABLE 9.1 Comparison of computational complexity of adaptive DFE algorithms.

Algorithm	Complex multiplications	Complex divisions
RAP	$2KN + K$	1 (Real)
FRLS	$20N + 5$	3
LMS	$2N + 1$	0
RLS	$2.5N^2 + 4.5N$	2

The numbers represent operations per input sample.

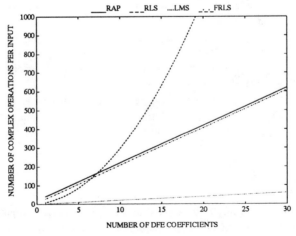

Figure 9.16 A comparison of the computational complexity of the RAP, RLS, LMS, and fast RLS algorithms.

was shown to have instability and noise-amplification properties that are traceable to the eigenvalue spread of the data-correlation matrix. The RAP algorithm does not suffer from these problems because it computes a regularized inverse solution of the DFE coefficients. The RAP algorithm was shown to provide improved performance over the conventional LMS and RLS algorithms for fractionally spaced DFE of a mobile communications channel. Simulation results show that the RAP algorithm has a lower average output MSE and a lower probability of error than either the LMS or RLS algorithm. The RAP algorithm also had computational complexity comparable to the fast RLS algorithm (FRLS), without exhibiting any of the numerical instability inherent in the FRLS algorithm.

ACKNOWLEDGMENTS

The author acknowledges the support of the Rutgers University Center for Computer Aids for Industrial Productivity (CAIP) and the Rutgers University Wireless Information Networks Laboratory (WINLAB) during portions of the preparation of this manuscript. The author was also supported through a Harpole–Pentair Faculty Fellowship while at Iowa State University.

REFERENCES

1. P. Monsen. Fading channel communications. *IEEE Commun. Soc. Mag.*, **18**(1):27–36, 1980.

2. L. B. Milstein. Interference suppression to aid acquisition in direct-sequence spread-spectrum communication. *IEEE Trans. Commun.*, **36**(11):1200–1207, 1988.

3. A. Fihel and H. Sari. Performance of reduced-bandwidth 16 QAM with decision-feedback equalization. *IEEE Trans. Commun.*, COM-35:715–723, 1987.

4. J. G. Proakis. *Digital Communications*, 2nd ed. New York: McGraw-Hill, 1989.

5. S. Haykin. *Adaptive Filter Theory*. Englewood Cliffs, N.J.: Prentice-Hall, 1986.

6. G. E. Bottomley and S. T. Alexander. A theoretical basis for the divergence of conventional recursive least squares filters. In *Proceedings of the IEEE International Conference on Acoustics, Speech, and Signal Processing*, 1989.

7. G. Long, F. Ling, and J. G. Proakis. Fractionally-spaced equalizers based on singular value decomposition. In *Proceedings of the IEEE International Conference on Acoustics, Speech, and Signal Processing*, pp. 1514–1517, 1988.

8. J. M. Wozencraft and I. M. Jacobs. *Principles of Communication Engineering*. New York: Wiley, 1965.

9. A. Papoulis. *Probability, Random Variables, and Stochastic Processes*, 2nd ed. New York: McGraw-Hill, 1984.

10. D. Hatzinakos and C. L. Nikias. Estimation of multipath channel response in frequency selective channels. *IEEE J. Selected Areas Commun.*, **7**(1):12–19, 1989.

11. E. Eleftheriou and D. D. Falconer. Adaptive equalization techniques for HF channels. *IEEE J. Selected Areas Commun.*, SAC-5(2):238–247, 1987.

12. E. A. Lee and D. G. Messerschmitt. *Digital Communication*. Kluwer Academic, 1988.

13. S. U. H. Qureshi. Adaptive equalization. *Proc. IEEE*, **73**(9), 1985.

14. J. Salz. Optimum mean-square decision feedback equalization. *Bell Syst. Techn. J.*, **52**(8):1341–1373, 1973.

15. J. E. Mazo. Analysis of decision directed convergence. *Bell Syst. Techn. J.*, **59**:1858–1876, 1980.

16. D. L. Duttweiler, J. E. Mazo, and D. G. Messerschmitt. Error propagation in decision-feedback equalizers. *IEEE Trans. Inform. Theory*, **IT-20**:490–497, 1974.

17. C. I. Podilchuk and R. J. Mammone. A comparison of projection techniques for image restoration. In *SPIE Proceedings on Electron Microscopy and Video Processing*, Los Angeles, January 1989.

18. J. F. Doherty and R. J. Mammone. A new fast method for channel estimation. In *Proceedings of the IEEE International Conference on Communications*, pp. 26.1.1–26.1.5, June 1989.

19. E. Kuisma et al. Signal processing requirements in Pan-European digital mobile communications. In *Proceedings of the IEEE International Symposium on Circuits and Systems*, pp. 1803–1810, 1988.

20. J. F. Doherty and R. J. Mammone. A row-action projection algorithm for adaptive filtering. In *Proceedings of the IEEE International Conference on Acoustics, Speech, and Signal Processing*, p. 20.D5.4, April 1990.

21. S. L. Gay and R. J. Mammone. Acoustic echo cancellation using POCS on the DSP16. In *Proceedings of the IEEE International Conference on Acoustics, Speech, and Signal Processing*, April 1990.

22. A. N. Tikhonov and V. Y. Arsenin. *Solutions to Ill-Posed Problems*. Washington, D.C.: V.H. Winston and Sons, 1977.

23. J. F. Doherty and R. J. Mammone. Regularized adaptive equalization for fading communications channels. Technical Report 3, Rutgers Wireless Information Networks Laboratory, November 1989.

24. J. F. Doherty and R. J. Mammone. A fast stable method of adaptive filtering. Submitted to *IEEE Trans. Acoust. Speech Signal Process.*, December 1989.

25. C. K. Rushforth. Signal restoration, functional analysis, and Fredholm integral equations of the first kind. In H. Stark, ed, *Image Recovery: Theory and Application*, Chapter 1. New York: Academic Press, 1987.

26. F. Ling and J. G. Proakis. Adaptive lattice decision feedback equalizers-their performance and application to time variant multipath channels. *IEEE Trans. Commun.*, **COM-33**(4):348–356, 1985.

27. D. D. Falconer and L. Ljung. Application of fast kalman estimation to adaptive equalization. *IEEE Trans. Commun.*, **COM-26**(10):1439–1446, 1978.

10

ADAPTIVE BEAMFORMING

Kevin Farrell

CAIP Center
Rutgers University
Piscataway, New Jersey 08855–1390
email: farrell@caip.rutgers.edu

10.1 INTRODUCTION

Beamforming is a method of coherently summing the outputs of spatially distributed sensors to improve signal reception and transmission in the presence of noise. Adaptive beamforming is used to improve signal reception by maintaining a narrow beamwidth directed towards the desired signal while adaptively steering sidelobes away from directional interference. This technique maximizes the desired signal response while minimizing the interference gain, thus improving the signal-to-noise ratio (SNR). Adaptive beamforming algorithms generally require *a priori* information regarding either the direction of arrival (DOA) of the desired signal, or a reference signal that is correlated with the desired signal. Adaptive beamforming algorithms do not generally require *a priori* information regarding the direction of the interference, but will adaptively cancel the effects of interference in the array output.

The problem of receiving a signal contaminated with noise, such as thermal noise or directional interference, is commonly encountered in sonar, radar, and geoscience applications. Adaptive beamforming provides an attractive solution for the signal-reception problem encountered in these applications. For example, military communication systems may be subject to jamming by electronic-countermeasures (ECM) equipment in a dynamically changing hostile environment. Adaptive beamforming may effectively be deployed in electronic counter-countermeasures (ECCM) to automatically cancel jammers while continuously tracking and adapting to changes in the environment. Another example may involve tracking a submarine with a sonar array. The sonar array may deploy adaptive beamforming by using the *a priori* information of the propeller frequency to track the submarine and cancel the effects of spurious signals that may come from nearby ships, ocean disturbances, etc. Adaptive arrays have also been suggested for microphone arrays to improve sound pick-up for conferencing applications [1].

This chapter develops a method of performing adaptive beamforming via the row-action projection (RAP) method. In the following two sections, the basic concepts of wave propagation and spatial frequency are explained to provide the necessary background for adaptive arrays. The motivation for adaptive beamforming is then provided through a comparison of phased arrays to dish antennas. Next, narrowband and broadband arrays and their basic structure are introduced. The RAP method for solving a set of linear equations is then introduced. The adaptive beamforming concepts are then presented as constrained optimization problems for the narrowband and broadband arrays and it is shown how the RAP method can be applied. Simulation results are provided for the narrowband and broadband models that show the RAP method to converge to the optimal Wiener solution. This chapter concludes with an algorithm and performance comparison to the well-known least mean squares (LMS) methods. It is shown that RAP, which is actually a form of *normalized* LMS, converges with comparable speed to the LMS algorithms, without requiring *a priori* information of the eigenvalue spread of the input autocorrelation matrix.

10.2 WAVE PROPAGATION

Adaptive beamforming uses adaptive arrays to enhance signal reception in the presence of noise. Adaptive arrays consist of a number of sensors typically configured in a line pattern that utilize the spatial characteristics of signals to improve the reception of a desired signal and/or cancellation of undesired signals. Signals may be spatially characterized by their angle of arrival with respect to the array. The angle of arrival of a signal is defined as the angle between the propagation path of the signal and the normal vector of the array. The angle of arrival of a signal is shown in Figure 10.1.

Note in Figure 10.1 that wavefronts emanating from a point source may be characterized by plane waves (i.e., the locus of constant phase falls on straight lines) when originating from the far-field (or Fraunhofer) region. The far-field approximation is valid for signals that satisfy the condition:

$$s \geq \frac{D^2}{\lambda},\tag{10.1}$$

where s is the distance between the signal source and the array, λ is the wavelength of the signal, and D is the length of the array. Wavefronts that originate closer than D^2/λ are considered to be in the near-field (or Fresnel) region. Wave-

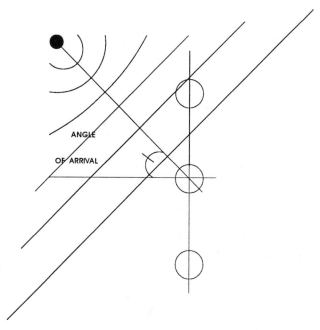

Figure 10.1 Angle of arrival.

fronts originating in the near field will be convex but not planar upon striking the array sensors. This chapter assumes that all sources are in the far-field region.

In Figure 10.1 it can be seen that the signal waveform will experience a time delay between crossing each sensor. This time delay corresponds to a shift in phase of the signal as observed by each sensor. This concept is illustrated in Figure 10.2, where θ is the angle of arrival.

The time delay, τ_1, of the waveform striking the first and then second sensors in Figure 10.2 may be calculated as

$$\tau_1 = \frac{d}{c}\sin\theta, \tag{10.2}$$

where d is the sensor spacing, c is the speed of propagation of the given waveform for a particular medium (i.e., 3×10^8 m/sec for electromagnetic waves through air, 1.5×10^3 m/sec for sound waves through water, etc.), and θ is the angle of arrival of the wavefront. The phase shift, or electrical angle, ϕ observed at each sensor to the angle of arrival of the wavefront may be found as

$$\phi = \frac{2\pi d}{\lambda_o}\sin\theta = \frac{\omega_o d}{c}\sin\theta \tag{10.3}$$

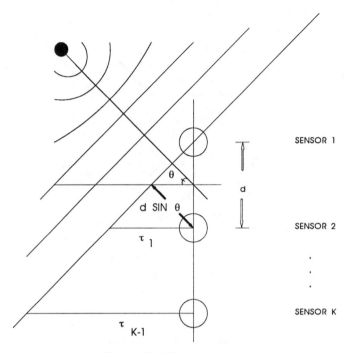

Figure 10.2 Wave propagation.

where λ_o is the wavelength of the signal at frequency f_o, as defined by

$$\lambda_o = \frac{c}{f_o}. \tag{10.4}$$

Therefore, a signal $x(k)$ crossing the sensor array at an angle of arrival θ may be characterized by the vector $X(k)$, where

$$X(K) = x(k) \begin{bmatrix} 1 \\ e^{-j\phi} \\ e^{-2j\phi} \\ \vdots \\ e^{-j(K-1)\phi} \end{bmatrix}. \tag{10.5}$$

Note that the phase-shift vector used for $X(k)$ assumes uniform sensor spacing.

Arrays that have a *visible region* of $-90°$ to $+90°$ (i.e., the azimuth range for signal reception) require that the sensor spacing satisfy

$$d \le \frac{\lambda}{2} \tag{10.6}$$

If the sensor spacing does not satisfy this relation, the array will be subject to grating lobes. Grating lobes are lobes other than the main lobe that appear in the visible region and can amplify signals from undesired directions. The above relation for sensor spacing is analogous to the Nyquist sampling rate for frequency domain analysis.

Example 10.2.1 It is desired to optimize the reception of a radio signal at 102.7 MHz using a five-sensor array (use $c = 3 \times 10^8$ m/sec).

1. Calculate the sensor spacing d in meters (assuming uniform spacing) if it is desired that the array length be equal to two wavelengths. If the array has a visible region of $-90°$ to $+90°$, will grating lobes appear?

2. The waveform described above arrives $30°$ above the normal of the array (see Figure 10.2 with $\theta = 30°$). Calculate the delays necessary at each sensor to have the waveform appear as if it were normal to the array. (Assume that the delay of the fifth sensor element is zero.)

3. Using the sensor spacing given in 1, calculate the distance in meters that is required for the far-field assumption to be valid.

Solution

1. Using (10.4) for wavelength, λ_o can be calculated as

$$\lambda_o = \frac{c}{f_o} = \frac{3 \times 10^8 \text{ m/s}}{102.7 \times 10^6 \text{ Hz}} = 2.92 \text{ m}, \qquad 2\lambda_o = 5.84 \text{ m}.$$

For a five-element array, the total sensor spacing will be four times d; therefore, the sensor spacing d may be found as

$$d = \frac{2\lambda_o}{4} = \frac{5.84 \text{ m}}{4} = 1.46 \text{ m}$$

Using the criteria stated in (10.6) for sensor spacing to avoid grating lobes:

$$d \le \frac{\lambda_o}{2} = 1.46 \text{ m} = d;$$

since d is exactly equal to $\lambda/2$, grating lobes will not appear.

2. In Figure 10.2, it can be seen that a wavefront arriving above the normal will first strike the first sensor, then the second, etc. Therefore, if the appropriate delays are applied directly after the sensors (with the largest delay following the first sensor, etc.), the wavefront may be made to appear normal to the array. The time constant τ may be calculated using (10.2)

as:

$$\tau = \frac{d}{c} \sin\theta = \frac{1.46 \text{ m}}{3 \times 10^8 \text{ m/sec}} \sin 30 = 2.4 \times 10^{-9} \text{ sec.}$$

Thus, delays of 4τ, 3τ, 2τ, and τ at the first through fourth sensors, respectively, will make the wavefront appear normal to the array.

3. The far-field assumption is valid for

$$\text{far-field distance} = \frac{D^2}{\lambda} = 11.68 \text{ m}$$

10.3   SPATIAL FREQUENCY VERSUS TEMPORAL FREQUENCY

The angle of arrival of a wavefront determines a quantity known as the *spatial frequency*. Beamforming uses information regarding the spatial frequency to suppress undesired directional signals. The spatial frequency is defined as the frequency observed across an array of sensors due to the phase shift of a signal arriving at some angle of arrival. Consider a sinusoid that is transmitted at a temporal frequency ω_c and is incident on the array at the three different angles of arrival shown in Figures 10.3(a–c).

Signals that arrive perpendicularly to the array (known as boresight) will create identical waveforms at each sensor. This case is illustrated in Figure 10.3(a). Signals that do not arrive perpendicular to the array will not create waveforms that are identical at each sensor, assuming that there is no spatial aliasing due to insufficiently spaced sensors. These waveforms will have a phase delay at each sensor that corresponds to the electrical angle given by (10.3) in the previous section. In Figures 10.3(a–c), it is assumed that the electrical angle and angle of arrival are equal in order to clarify the concept of spatial frequency. At some instant in time, hypothetical values measured at each sensor could be those values in parentheses shown in Figures 10.3(a–c). In the array in Figure 10.3(a) the same value (1.0) appears at each sensor. Since the values at each sensor

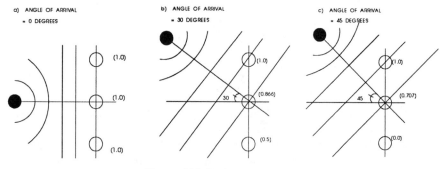

Figure 10.3 Spatial configurations.

are equal, there are no observed periodicities. In the array in Figure 10.3(b), the phase shift at each sensor due to the electrical angle results in the observed values of (1.0), (0.866), and (0.5) at each sensor. If each sensor observes a 30° phase shift of the waveform, the waveform will complete a full period in 12 sensors. Similarly, in the array in Figure 10.3(c), the waveform completes a full period in eight sensors. The periodicity of the three waveforms arriving at the different angles of arrival in Figures 10.3(a–c) is illustrated in Figures 10.4(a–c), respectively.

The periodicity of the three waveforms shown in Figures 10.4(a–c) corresponds to the spatial frequency of Figures 10.3(a–c), respectively. The unit of measure for spatial frequency is cycles/m as opposed to cycles/sec for temporal frequency. To simplify the remaining discussion, it is assumed that the total distance between the first and last sensors is one meter. Therefore, since the array response shown in Figure 10.4(a) does not observe a periodic response, the spatial frequency is 0 cycles/m. The array response in Figure 10.4(b) has a spatial frequency of one period per 12 sensors, which corresponds to 1 cycle/m. The array response in Figure 10.4(c) has a spatial frequency of one period per eight sensors, which corresponds to a spatial frequency of 1.5 cycles/m. It can be seen

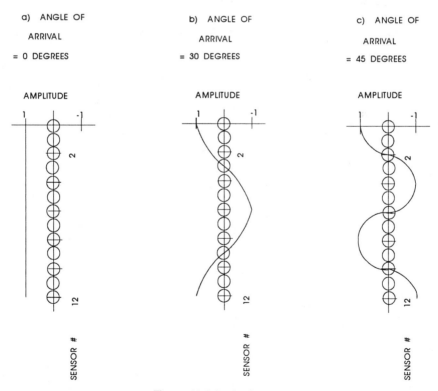

Figure 10.4 Spatial frequency.

from Figures 10.3(a–c) and 10.4(a–c) that, as the angle of arrival increases, so does the spatial frequency. It can also be deduced that retaining signals having an angle of arrival equal to 0° while suppressing signals from other directions is equivalent to low-pass filtering the spatial frequency. This will maintain the boresight signal (which has a spatial frequency of 0 cycles/m) while suppressing signals arriving from other directions that have a spatial frequency greater than zero. Adaptive beamforming uses this principle of spatial filtering to suppress directional signals.

10.4   PHASED ARRAY VERSUS DISH ANTENNA

Narrow-width beam generation can be realized by a large-aperture antenna or an array of small antennas. Without loss of generality, we shall consider a rectangular dish antenna of aperture length L that receives a wavefront originating in the far field. The relationships found for the rectangular aperture are similar to those for circular and other geometries [2]. Wavefronts arriving from the far field will create uniform waveforms on the antenna. Thus, the waveform incident on the rectangular antenna will appear as shown in Figure 10.5.

The resulting radiation pattern of the antenna is proportional to the Fourier transform of the incident waveform. Taking the Fourier transform of the rectangle shown in Figure 10.5 results in a sinc function as follows (where a rectangle height of $1/L$ is used for simplicity):

$$F[\text{rect}(L)] = \int_{-L/2}^{L/2} (1/L)e^{-jut}\, dt = \frac{\sin((L/2)u)}{(L/2)u} = \text{sinc}\left(\frac{L}{2}u\right). \qquad (10.7)$$

In (10.7), u represents the normalized spatial frequency and is defined as

$$u = \sin\theta. \qquad (10.8)$$

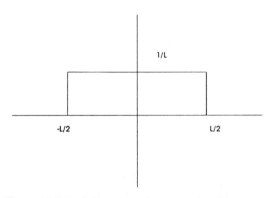

Figure 10.5 Far-field pattern of a rectangular dish antenna.

The excitation of this pattern by a uniform current field from a wavefront having a wavelength λ will appear as

$$\mathrm{sinc}(\pi \frac{L}{\lambda} u).$$

The pattern for this sinc function is shown in Figure 10.6.

In Figure 10.6 note that, as the visible region varies from -90° to 90°, the normalized spatial frequency varies from -1 to 1. The width of the mainlobe shown in Figure 10.6 represents the angular width of the beam formed by the rectangular dish antenna. This angular width is inversely proportional to the aperture length L. Therefore, as the required beamwidth narrows, the required aperture length increases. Since the size of a dish antenna is physically limited, a stringent beamwidth requirement can result in an antenna specification that is physically unrealizable. This is a fundamental limitation of single-element dish antennas.

Now consider a phased array that is receiving a wavefront originating in the far field. A phased array that has elements spaced a distance d apart, with an overall array length D and individual aperture responses characterized by rectangles of length M, will have the following response to the wavefront:

$$[\mathrm{rect}(D)\,\mathrm{comb}(d)] * \mathrm{rect}(M),$$

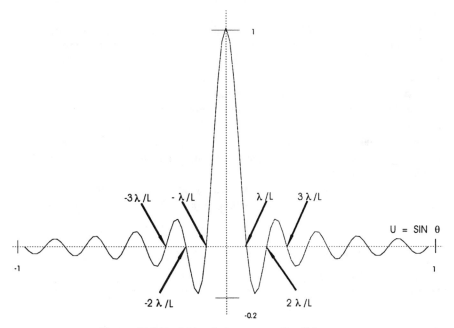

Figure 10.6 Far-field radiation pattern of a dish antenna.

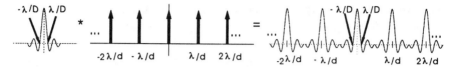

Figure 10.7 Fourier transform of a sampled aperture.

where the * denotes convolution. The resulting radiation pattern of the antenna can be found by taking the Fourier transform of the above response:

$$F[\text{rect}(D)\,\text{comb}(d) * \text{rect}(M)] = (F[\text{rect}(D)] * F[\text{comb}(d)])F[\text{rect}(M)].$$

The resulting pattern of the first enclosed term will appear as shown in the following equation and Figure 10.7.

$$F[\text{rect}(D)] * F[\text{comb}(d)] = F[\text{rect}(D)] * F\left[\sum_{n=0}^{\infty} \delta(x - nd)\right] \tag{10.9}$$

$$= \sum_{m=0}^{\infty} \frac{\sin(\pi D/\lambda)(u - m\lambda/d)}{(\pi D/\lambda)(u - m\lambda/d)}.$$

Multiplying the train of sinc functions with the sinc function response for each individual element results in the far-field radiation pattern of the phased array shown in Figure 10.8.

Note that the beamwidth for the phased array is inversely proportional to the array width instead of being dependent on an individual element pattern. Narrow beamwidths can therefore be realized by increasing the array width and the number of elements without being limited by the physical constraints of single antennas. This characteristic is also advantageous for airborne applications, where a phased array may be more convenient to mount on a fuselage than a large dish antenna.

The advantage of phased arrays as described above and in [3, 4] provides the major incentive of multielement arrays. Additional gains may be obtained from multielement arrays by processing the outputs of each individual antenna followed by a coherent summation. This is known as adaptive beamforming. Adaptive beamforming distinguishes signals arriving from different directions and provides

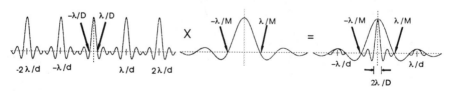

Figure 10.8 Fourier transform of a sampled aperture.

a spatial filter for signals that arrive from a desired direction. Adaptive beam-forming is implemented on adaptive arrays that can be narrowband or broadband, depending on the bandwidth of the signals in the target environment. Narrowband and broadband arrays will be discussed in the remainder of this chapter.

10.5 NARROWBAND ARRAYS

Narrowband adaptive arrays are used in applications involving signals that may be characterized by a single frequency and thus occupy a relatively narrow bandwidth. Signals whose envelope does not change during the time their wavefront is incident on the antenna elements are considered to be narrowband. A narrowband adaptive array consists of an array of sensors followed by a set of adjustable weights from which the outputs of the multipliers are summed to produce the array output.

A narrowband array is shown in Figure 10.9.

The input vector $X(k)$ consists of the sum of the desired signal $S(k)$ and noise $N(k)$:

$$X(k) = S(k) + N(k), \tag{10.10}$$

$$X^T(k) = [x_1, x_2, \ldots, x_K] \; S^T(k) = [s_1, s_2, \ldots, s_K] \; N^T(k) = [n_1, n_2, \ldots, n_K],$$

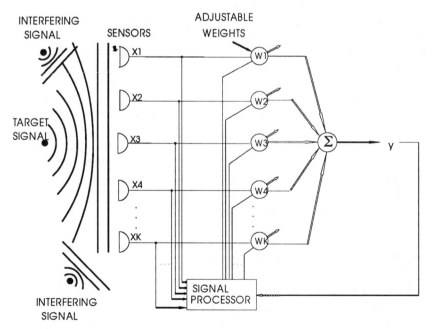

Figure 10.9 Narrowband array.

where k denotes the time instant of the input vector, K denotes the length of the array, and the superscript T denotes the transpose.

The noise vector $N(k)$ will generally consist of thermal noise and directional interference. At each time instant, the input vector is multiplied with the weight vector to obtain the array output, which is given as

$$y = \sum_{i=1}^{K} x_i * w_i = X^T(k)W = W^T X(k) \tag{10.11}$$

$$W^T = [w_1, w_2, \ldots, w_K]$$

The array output is then passed to the signal processor, which uses the previous value of the output and the current values of the inputs to determine the adjustment to make to the weights. The weights are then adjusted and multiplied with the new input vector to obtain the next output, etc. The output feedback loop allows the weights to be adjusted adaptively, thus accommodating nonstationary environments.

The output signal y will converge to the value of the desired signal as the interfering signals are canceled. The convergence is in the least squares sense; that is, the quantity $(y_{\text{true}} - y)^2 = (y_{\text{true}} - W^T X)^2$ is minimized as time increases. The criterion that allows for the output signal y to approximate the target signal while suppressing the directional interference is based on the selection of values for the weight vector W. Thus, the array-processing problem basically comes down to finding the optimal weight vector W.

Narrowband arrays rely on the assumption that wavefronts normal to the array will create identical waveforms at each sensor and wavefronts arriving at angles not normal to the array will create a linear phase shift at each sensor. Signals that occupy a large bandwidth and do not arrive normal to the array violate this assumption, since the phase shift is a function of f_o and varying frequency will cause a varying phase shift. Broadband target signals, which thus far have been assumed to arrive normal to the array, are not subject to frequency-dependent phase shifts at each sensor. This is attributed to the coherent summation of the target signal at each sensor, where the phase shift is a uniform random variable with zero mean. A modified array structure, however, is necessary to compensate the interference waveform inconsistencies that are caused by variations about the center frequency. This modified array structure is known as a broadband array and is discussed in the following section.

10.6 BROADBAND ARRAYS

Broadband adaptive arrays are used in applications involving signals that may not be sufficiently characterized by a center frequency and thus occupy a relatively large bandwidth. Broadband adaptive arrays consist of an array of sensors followed by tapped delay lines, which is the major implementation difference

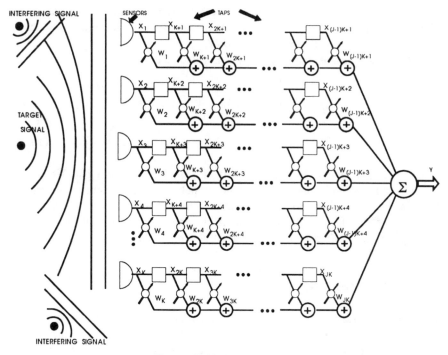

Figure 10.10 Broadband array.

between a broadband and narrowband array. A broadband array is shown in Figure 10.10.

Consider the transfer functions for a given sensor of the narrowband and broadband arrays, shown by (10.12) and (10.13), respectively.

$$H_{\mathrm{narrow}}(w) = w_1, \tag{10.12}$$

$$H_{\mathrm{broad}}(w) = w_1 + w_2 e^{-jwT} + w_3 e^{-2jwT} + \cdots + w_J e^{-j(J-1)wT}. \tag{10.13}$$

The narrowband transfer function has only a single weight that is constant with frequency. However, the broadband transfer function, which is actually a Fourier series expansion, is frequency dependent and allows for choosing a weight vector that may compensate phase variations due to signal bandwidth. This property of tapped delay lines provides the necessary flexibility for processing broadband signals. Note that, typically, four or five taps will be sufficient to compensate most bandwidth variances. Using more than four or five taps will, in general, add uncorrelated noise to the output and require increased processing resources.

The broadband array shown in Figure 10.10 obtains values at each sensor and then propagates these values through the array at each time interval. Therefore, if the values x_1 through x_K are input at time instant 1, then at time instant 2, x_{K+1} through x_{2K} will have the values previously held by x_1 through x_K, x_{2K+1}

through x_{3K} will have the values previously held by x_{K+1} through x_{2K}, etc. Also, at each time instant, a scalar value y will be calculated as the inner product of the input vector X and the weight vector W.

At each time instant, the array output is calculated as

$$y = \sum_{i=1}^{JK} x_i * w_i = X^T(k)W = W^T X(k), \tag{10.14}$$

$$X^T(k) = [x_1, x_2, \ldots, x_{JK}], \qquad W^T(k) = [w_1, w_2, \ldots, w_{JK}]. \tag{10.15}$$

Although not shown in Figure 10.10, a signal processor exists as in the narrowband array, which uses the previous output and current inputs to determine the adjustments to make to the W vector. The output signal y will approach the value of the desired signal as the interfering signals are canceled until it converges to the desired signal in the least squares sense.

10.7 NULL STEERING

Null steering consists of adjusting the lobe pattern of an antenna to have minimum gain in the direction of interference. This is equivalent to the effects of adjusting the "rabbit ear" antennas of a television to optimize reception via minimization of interference that may be generated by some other appliance. A possible antenna lobe pattern before and after adjustment of the antennas of a television is shown in Figure 10.11.

Figure 10.11 demonstrates the suppression of interference by minimizing the antenna gain in the direction of the interference. Antenna arrays with multi-

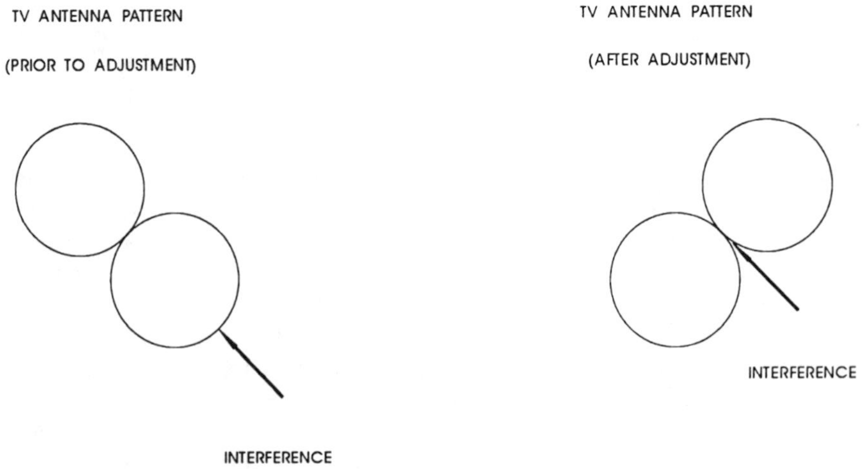

TV ANTENNA PATTERN

(PRIOR TO ADJUSTMENT)

TV ANTENNA PATTERN

(AFTER ADJUSTMENT)

INTERFERENCE

INTERFERENCE

Figure 10.11 Television antenna patterns.

ple elements use the same principles as a two-element array, such as television antennas, but have more complicated lobe patterns. For example, a communication system utilizing a multielement array may be subject to jamming within its sidelobes. This concept is illustrated in Figure 10.12.

The sidelobes may be steered to avoid the interference and minimize the gain in the direction of the interference. This will minimize the effects of the interference on the desired signal, which is contained in the main lobe. A possible sidelobe adjustment to avoid directional interference is shown in the second antenna lobe pattern of Figure 10.12. Note that, as the number of sensors increases, so does the number of sidelobes, and, hence, the complexity of the lobe pattern. Therefore, complicated lobe patterns will generally require a computer to calculate the lobe adjustments, while a two-element lobe pattern may be adjusted manually.

Another important criterion for antenna lobe patterns is known as *degrees of freedom*. The degrees of freedom of an array of sensors is defined as the number of signals that the array can minimize or maximize. A degree of freedom is consumed by each interference source (arriving from a different direction) to be nulled. A degree of freedom is also consumed by any main lobe constraint, such as directing the main lobe to be normal to the array. More detailed explanations of degrees of freedom can be found in [5, 6].

Consider a scenario with two interference sources at different directions and a target signal arriving normal to the array. Canceling the interference sources while maintaining a main lobe on the desired signal will require three degrees of freedom. The number of degrees of freedom for a given sensor array is given by the number of sensors minus one. Therefore, it may be stated that, in the

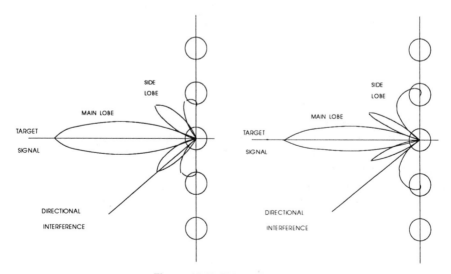

Figure 10.12 Sidelobe jamming.

TV ANTENNA PATTERN

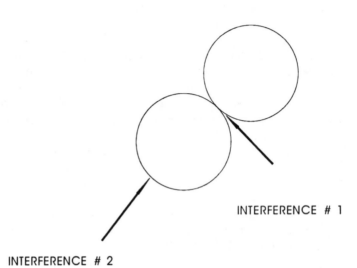

INTERFERENCE # 1

INTERFERENCE # 2

Figure 10.13 TV antennapPattern with two interference sources.

television antenna example considered above, because there are two antennas there is one degree of freedom. This may be intuitively seen in Figure 10.13.

Note that there is no antenna adjustment that will be capable of nulling both interference sources. A third antenna would be necessary to provide the two degrees of freedom that are necessary to cancel both interference sources.

10.8   NULL PROJECTING

Null projecting is considered in this chapter to be the linear algebra equivalent of null steering. Maintaining a mainlobe directed towards a target signal while steering the sidelobes away from interference is equivalent to projecting the target signal onto the range space of the weight vector W while projecting the interference onto the null space. The range space is defined here as the component of W that will retain the information of a signal following its projection onto W. The null space is defined as the component of W that will remove the information of a signal following its projection onto W. The beamforming vector will be defined as the weight vector satisfying the projection requirements. Choosing a weight vector W that satisfies the range-space and null-space projection requirements will form an antenna lobe pattern that minimizes the gain in the direction of interference while allowing the target signal to pass unaltered. For example, consider a three-sensor narrowband array with a 1 kHz tone arriving at boresight and a 2 kHz tone arriving at $30°$ off boresight. The composite signal will appear at each of the narrowband sensors, as shown in Figure 10.14.

Numerical values for the corresponding input vectors could appear as

$$X(k) = S(k) + N(k),$$

$$X(k) = \begin{bmatrix} 1.707 \\ 1.573 \\ 1.207 \end{bmatrix} \quad S(k) = \begin{bmatrix} 0.707 \\ 0.707 \\ 0.707 \end{bmatrix} \quad N(k) = \begin{bmatrix} 1.000 \\ 0.866 \\ 0.500 \end{bmatrix}.$$

We shall define the optimal weight vector W as that which projects $S(k)$ onto the range space of W while projecting $X(k)$ onto the null space. The W vector for this example may be found as

$$W_{\text{opt}} = \begin{bmatrix} 3.732 \\ -6.464 \\ 3.732 \end{bmatrix}.$$

Note that the projection requirements are satisfied as

$$W^T S(k) = s(k), W^T N(k) = 0,$$

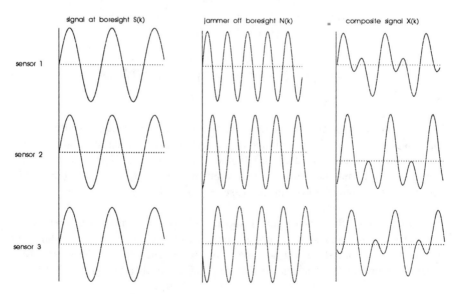

Figure 10.14 Composite narrowband signal.

hence,

$$W^T X(k) = S^T[S(k) + N(k)] = W^T S(k) = s(k),$$

where $s(k)$ is the scalar value of the target signal at time instant k; in this example, 0.707. This weight vector will cancel the effects on the output of any signal arriving at $30°$. This weight vector does not change until there is a change in the environment (i.e., the interference changes location). For example, consider the next three time instances of the above example (assuming the sampling rate to be 8 kHz):

$$X(k) = \begin{bmatrix} 0.924 \\ 0.924 \\ 0.924 \end{bmatrix} + \begin{bmatrix} 0.708 \\ 0.260 \\ -0.258 \end{bmatrix} = \begin{bmatrix} 1.631 \\ 1.183 \\ 0.665 \end{bmatrix},$$

$$W^T X(k) = [3.732 \ -6.464 \ 3.732] \begin{bmatrix} 1.631 \\ 1.183 \\ 0.665 \end{bmatrix} = 0.923,$$

$$X(k) = \begin{bmatrix} 1.000 \\ 1.000 \\ 1.000 \end{bmatrix} + \begin{bmatrix} 0.000 \\ -0.500 \\ -0.866 \end{bmatrix} = \begin{bmatrix} 1.000 \\ 0.500 \\ 0.134 \end{bmatrix},$$

$$W^T X(k) = [3.732 \ -6.464 \ 3.732] \begin{bmatrix} 1.000 \\ 0.500 \\ 0.134 \end{bmatrix} = 1.0,$$

$$X(k) = \begin{bmatrix} 0.924 \\ 0.924 \\ 0.924 \end{bmatrix} + \begin{bmatrix} -0.707 \\ -0.966 \\ -0.966 \end{bmatrix} = \begin{bmatrix} 0.217 \\ -0.042 \\ -0.042 \end{bmatrix},$$

$$W^T X(k) = [3.732 \ -6.464 \ 3.732] \begin{bmatrix} 0.217 \\ -0.042 \\ -0.042 \end{bmatrix} = 0.925. \tag{10.16}$$

It can be seen that, at each time instant, the W vector successfully cancels the interference waveform arriving at $30°$ while maintaining the value of the target signal. Again note that this method of signal cancellation is equivalent to adjusting the sidelobe patterns of an antenna array to have zero gain in the direction of the interference (in this case, at $30°$).

Example 10.8.1 It is given that a signal $s(k)$ arrives normal to a three-sensor narrowband array and an interfering signal arrives at an angle of incidence of $45°$ above the normal. These are the only signals incident on the array. At a given time instant k_1, it is known that $s(k_1) = 0.5$ and $X^T(k_1) = [1.5 \ 1.207 \ 0.5]$.

1. Find the noise vector $N(k)$ at time instant $k = k_1$.

2. Find the optimal weight vector W which will cancel the interfering signal N while allowing the target signal $s(k)$ to pass with unity gain.

3. Find the optimal weight vector W that will cancel the target signal $s(k)$ while allowing the interfering signal N to pass with unity gain (assume that the value of N at time instant k_1 is 1.0).

(Hint: assume that $w_1 = w_3$ for b) and c).)

Solution

1. The input vector consists of the sum of the signal and noise. Therefore

$$N(k_1) = X(k_1) - S(k_1) = \begin{bmatrix} 1.5 \\ 1.207 \\ 0.5 \end{bmatrix} - \begin{bmatrix} 0.5 \\ 0.5 \\ 0.5 \end{bmatrix} = \begin{bmatrix} 1.0 \\ 0.707 \\ 0.0 \end{bmatrix}$$

2. Since it is known that an optimal weight vector W will satisfy the conditions

$$W^T X(k_1) = [w_1 \ w_2 \ w_3] \begin{bmatrix} 1.5 \\ 1.207 \\ 0.5 \end{bmatrix} = s(k_1) = 0.5,$$

therefore,

$$1.5w_1 + 1.207w_2 + 0.5w_3 = 0.5;$$

and

$$W^T N(k_1) = [w_1 \ w_2 \ w_3] \begin{bmatrix} 1.000 \\ 0.707 \\ 0.000 \end{bmatrix} = 0.0,$$

therefore,

$$w_1 + 0.707w_2 = 0.0;$$

and, using the hint,

$$w_1 = w_3, \qquad w_1 - w_3 = 0.$$

This problem has now been reduced to a system with three equations and three unknowns and may be written in a matrix-vector format:

$$AW = \begin{bmatrix} 1.500 & 1.207 & 0.500 \\ 1.000 & 0.707 & 0.000 \\ 1.000 & 0.000 & -1.000 \end{bmatrix} \begin{bmatrix} w_1 \\ w_2 \\ w_3 \end{bmatrix} = \begin{bmatrix} 0.500 \\ 0.000 \\ 0.000 \end{bmatrix}$$

$$A = \begin{bmatrix} 1.500 & 1.207 & 0.500 \\ 1.000 & 0.707 & 0.000 \\ 1.000 & 0.000 & -1.000 \end{bmatrix}, B = \begin{bmatrix} 0.500 \\ 0.000 \\ 0.000 \end{bmatrix}$$

The solution may now be determined by finding the inverse of the A matrix:

$$W = A^{-1}B = \begin{bmatrix} 3.413 & -5.835 & 1.707 \\ -4.827 & 9.666 & -2.414 \\ 3.413 & -5.830 & 0.707 \end{bmatrix} \begin{bmatrix} 0.500 \\ 0.000 \\ 0.000 \end{bmatrix} = \begin{bmatrix} 1.707 \\ -2.414 \\ 1.707 \end{bmatrix}.$$

Note that the hint—the assumption that the weights are symmetric—is not always true. This example was specifically chosen to satisfy this condition to provide the third equation necessary to solve for the three unknowns.

3. The conditions for canceling the target signal while maintaining the interference signal are

$$W^T X(k_1) = [w_1 \ w_2 \ w_3] \begin{bmatrix} 1.500 \\ 1.207 \\ 0.500 \end{bmatrix} = n(k_1) = 1.0,$$

therefore,

$$1.5w_1 + 1.207w_2 + 0.5w_3 = 1.0,$$

and

$$W^T N(k_1) = [w_1 \ w_2 \ w_3] \begin{bmatrix} 1.000 \\ 0.707 \\ 0.000 \end{bmatrix} = 1.0,$$

therefore,

$$w_1 + 0.707w_2 = 1.0.$$

Using the hint that $w_1 = w_3$, and realizing that the only change from part 1 is in the B vector, since the A matrix has not changed, the problem may be solved as follows:

$$AW = B,$$

$$W = A^{-1}B = \begin{bmatrix} 3.413 & -5.835 & 1.707 \\ -4.827 & 9.666 & -2.414 \\ 3.413 & -5.830 & 0.707 \end{bmatrix} \begin{bmatrix} 1.000 \\ 1.000 \\ 0.000 \end{bmatrix} = \begin{bmatrix} -2.422 \\ 4.839 \\ -2.417 \end{bmatrix}.$$

Note that Example 10.8.1, while illustrating the concept of optimal weight vectors, assumed that the direction of interference was known *a priori*. This is rarely the case in practical applications. The remainder of this chapter will consider applications where the interference directions are unknown and must be learned adaptively.

10.9 THE ROW-ACTION PROJECTION METHOD

The row-action projection (RAP) method is an iterative technique for solving a system of linear equations. The RAP method operates by creating orthogonal projections in the space defined by the RAP matrix. The RAP matrix is defined here as the matrix containing the equations for a linear system. The concept of row-action projection may be observed in Figure 10.15.

Following Figure 10.15, first RAP chooses a solution that satisfies the first equation contained in the RAP matrix (i.e., the first row). RAP then creates an orthogonal projection (i.e., perpendicular) to the second equation and uses this intersection (indicated by the circle) as its next solution. Then RAP creates an orthogonal projection from the point on the second equation to a point on the third equation. After RAP has finished projecting onto all constraints, it then projects back onto the cost function and repeats this process until the best solution is reached (i.e., the intersection of the three lines in Figure 10.15). This solution is typically "best" in the least squares sense.

The RAP method, as described in Chapter 2, forms its orthogonal projections by calculating the error obtained at each iteration and then using this error to form the projection for the next row. Given a system of equations in the form

$$AX = B, \tag{10.17}$$

where it is desired to find the vector X, RAP is implemented using the equations

$$X^{(k+1)} = X^{(k)} + \lambda \frac{\epsilon_i}{|A_i|} \frac{A_i^T}{|A_i|}, \qquad \epsilon_i = B_i - A_i X^{(k)}. \tag{10.18}$$

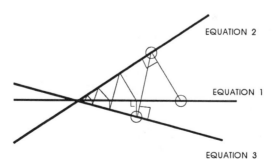

Figure 10.15 Row-action projections.

In (10.18), the superscript k denotes the iteration, the subscript i refers to the row number of the matrix or vector, ϵ represents the current error value, and λ is a relaxation parameter, which in general is chosen between zero and two. The choice of λ is important for performance characteristics and has the tradeoff that a large λ will provide faster convergence, while a small λ will provide greater accuracy. Also, note that choosing λ between one and two may, in some instances, prevent convergence.

The choice of different values of λ is illustrated in Figure 10.16.

Choosing $\lambda = 1$ will satisfy the next projection exactly, while choosing $\lambda = 0.5$ and $\lambda = 2$ will project half the distance and twice the distance to the hyperplane given by the next equation, respectively.

Example 10.9.1 Given the two equations $x_1 + x_2 = 4$ and $-x_1 + 3x_2 = -12$, find the solution for x_1 and x_2 using the RAP method. Use $x_1 = 0$ and $x_2 = 0$ as the initial conditions and assume that $\lambda = 1$.

Solution First, these equations must be put into the matrix-vector form:

$$AX = B, \qquad \begin{bmatrix} 1 & 1 \\ -1 & 3 \end{bmatrix} \begin{bmatrix} x_1 \\ x_2 \end{bmatrix} = \begin{bmatrix} 4 \\ -12 \end{bmatrix},$$

$$A = \begin{bmatrix} 1 & 1 \\ -1 & 3 \end{bmatrix}, \qquad B = \begin{bmatrix} 4 \\ -12 \end{bmatrix}.$$

Now, starting with $k = 0$ and $i = 1$,

$$\epsilon_1 = b_1 - A_1 X^{(0)} = 4 - \begin{bmatrix} 1 & 1 \end{bmatrix} \begin{bmatrix} 0 \\ 0 \end{bmatrix} = 4,$$

$$X^{(1)} = X^{(0)} + \lambda \frac{\epsilon_1}{|A_1|} \frac{A_1^T}{|A_1|} = \begin{bmatrix} 0 \\ 0 \end{bmatrix} + (1)\frac{4}{\sqrt{1^2 + 1^2}^2} \begin{bmatrix} 1 \\ 1 \end{bmatrix} = \begin{bmatrix} 2 \\ 2 \end{bmatrix}.$$

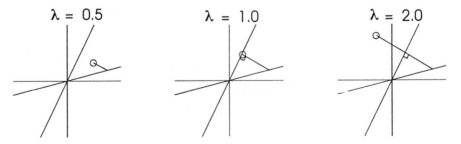

$$\lambda = 0.5 \qquad\qquad \lambda = 1.0 \qquad\qquad \lambda = 2.0$$

Figure 10.16 Row-action projections.

Next, for $k = 1$ and $i = 2$,

$$\epsilon_2 = b_2 - A_2 X^{(1)} = -12 - \begin{bmatrix} -1 & 3 \end{bmatrix} \begin{bmatrix} 2 \\ 2 \end{bmatrix} = -16,$$

$$X^{(2)} = X^{(1)} + \lambda \frac{\epsilon_2}{|A_2|} \frac{A_2^T}{|A_2|} = \begin{bmatrix} 2 \\ 2 \end{bmatrix} + (1) \frac{-16}{\sqrt{-1^2 + 3^2}} \begin{bmatrix} -1 \\ 3 \end{bmatrix} = \begin{bmatrix} 3.6 \\ -2.8 \end{bmatrix}.$$

For $k = 2$ and $i = 1$,

$$\epsilon_1 = b_1 - A_1 X^{(2)} = 4 - \begin{bmatrix} 1 & 1 \end{bmatrix} \begin{bmatrix} 3.6 \\ -2.8 \end{bmatrix} = 3.2,$$

$$X^{(3)} = X^{(2)} + \lambda \frac{\epsilon_1}{|A_1|^T} \frac{A_1}{|A_1|} = \begin{bmatrix} 3.6 \\ -2.8 \end{bmatrix} + (1) \frac{3.2}{\sqrt{1^2 + 1^2}} \begin{bmatrix} 1 \\ 1 \end{bmatrix} = \begin{bmatrix} 5.2 \\ -1.2 \end{bmatrix}.$$

For $k = 3$ and $i = 2$,

$$\epsilon_2 = b_2 - A_2 X^{(3)} = -12 - \begin{bmatrix} -1 & 3 \end{bmatrix} \begin{bmatrix} 5.2 \\ -3.2 \end{bmatrix} = -3.2$$

$$X^{(4)} = X^{(3)} + \lambda \frac{\epsilon_2}{|A_2|} \frac{A_2^T}{|A_2|} = \begin{bmatrix} 5.2 \\ -3.2 \end{bmatrix} + (1) \frac{-3.2}{\sqrt{-1^2 + 3^2}} \begin{bmatrix} -1 \\ 3 \end{bmatrix} = \begin{bmatrix} 5.52 \\ -2.16 \end{bmatrix}.$$

By continuing this iterative process, the vector X^T will be found to converge to $[6 \; -2]$.

A graphical representation of the iterations of Example 10.9.1 can be seen in Figure 10.17.

It can be seen from Figure 10.17 that RAP will converge to the intersection of the two lines via orthogonal projections. The RAP method has found numerous applications in digital signal processing. These applications have included image processing [7, 8], adaptive filtering [9], echo cancellation [10], spectrum estimation [11], and numerous other applications. This chapter will apply the RAP method to adaptive beamforming.

10.10   NARROWBAND PROBLEM FORMULATION

For the narrowband array shown in Figure 10.9, an algorithm is desired that will spatially filter directional signals. This algorithm may be realized by projecting the signal vector onto the range space of W while projecting the noise vector onto the null space.

The null-space projection may be realized by driving the input vector to zero, as shown in the following equation:

$$X^T(k)W = 0. \tag{10.19}$$

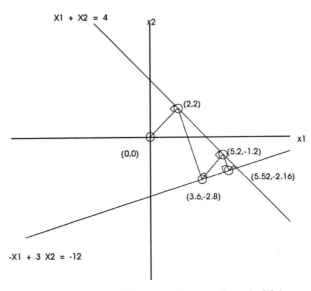

Figure 10.17 Row-action projections for Example 3.8.1.

Note that (10.19) projects the composite input vector onto a null space, not just the noise component. However, this projection is still valid because the range-space projection will maintain the target signal. The range-space projection may be realized as follows:

$$\begin{bmatrix} 1 & 1 & 1 & \cdots & 1 \end{bmatrix} W = 1. \tag{10.20}$$

This condition attempts to have the weight vector equal $[1/K, 1/K, \ldots, 1/K]$ (K is the number of sensors) by imposing the constraint that the sum of the elements in the weight vector column be equal to one.

Therefore, a signal that creates identical waveforms at each sensor will have unity gain in the output. For example,

$$X^T(k)W = S^T(k)W = \begin{bmatrix} s(k) & s(k) & \cdots & s(k) \end{bmatrix} \begin{bmatrix} w_1 \\ w_2 \\ \vdots \\ w_K \end{bmatrix}$$

$$= w_1 s(k) + w_2 s(k) + \cdots + w_K s(k) = (w_1 + w_2 + \cdots + w_K)s(k) = s(k)$$

The two constraints shown in (10.19) and (10.20) may now be put into a matrix form that is suitable for the RAP method. This matrix is shown in the following equation:

$$
\begin{bmatrix} x_1(k) & x_2(k) & \cdots & x_K(k) \\ 1 & 1 & \cdots & 1 \end{bmatrix}
\begin{bmatrix} w_1 \\ w_2 \\ \vdots \\ w_K \end{bmatrix}
= \begin{bmatrix} 0 \\ 1 \end{bmatrix} \tag{10.19}
$$

The system shown in (10.19) is the RAP solution for narrowband array processing. The RAP method will create orthogonal projections onto both rows of the above matrix until it converges to the best solution (refer to Chapter 2 for types of convergence). This solution will correspond to the optimal weight vector W that will allow the boresight signal to pass with unity gain while canceling signals that arrive from directions other than boresight. Simulation results for the RAP method, as applied to narrowband array processing, are provided in the following section.

10.11   RAP NARROWBAND SIMULATION RESULTS

The RAP method for narrowband array processing described in the previous section was applied to a scenario for which the performance curves are provided in this section. The narrowband array used for this scenario consisted of three equispaced sensors. The signal environment consisted of a boresight signal of frequency f_o and an interference signal $30°$ above boresight with frequency $2f_o$. The input was also contaminated with Gaussian white noise to give an input SNR of 37 dB. The input spectrum appeared as shown in Figure 10.18.

The RAP method was applied to this scenario with a "scaled" λ. The value of λ was initialized to 0.5 and linearly decreased with respect to the number of samples. The equation used to scale λ was

$$
\lambda = 0.5 \frac{N - i}{N}, \tag{10.20}
$$

where N is the total number of iterations and i is the current iteration number. This technique takes advantage of the speed of convergence in the initial stages while obtaining accuracy in the final stages. The graph for the output error squared is shown in Figure 10.19, where the output error squared is considered to be

$$
\text{output error squared} = (y_{\text{true}} - W^T X)^2. \tag{10.21}
$$

The output spectrum can be seen in Figure 10.20.

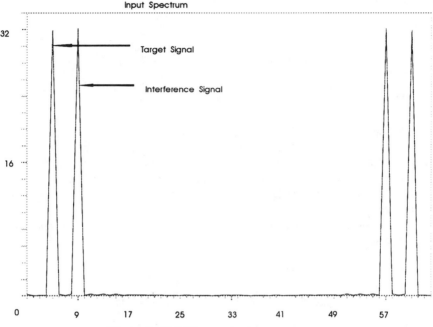

Figure 10.18 RAP narrowband input spectrum.

10.12 BROADBAND PROBLEM FORMULATION

Broadband arrays have been analyzed by Widrow [12], Griffiths [13, 14], and Frost [15]. Widrow proposed an LMS algorithm that minimized the square of the difference between the observed output and the expected output, which was estimated with a training signal. This approach assumes that the direction of arrival (DOA) and a training signal are known *a priori*. Griffiths proposed an LMS algorithm that minimized the difference between the autocorrelation and cross-correlation matrices obtained from the observed inputs and outputs. This method assumes that the DOA and second-order signal statistics are known *a priori*. The methods proposed by Widrow and Griffiths were forms of *unconstrained* optimization. Frost [15] proposed a *constrained* LMS algorithm that assumes *a priori* knowledge of the DOA and the frequency band of interest. The RAP broadband algorithm will be formulated as a *constrained* optimization problem that uses the same *a priori* information as the Frost algorithm.

For the broadband array shown in Figure 10.10, waveforms propagating normal to the array (i.e., the target signal) may be considered to create identical waveforms at each sensor. Target signals that are not normal to the array may be made to appear normal by applying delays directly after the sensors to compensate for the phase shift. This technique was demonstrated in Example 10.2.1. The addition of delays essentially has the effect of a *steering vector*, which redi-

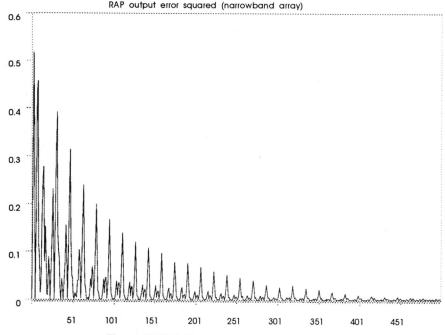

Figure 10.19 RAP narrowband output error squared.

rects the array response, or main lobe, towards the angle of arrival of the target signal.

Since the steered target signal will tend to create identical waveforms at each sensor, the output of each column of delays will be identical and the signal will appear to propagate through the array as a wavefront perpendicular to the look direction. Since the taps in each column see the same signal, this array may be collapsed to a single-sensor finite impulse response (FIR) filter, as shown in Figure 10.21.

By constraining the column vectors of the broadband adaptive array (as shown in Figure 10.21) to be equal in sum to the coefficients of a FIR filter, the desired frequency response in the look direction may be obtained. For example,

$$w_1 + w_2 + \cdots + w_K = h_1,$$

$$w_{K+1} + w_{K+2} + \cdots + w_{2K} = h_2,$$

$$\vdots$$

$$w_{(J-1)K+1} + w_{(J-1)K+2} + \cdots + w_{JK} = h_J,$$

where h_1, h_2, etc. are the FIR filter coefficients defining the desired frequency response of the target signal.

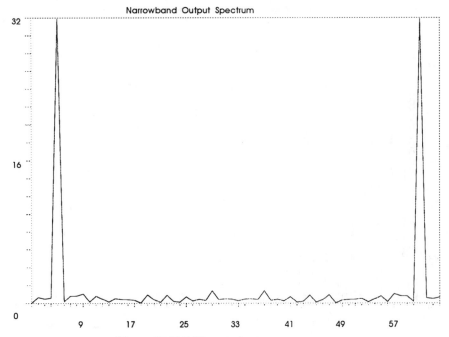

Figure 10.20 RAP narrowband output spectrum.

Minimizing the output power of the signals observed by the broadband array shown in Figure 10.10 will minimize the effects of signals that are not normal to the sensor array. Therefore, minimizing the power of the composite signals observed at each sensor will tend to remove directional interference, while in-phase signals will add coherently and thus be maintained. This observation was made by Frost [15] and is summarized below.

The expectation of the second moments of the input vectors $X(k)$, $S(k)$, and $N(k)$ are defined by the autocorrelation matrices R_{xx}, R_{ss}, and R_{nn}, respectively:

$$E[X(k)X^T(k)] = R_{xx}, \qquad E[S(k)S^T(k)] = R_{ss}, \qquad E[N(k)N^T(k)] = R_{nn}.$$

The expected output power may be found as

$$E[y^2(k)] = E[W^T X(k)X^T W] = W^T R_{xx} W. \tag{10.22}$$

For interference cancellation, it is assumed that $S(k)$ and $N(k)$ are uncorrelated,

$$E[S^T(k)N(k)] = E[N^T(k)S(k)] = 0.$$

If this condition is not met, then the array will cancel that part of $S(k)$ which is correlated with $N(k)$. Using the condition that $S(k)$ and $N(k)$ are uncorrelated,

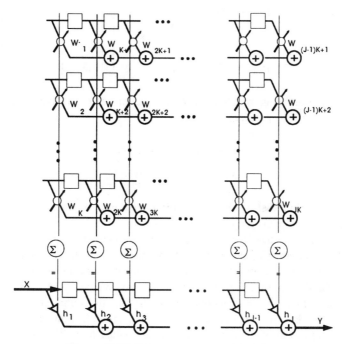

Figure 10.21 Column summation to FIR filter.

the expected output power can be found to reduce to

$$W^T R_{xx} W = E[W^T X(k) X^T(k) W]$$
$$= E[W^T (S(k) + N(k))(S(k) + N(k))^T W] = W^T (R_{ss} + R_{nn}) W.$$

Therefore, the optimal weight vector W will minimize the noise power contained in R_{xx} so that only the signal power will remain:

$$W^T (R_{ss} + R_{nn}) W = W^T R_{ss} W + W^T R_{nn} W = W^T R_{ss} W, \quad W^T R_{nn} W = 0.$$

Implementing the constraint that the columns of the broadband array sum to the values of the vector H which contains the tap values for the desired FIR filter response,

$$H = \begin{bmatrix} h_1 & h_2 & \cdots & h_J \end{bmatrix}, \tag{10.23}$$

may be realized through the multiplication of a projection matrix C^T. This is shown as follows:

$$C^T W = H, \tag{10.24}$$

$$
\begin{bmatrix}
1 & \cdots & 1 & 0 & \cdots & 0 & \cdots & 0 & \cdots & 0 \\
0 & \cdots & 0 & 1 & \cdots & 1 & \cdots & 0 & \cdots & 0 \\
\vdots & & & & & & & & & \\
0 & \cdots & 0 & 0 & \cdots & 0 & \cdots & 1 & \cdots & 1
\end{bmatrix}
\begin{bmatrix}
w_1 \\ w_2 \\ \vdots \\ w_{JK}
\end{bmatrix}
=
\begin{bmatrix}
h_1 \\ h_2 \\ \vdots \\ h_J
\end{bmatrix}.
$$

The array-processing problem may now be presented as the following constrained optimization problem:

$$\text{minimize} \quad W^T R_{xx} W, \tag{10.25}$$

$$\text{subject to} \quad C^T W = H, \tag{10.26}$$

where the criterion that $W^T R_{xx} W$ be minimized will be the cost function and $C^T W = H$ will comprise the constraint equations.

RAP may be used to solve the adaptive array-processing problem by constructing a matrix that contains the cost-function equation and the constraint equations. The cost function was previously derived to be the minimization of the output power $W^T R_{xx} W$. The output power may be approximated by y^2. Therefore, minimizing the output power is equivalent to driving the output power, or y^2, to zero, since this is a nonnegative quantity. Also, minimization of the quantity y^2 implies minimization of the quantity y, which is equal to $X^T W$. The minimization of the output power $W^T R_{xx} W$ may therefore be represented by the linear equation $X^T W = 0$. The constraint equations given by $C^T W = H$ are already in a form that RAP can utilize. Therefore, adding a row representing X^T to the constraint matrix C and then adding a zero to the solution vector H, we may form the matrix-vector problem for adaptive broadband-array processing as follows:

$$
\begin{bmatrix}
x_1 & \cdots & x_K & x_{K+1} & \cdots & x_{2K} & \cdots & x_{(J-1)K+1} & \cdots & x_{JK} \\
1 & \cdots & 1 & 0 & \cdots & 0 & \cdots & 0 & \cdots & 0 \\
0 & \cdots & 0 & 1 & \cdots & 1 & \cdots & 0 & \cdots & 0 \\
\vdots & & & & & & & & & \\
0 & \cdots & 0 & 0 & \cdots & 0 & \cdots & 1 & \cdots & 1
\end{bmatrix}
$$

$$
*
\begin{bmatrix}
w_1 \\ w_2 \\ \vdots \\ w_{JK}
\end{bmatrix}
=
\begin{bmatrix}
0 \\ h_1 \\ \vdots \\ h_J
\end{bmatrix}. \tag{10.27}
$$

Note that an additional adjustment must be made to the broadband RAP algorithm. The normalized projection of the first row (i.e., $X^T(k)W$) requires that the magnitude of the input vector $X(k)$ be obtained. Therefore, the $X(k)$ vector must be squared at some point in the algorithm. Squaring the $X(k)$ vector is equivalent to doubling the input signal bandwidth. Doubling the input signal bandwidth introduces high-frequency terms that degrade the performance of the

RAP algorithm. An intuitive approach to alleviating this problem would be to low-pass filter the squared input (or magnitude) term in the RAP algorithm. A low-pass filter may be conveniently applied to the magnitude by using a *forgetting factor*. For example, the RAP update for the first row of the matrix in (10.27), without using the forgetting factor, would appear as

$$W^{k+1} = W^k + \lambda \frac{\epsilon}{|X|} \frac{X}{|X|}, \qquad \epsilon = 0 - X^T W^k. \tag{10.28}$$

The RAP update for the first row of the matrix in (10.27) with the forgetting factor would appear as

$$W^{k+1} = W^k + \lambda \frac{\epsilon X}{\text{mag}}$$

$$\epsilon = 0 - X^T W^k, \qquad \text{mag} = \gamma \, \text{mag} + (1 - \gamma)|X||X|, \tag{10.29}$$

where γ is the forgetting factor, which has a value of $\gamma = 0.9$ for this application. Note that the use of a forgetting factor, as in (10.29), is equivalent to applying an infinite impulse response (IIR) filter to the magnitude. The RAP algorithm presented in (10.27) and (10.29) may now be applied to broadband-array processing.

10.13 RAP BROADBAND SIMULATION RESULTS

The RAP method for broadband array processing described in the previous section was applied to a scenario for which the performance curves are provided in this section. The broadband array for this scenario consisted of five equispaced sensors, each followed by five taps. The signal environment is represented by the following table:

Signal	$BW(\Delta f/f_s)$	Frequency	Azimuth
Target signal	0.0625	f_o	$0°$
Interference 1	0.09375	$0.4f_o$	$30°$
Interference 2	0.0625	$1.5f_o$	$-47°$
White noise	∞	—	—

The white noise consisted of a zero-mean, unit-variance Gaussian random variable. The input SNR resulting from the additive white noise was 17 dB. The input spectrum for the broadband scenario is shown in Figure 10.22. (Note that the RAP broadband array is using a scaled λ, as given by (10.20) for the RAP narrowband array.)

The output spectrum for the RAP broadband array is shown in Figure 10.23.

The output error squared for the RAP broadband array is shown in Figure 10.24.

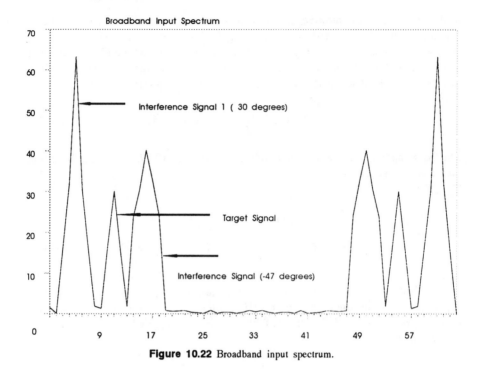

Figure 10.22 Broadband input spectrum.

A description of the constrained LMS algorithm derived by Frost will be given in the following section, along with performance curves for the identical broadband example.

10.14   CONSTRAINED LMS AND RAP COMPARISON

Frost [15] derived the constrained LMS algorithm for broadband array processing using Lagrange multipliers. Define the cost function as

$$H(W) = \frac{1}{2} W^T R_{xx} W + \lambda^T (C^T W - H), \qquad (10.30)$$

where λ is a Lagrange multiplier and H is a vector representative of the desired frequency response. Minimizing the function $H(W)$ with respect to W will obtain the recursive equation

$$W(k + 1) = P[W(k) - \mu R_{xx} W(k)] + C(C^T C)^{-1} H, \qquad (10.31)$$

where

$$P = I - C(C^T C)^{-1} C^T$$

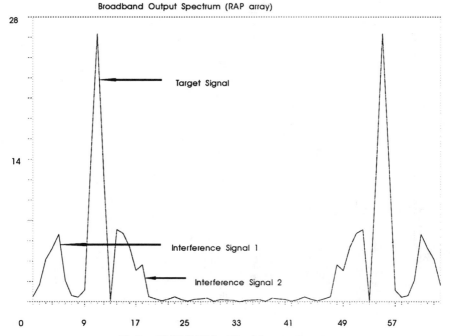

Figure 10.23 RAP broadband output spectrum.

$$W(0) = C(C^T C)^{-1} H$$

$$H = \begin{bmatrix} h_1 & h_2 & \cdots & h_J \end{bmatrix}.$$

In (10.31), μ is a scaling parameter that controls the speed of convergence and the accuracy of the estimate for the optimal weight vector W. To ensure convergence, μ must be less than the inverse of the largest eigenvalue of the autocorrelation matrix formed by the input vector X, as is elaborated upon in [16]:

$$0 < \mu < \frac{1}{\lambda_{\max}}. \tag{10.32}$$

The Frost algorithm was applied to the same scenario to which the RAP broadband method was applied in the previous section. (A value of $\mu = 0.0025$ was used for the Frost algorithm.) The output spectrum resulting from the Frost algorithm is shown in Figure 10.25.

The signal-to-jammer ratio in the output spectrum for the Frost algorithm was roughly 2.2. The signal-to-jammer ratio in the output spectrum for the RAP algorithm, as shown in Figure 10.23, was roughly 4.5. The output error squared for the Frost algorithm is shown in Figure 10.26. The output error squared plots for the RAP and Frost algorithms were similar.

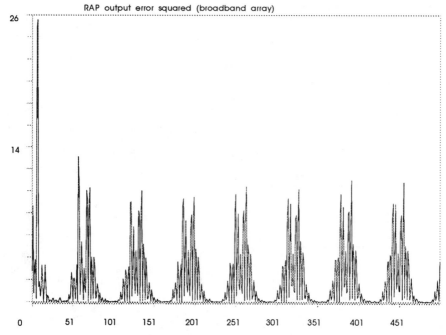

Figure 10.24 RAP broadband output error squared.

10.15 CONCLUSION

A new method for adaptive beamforming was presented in this chapter. The algorithm was shown to converge to the least mean square solution for both narrowband and broadband arrays. The row-action projection method is a simple technique for creating orthogonal projections within a space defined by a set of hyperplanes. The row-action projection method can easily be applied to both unconstrained and constrained optimization problems whose solution lies in a convex set (i.e., no local maxima or minima). Adaptive beamforming falls into this category, and it has been shown that the row-action projection method is a viable solution for this application.

Adaptive beamforming can be implemented with narrowband or broadband arrays. In this chapter, the row-action projection method was successfully formulated for both implementations. Both formulations assume that the angle of arrival of the desired signal is known, but do not assume any knowledge of the interfering sources. The narrowband array of Section 10.10 uses range-space and null-space projections of the signal and noise components to adaptively update the narrowband array weights. The broadband array of Section (10.12) uses the criteria for output power minimization, along with a frequency-response constraint in the direction of the desired signal, to update the broadband array

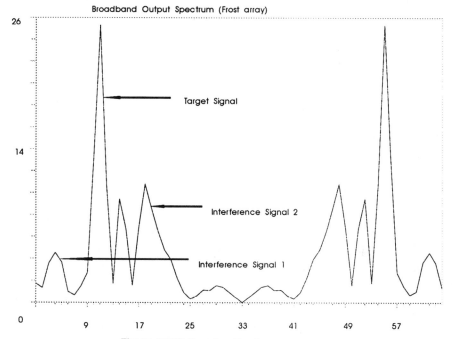

Figure 10.25 Frost broadband output spectrum.

weights. Simulation results were provided for the narrowband and broadband arrays in Sections 10.11 and 10.13, respectively.

A simulation example was provided that compared the row-action projection method to a constrained LMS method for broadband arrays. An advantage of the row-action projection method is the fact that *a priori* information regarding the eigenvalues of the input autocorrelation matrix is not required. The only adjustable parameters of the row-action projection method are the λ and γ parameters. Convergence is guaranteed for λ if $0 < \lambda < 1$. The γ parameter controls the speed and accuracy of the low-pass filter response. Values for γ that are close to one will provide slow but accurate responses. Values of γ near zero will provide fast responses while compromising accuracy. It is recommended that γ be chosen to be between 0.9 and 0.95. Regarding the simulation results, the row-action projection method was more successful in suppressing the directional interference than the constrained LMS method. Based on the output spectra in Figures 10.23 and 10.25, the RAP method showed a larger output signal-to-jammer ratio. Though conclusions should not be based upon a single example, the initial results of the row-action projection method show it to be a potential alternative to the constrained LMS method.

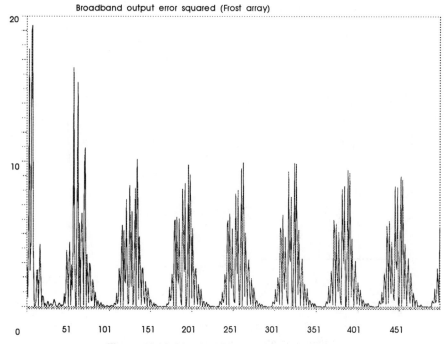

Figure 10.26 Frost broadband output error squared.

ACKNOWLEDGMENTS

This research reported here was made possible through the support of the New Jersey Commission on Science and Technology and the Center for Computer Aids for Industrial Productivity at Rutgers University.

REFERENCES

1. J. L. Flanagan, J. D. Johnston, R. Zahn, and G. W. Elko. Computer-stered microphone arrays for sound transduction in large rooms. *J. Acoust. Soc. Am.* **78**(11):1508–1518, 1985.

2. J. Goodman. *Introduction to Fourier Optics.* New York: McGraw Hill, 1968.

3. A. Macovski. *Medical Imaging Systems.* Englewood Cliffs, N.J.: Prentice Hall, 1983.

4. B. Steinberg. *Principles of Aperture and Array System Design.* New York: Wiley, 1976.

5. R. T. Compton. *Adaptive Antennas.* Englewood Cliffs, N.J.: Prentice Hall, 1988.

6. J. E. Hudson. *Adaptive Array Principles.* Institution of Electrical Engineers, 1981.

7. C. I. Podilchuk and R. J. Mammone. Image recovery by convex projectionns using a least squares constraint. *J. Opt. Soc. Am.*, March 1990.

8. C. I. Podilchuk and R. J. Mammone. Row and block action projection techniques for image restoration. In *Signal Recovery and Synthesis III*, Cape Cod, Mass.: Optical Society of America, June 1989.

9. J. F. Doherty and R. J. Mammone. A new fast method for channel estimation. In *Proceedings of the IEEE ICC 89*, Boston, June 1989.

10. S. L. Gay and R. J. Mammone. Acoustic echo cancellation using POCS on the DSP16. In *Proceedings of the IEEE International Conference on Acoustics, Speech, and Signal Processing 1990*, Albuquerque, N.M., April 1990.

11. A. B. Fineberg and R. J. Mammone. An adaptive technique for high resolution time-varying spectral estimation. In *Proceedings of the IEEE International Conference on Acoustics, Speech, and Signal Processing 1991*, Toronto, Ontario, May 1991.

12. B. Widrow and S. D. Stearns. *Adaptive Signal Processing*. Englewood Cliffs, N.J.: Prentice-Hall, 1985.

13. L. J. Griffiths. A simple adaptive algorithm for real-time processing in antenna arrays. *Proc. IEEE*, **57**(10):1696–1704, 1969.

14. L. J. Griffiths and C. W. Jim. An alternative approach to linearly constrained adaptive beamforming. *IEEE Trans. Antennas Propagation*, **AP-30**(1):27–34, 1982.

15. O. L. Frost III. An algorithm for linearly constrained adaptive array processing. *Proc. IEEE*, **60**(8):926–935, 1972.

16. S. J. Orfanidis. *Optimum Signal Processing*, 2nd ed. New York: Macmillan, 1988.

17. K. Takao, M. Fujita, and T. Nishi. An adaptive antenna array under directional constraint. *IEEE Trans. Antennas Propagation*, **AP-24**(9):662–669, 1976.

18. L. J. Griffiths. Linearly-constrained adaptive signal processing methods. In *Advanced Algorithms and Architectures for Signal Processing II*, pp. 96–100. SPIE, 1987.

19. S. P. Applebaum. Adaptive arrays. *IEEE Trans. Antennas Propagation*, **AP-24**:585–598, 1976.

20. Y. Censor. Row-action techniques for huge and sparse systems and their applications. *SIAM Rev.*, **23**(4), 1981.

21. R. Monzingo. *Introdution to Adaptive Arrays*. New York: Wiley, 1980.

22. A. Giordano and F. Hsu. *Least Square Estimation with Applications to Digital Signal Processing*. New York: Wiley, 1985.

23. B. Noble and J. W. Daniel. *Applied Linear Algebra*. Englewood Cliffs, N.J.: Prentice Hall, 1988.

24. A. Papoulis. *Probability, Random Variables, and Stochastic Processes*. New York: McGraw Hill, 1984.

25. D. DeFatta, J. Lucas, and W. Hodgkiss. *Digital Signal Processing: A System Design Approach*. New York: Wiley, 1988.

26. W. Haykin. *Adaptive Filter Theory*. Englewood Cliffs, N.J.: Prentice Hall, 1991.

PART 4

SIGNAL RECOGNITION

11

NEURAL NETWORKS

Ananth Sankar[1]
AT&T Bell Laboratories
600 Mountain Avenue
Murray Hill, New Jersey 07947
email: sankar@research.att.com

11.1 INTRODUCTION

Neural networks have recently been found to offer an attractive approach to pattern recognition [1–5]. Unlike methods based on Bayes' classification, no assumptions need to be made as to the probability density functions of the different classes. Rather, an assumption is made as to the form of the decision region separating the classes. This assumption is implicitly made by the configuration of the neural network. The parameters of the network are then adjusted by one of many learning algorithms.

Pattern recognition is the process of assigning a class label to a given observation. Figure 11.1 shows a block diagram of a pattern-recognition system. This is comprised of two stages. The first stage is feature extraction. Feature extraction is the process of generating features that are representative of the different classes. For example, suppose the system is required to recognize squares and rectangles. Possible features to use for this problem could be the lengths of the two adjacent sides of the given figure. This results in a two-dimensional feature space. In this feature space, all squares lie on the bisector of the first quadrant of the feature space, as shown in Figure 11.2. Any other point corresponds to a rectangle. Feature extraction is followed by pattern classification. In this stage, the feature vectors that have already been generated are classified. Again returning to the above example, the classifier is required to recognize whether or

---

[1] This research was done when Dr. Sankar was a Ph.D. candidate at Rutgers University.

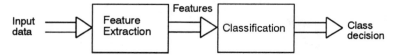

Figure 11.1 Block diagram of pattern-recognition system.

not a given point in the feature space lies on the bisector of the first quadrant. If it does, then the classifier decides the point corresponds to a square. If not, then the point corresponds to a rectangle.

Neural networks are typically used in pattern recognition as classifiers. It is assumed that feature vectors are available as the result of some appropriate preprocessing. The basic element of a neural network is the neuron. In the 1960s, a neuron-like element called the perceptron was the subject of much research [6, 7]. Essentially, perceptrons can implement a hyperplane in feature space and, thus, can implement a linear decision region. If the classes are separable by a hyperplane—i.e., if they are linearly separable—then a perceptron can be used for the problem. A learning algorithm for the perceptron was given by Rosenblat [6]. This algorithm finds an optimum hyperplane if the classes are linearly separable. However, if the classes are not linearly separable, as is the case in most practical applications, a perceptron does not properly classify the

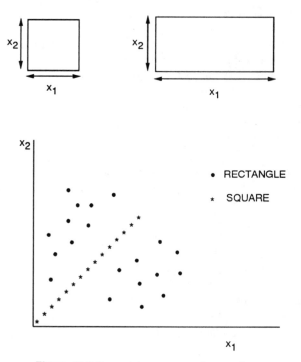

Figure 11.2 Recognizing squares and rectangles.

feature vectors. Nilsson [8] has discussed the use of layered machines, where each layer has many perceptrons. Such machines can solve non–linearly separable problems. However, there were no general training algorithms for these systems until the recent discovery of the backpropagation algorithm [1, 9, 10]. These layered perceptron machines are also called multilayer perceptrons (MLP).

Section 11.2 describes a single neuron, the basic processing element in a neural network. In Section 11.3, some learning algorithms are given to train a neuron for two-class problems. The two-class problem is extended to the multiclass case in Section 11.4. The MLP architecture is described in Section 11.5. Section 11.6 describes the use of the backpropagation algorithm to train MLPs.

11.2 THE NEURON

Let us consider a simple two-class pattern classification problem, where the feature vectors belonging to the two classes are linearly separable. Figure 11.3 illustrates this problem for a 2-dimensional feature space. As can be seen from the figure, a line can be drawn so that the two classes are on the two sides of this line. The equation of this line can be written as

$$w_1 x_1 + w_2 x_2 = \theta, \tag{11.1}$$

where w_1 and w_2 are the components of the normal vector to the line and θ is a bias term. The bias term shifts the position of the line along the direction of the normal vector, $\mathbf{w} = (w_1, w_2)$. In vector notation we may write (11.1) as

$$\mathbf{w} \cdot \mathbf{x} = \theta, \tag{11.2}$$

where $\mathbf{x} = (x_1, x_2)$ is an input feature vector that lies on the line. In order to simplify the analysis, we may treat the bias term, θ, as an additional weight to get the augmented 3-dimensional weight vector given by

$$\mathbf{w} = (w_1, w_2, -\theta). \tag{11.3}$$

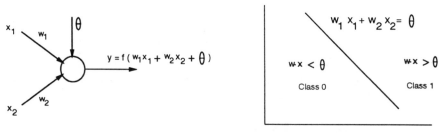

Figure 11.3 Linearly separable problem solved by single neuron.

The input vector, $\mathbf{x}$, is also augmented with a constant input of 1 to get

$$\mathbf{x} = (x_1, x_2, 1).$$ (11.4)

Now (11.2) can be written more simply as

$$\mathbf{w} \cdot \mathbf{x} = 0.$$ (11.5)

The decision rule used to classify an input feature vector is given by

$$\begin{aligned} \mathbf{x} &\in \text{class } 0, \quad \text{if } \mathbf{w} \cdot \mathbf{x} \le 0, \\ \mathbf{x} &\in \text{class } 1, \quad \text{if } \mathbf{w} \cdot \mathbf{x} > 0. \end{aligned}$$ (11.6)

This decision rule can be implemented by a neuron, as shown in Figure 11.3. The neuron calculates a weighted sum of its inputs and adds a bias term to the result. This value is then passed through a nonlinear function that implements the line. For example, suppose the nonlinear function used is the hardlimiter, given by

$$f(x) = \begin{cases} 1, & \text{if } x > 0, \\ 0, & \text{if } x \le 0. \end{cases}$$ (11.7)

The input to the hardlimiter is the weighted sum given by

$$s = w_1 x_1 + w_2 x_2 - \theta.$$ (11.8)

If $s < 0$, then $\mathbf{x}$ belongs to class 0 and the output of the neuron, $f(s)$, is 0. If $s > 0$, then $\mathbf{x}$ belongs to class 1 and the output of the neuron is 1. Thus, the neuron implements the following decision rule:

$$\begin{aligned} \mathbf{x} &\in \text{class } 0, \quad \text{if } f(s) = 0, \\ \mathbf{x} &\in \text{class } 1, \quad \text{if } f(s) = 1. \end{aligned}$$ (11.9)

The hardlimiter is not the only nonlinearity that is used. Another commonly used function is the sigmoid function given by

$$f(x) = \frac{1}{1 + e^{-x}}.$$ (11.10)

The essential property that must be satisfied by the nonlinear function is that it be monotonically increasing. In this case, some threshold ϕ can be set up so that the following decision rule can be implemented:

$$\begin{aligned} \mathbf{x} &\in \text{class } 0, \quad \text{if } f(s) < \phi, \\ \mathbf{x} &\in \text{class } 1, \quad \text{if } f(s) > \phi. \end{aligned}$$ (11.11)

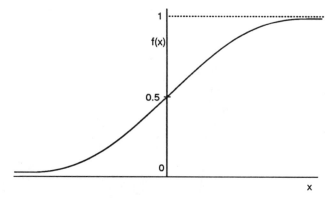

Figure 11.4 Sigmoid function.

For example, in the case of the sigmoid function shown in Figure 11.4, the threshold, ϕ, is 0.5. Both the hardlimiter and the sigmoid are monotonically increasing. In addition, the sigmoid is also differentiable, a property that is required for some learning algorithms, such as the backpropagation algorithm (see Section 11.6).

Figure 11.3 showed a linearly separable problem in a two-dimensional feature space. In this case, a line is used to separate the classes. If a higher-dimensional feature space is used, then a hyperplane is used to separate the classes. In general, in a d-dimensional feature space, a $(d-1)$-dimensional hyperplane is used.

We have discussed how a neuron can implement a hyperplane in feature space. We now give some details concerning the power, or capacity, of a single neuron.

Consider a d-dimensional feature space. Assume that there are N training data in general position.[2] Cover [11] showed that the probability of a randomly picked dichotomy being linearly separable is given by

$$P(N,d) = \begin{cases} \dfrac{1}{2^{N-1}} \displaystyle\sum_{i=0}^{d} \binom{N-1}{i}, & \text{for } N > d, \\[2ex] 1, & \text{for } N \leq d. \end{cases} \tag{11.12}$$

The above probability gives a measure of the effectiveness of a hyperplane. In Figure 11.5, we plot the probability, $P(N,d)$, against $N/(d+1)$ for different values of d. It is seen that, for very large dimensional feature vectors, if the number of feature vectors is less than $2(d+1)$, then the probability of a dichotomy being linearly separable is almost 1; whereas, if the number of feature vectors is more

---

[2]N points are in general position in d-dimensional space if no $d+1$ points lie on a $(d-1)$-dimensional hyperplane.

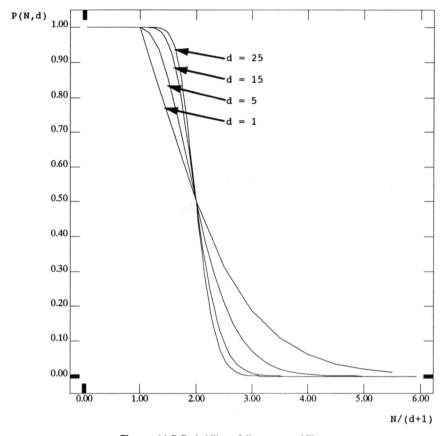

Figure 11.5 Probability of linear separability.

than $2(d + 1)$, this probability is almost 0. This leads to the definition of the capacity, C, of the hyperplane as

$$C = 2(d + 1). \tag{11.13}$$

Thus, the capacity of a hyperplane is equal to twice the number of weights. This definition is due to Cover [11] . In fact, Cover [11] also showed that the capacity of any discriminant function with m free variables is $2m$. Table 11.1 tabulates the capacity of some commonly used discriminant functions.

11.3 SINGLE-NEURON LEARNING ALGORITHMS

Historically there have been two main iterative techniques to update the weights of a neuron. They are the *error-correcting algorithms* and the *minimum square*

TABLE 11.1 Capacity of Discriminant Functions.

Discriminant Function	Number of Free Variables	Capacity
Hyperplane	$d+1$	$2(d+1)$
Hypersphere	$d+2$	$2(d+2)$
Hypercone	$d+1$	$2(d+1)$
rth order polynomial	$\dbinom{d+r}{r}$	$2\dbinom{d+r}{r}$

error algorithms. These techniques differ in that error-correcting techniques update the weight vector only when a pattern is misclassified, whereas minimum square error techniques update the weight vector for every pattern presentation. These two techniques are reviewed in the following sections.

11.3.1 Error-Correcting Algorithms

In the two-class case, recall that the linear discriminant makes class decisions according to the following rule:

$$
\begin{aligned}
\mathbf{x} \in \text{class } 0, &\quad \text{if } \mathbf{w}\cdot\mathbf{x} \leq 0, \\
\mathbf{x} \in \text{class } 1, &\quad \text{if } \mathbf{w}\cdot\mathbf{x} > 0.
\end{aligned}
\tag{11.14}
$$

Figure 11.6 shows a processor called a *perceptron* that implements (11.14). The perceptron is essentially a neuron that uses the hardlimiter function, f, to operate on the weighted sum of its inputs, $\mathbf{w}\cdot\mathbf{x}$. Thus, the perceptron puts $\mathbf{x}$ in class 0 if $f(\mathbf{w}\cdot\mathbf{x}) = 0$ and in class 1 if $f(\mathbf{w}\cdot\mathbf{x}) = 1$.

In error-correcting methods, the weight vector, $\mathbf{w}$, of the perceptron is updated only for those patterns, $\mathbf{x}$, that are presently misclassified, i.e., if $\mathbf{x} \in$ class 0 and $\mathbf{w}\cdot\mathbf{x} > 0$ or if $\mathbf{x} \in$ class 1 and $\mathbf{w}\cdot\mathbf{x} \leq 0$. The best known of these methods is the *perceptron learning rule* [6]. The weight update rule for the perceptron learning

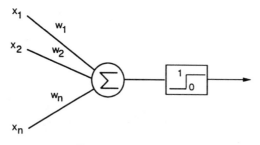

Figure 11.6 Perceptron.

rule is given by

$$\begin{aligned}
\mathbf{w}^{n+1} &= \mathbf{w}^n + \eta\mathbf{x}, && \text{if } \mathbf{x} \in \text{class 1 and } \mathbf{w}\cdot\mathbf{x} \le 0, \\
\mathbf{w}^{n+1} &= \mathbf{w}^n - \eta\mathbf{x}, && \text{if } \mathbf{x} \in \text{class 0 and } \mathbf{w}\cdot\mathbf{x} > 0, \qquad (11.15) \\
\mathbf{w}^{n+1} &= \mathbf{w}^n && \text{otherwise,}
\end{aligned}$$

where the superscript denotes the nth iteration and η is a positive constant. The input patterns are presented cyclically and the weights updated.

The following simple 2-dimensional pattern-classification problem shows the working of the perceptron algorithm. Figure 11.7 shows the feature space with the points $(0,0)$ and $(0,1)$ in class 0 and the points $(1,0)$ and $(1,1)$ in class 1. Let the required output of the perceptron for class-0 features be 0 and the required output for class-1 features be 1. Let the initial weights be given by

$$w_1 = 0.5,$$

$$w_2 = 0.5,$$

$$\theta = 0.0,$$

and the step size be 1. We give four iterations of the algorithm below.

First iteration

$$\mathbf{x} = (0, 0, 1),$$

$$\mathbf{w} = (0.5, 0.5, 0),$$

$$\mathbf{w}\cdot\mathbf{x} = 0.$$

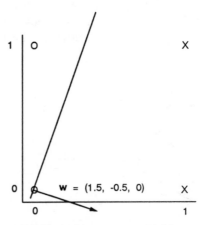

Figure 11.7 Figure for perceptron algorithm example.

Since $\mathbf{w}\cdot\mathbf{x} = 0$ and $\mathbf{x}$ is in class 0, $\mathbf{x}$ is correctly classified and no weights are updated.

Second iteration

$$\mathbf{x} = (0, 1, 1),$$
$$\mathbf{w} = (0.5, 0.5, 0),$$
$$\mathbf{w}\cdot\mathbf{x} = 0.5.$$

Since $\mathbf{w}\cdot\mathbf{x} = 0.5$ and $\mathbf{x}$ is in class 0, $\mathbf{x}$ is misclassified and the weights are updated:

$$\mathbf{w} = \mathbf{w} - (0, 1, 1) = (0.5, -0.5, -1).$$

Third iteration

$$\mathbf{x} = (1, 0, 1),$$
$$\mathbf{w} = (0.5, -0.5, -1),$$
$$\mathbf{w}\cdot\mathbf{x} = -0.5.$$

Since $\mathbf{w}\cdot\mathbf{x} = -0.5$ and $\mathbf{x}$ is in class 1, $\mathbf{x}$ is misclassified and the weights are updated:

$$\mathbf{w} = \mathbf{w} + (1, 0, 1) = (1.5, -0.5, 0).$$

Fourth iteration

$$\mathbf{x} = (1, 1, 1),$$
$$\mathbf{w} = (1.5, -0.5, 0),$$
$$\mathbf{w}\cdot\mathbf{x} = 1.$$

Since $\mathbf{w}\cdot\mathbf{x} = 1$ and $\mathbf{x}$ is in class 1, $\mathbf{x}$ is correctly classified and no weights are updated. It can easily be verified that this is a solution weight vector and so no further weight updates will be made. This solution hyperplane is shown in Figure 11.7.

It can be shown that, if the patterns are linearly separable, the perceptron learning rule will, in finite time, produce a solution weight vector that correctly classifies all the training data. This statement is the celebrated *perceptron convergence theorem* [6–8, 12].

Theorem 11.1 *Let the training set,* $\mathcal{X}$*, be linearly separable so that there exists a solution weight vector,* **w**, *such that*

$$\mathbf{w} \cdot \mathbf{x} \leq 0, \quad \text{for } \mathbf{x} \in \text{class } 0, \tag{11.16}$$

$$\mathbf{w} \cdot \mathbf{x} > 0 \quad \text{for } \mathbf{x} \in \text{class } 1. \tag{11.17}$$

Then the perceptron learning rule will produce, in finite time, a solution weight vector, **w***. *Note that there may be more than one solution vector. The theorem only guarantees that one of them will be found.*

In the example given above, a solution weight vector was found in 4 iterations for the linearly separable classification problem shown in Figure 11.7. Proofs of the perceptron convergence theorem can be found in Duda and Hart [12], Nilsson [8] and Minsky [7]. The proof given here follows the one given in Duda and Hart [12].

Proof The problem is first converted to one of solving a set of linear inequalities. This can be done by simply multiplying the class-0 input vectors by –1. The problem is now transformed into one of finding a weight vector that satisfies

$$\mathbf{w} \cdot \mathbf{x} \geq 0 \quad \forall \ \mathbf{x} \in \mathcal{X}. \tag{11.18}$$

The perceptron learning rule now becomes

$$\begin{aligned} \mathbf{w}^{n+1} &= \mathbf{w}^n + \eta \mathbf{x}, &&\text{if } \mathbf{w}^n \cdot \mathbf{x} < 0, \\ \mathbf{w}^{n+1} &= \mathbf{w}^n, &&\text{otherwise.} \end{aligned} \tag{11.19}$$

Since the problem is assumed to be linearly separable, there exists at least one weight vector which satisfies (11.18). Let $\hat{\mathbf{w}}$ be one such vector and α be a positive scale factor. From (11.19),

$$\mathbf{w}^{k+1} - \alpha \hat{\mathbf{w}} = \mathbf{w}^k - \alpha \hat{\mathbf{w}} + \eta \mathbf{x}^k, \tag{11.20}$$

where $\mathbf{x}^k$ is the kth misclassified pattern in the sequence of patterns presented to the perceptron. By taking the squared magnitudes of the vectors on both sides of (11.20), we get

$$\|\mathbf{w}^{k+1} - \alpha \hat{\mathbf{w}}\|^2 = \|\mathbf{w}^k - \alpha \hat{\mathbf{w}}\|^2 + 2\eta(\mathbf{w}^k - \alpha \hat{\mathbf{w}}) \cdot \mathbf{x}^k + \eta^2 \|\mathbf{x}^k\|^2. \tag{11.21}$$

Since $\mathbf{x}^k$ is misclassified, $\mathbf{w}^k \cdot \mathbf{x}^k < 0$; hence,

$$\|\mathbf{w}^{k+1} - \alpha \hat{\mathbf{w}}\|^2 < \|\mathbf{w}^k - \alpha \hat{\mathbf{w}}\|^2 - 2\alpha\eta \hat{\mathbf{w}} \cdot \mathbf{x}^k + \|\mathbf{x}^k\|^2 \eta^2. \tag{11.22}$$

If

$$\beta^2 = \max_i \|x^i\|^2$$

and

$$\gamma = \min_i \hat{w} \cdot x^i \geq 0,$$

then

$$\|w^{k+1} - \alpha\hat{w}\|^2 < \|w^k - \alpha\hat{w}\|^2 - 2\alpha\eta\gamma + \eta^2\beta^2. \qquad (11.23)$$

If we set

$$\alpha = \frac{\beta^2\eta}{\gamma},$$

then

$$\|w^{k+1} - \alpha\hat{w}\|^2 < \|w^k - \alpha\hat{w}\|^2 - \eta^2\beta^2. \qquad (11.24)$$

Thus, the squared distance between w^k and $\alpha\hat{w}$ is decreased by at least $\beta^2\eta^2$ with each weight update. If the initial weight vector is w^1, then, after k updates,

$$\|w^{k+1} - \alpha\hat{w}\|^2 < \|w^1 - \alpha\hat{w}\|^2 - k\eta^2\beta^2. \qquad (11.25)$$

Since the squared distance can never be negative, (11.25) provides an upper bound on the number of weight updates, k_{max}, where

$$k_{max} = \frac{\|w^1 - \alpha\hat{w}\|^2}{\eta^2\beta^2}. \qquad (11.26)$$

Since every pattern is presented an arbitrarily large number of times, and there is a bound on the number of times the weight vector can be updated (11.26), it follows that a solution is found in at most k_{max} weight updates. This ends the proof of Theorem 11.17.

The bound given in (11.26) requires knowledge of a solution vector and hence does not really help in finding the number of updates needed to solve the problem. All that can be said is that, in the linearly separable case, a solution will be found in a finite number of steps. In the case where the patterns are not linearly separable, the perceptron algorithm is unstable. Nothing can be said about the separating power of any weight vector in the sequence of updated vectors. However, in the non–linearly separable case, we have the *perceptron cycling theorem*, which states that the magnitude of the weights is bounded [7].

In the case where the weights are restricted to be integers, this results in a finite number of vectors in the sequence of updated weight vectors.

As mentioned before, the problem of finding a separating weight vector can be treated as the problem of solving the set of inequalities given in (11.18). Iterative relaxation methods for finding a weight vector that satisfies (11.18) have been given by Agmon [13] and Motzkin and Schoenberg [14]. The weight update rule in these methods is given by

$$
\begin{array}{ll}
\mathbf{w}^{n+1} = \mathbf{w}^n, & \text{if } \mathbf{w} \text{ satisfies (11.18),} \\
\mathbf{w}^{n+1} = \mathbf{w}^n - \lambda \frac{\mathbf{w}^n \cdot \mathbf{x}^*}{\|\mathbf{x}^*\|^2} \mathbf{x}^*, & \text{otherwise,}
\end{array}
\tag{11.27}
$$

where

$$
\mathbf{w}^n \cdot \mathbf{x}^* = \max_j \mathbf{w}^n \cdot \mathbf{x}_j.
$$

In other words, if (11.18) is not satisfied, then $\mathbf{w}^{n+1}$ is the projection of $\mathbf{w}^n$ on the farthest hyperplane, $\mathbf{x}^*$. It is shown in [13, 14] that this algorithm will converge to a solution if $0 < \lambda \leq 2$. It is worth noting that these relaxation methods predate the perceptron learning rule.

11.3.2 Minimum Squared Error Algorithms

Perhaps the best-known algorithm of this type is the LMS algorithm due to Widrow and Hoff [15]. In this method, the inequalities in (11.18) are transformed to equalities by requiring that

$$
\mathbf{w} \cdot \mathbf{x}_i = b_i \qquad \forall \ \mathbf{x}_i \in \mathcal{X},
\tag{11.28}
$$

where $\mathcal{X}$ is the entire training set and b_i is an arbitrarily specified positive number.

The LMS algorithm uses a gradient-descent method to minimize the mean square error given by

$$
E = \sum_{\mathbf{x}_i \in \mathcal{X}} (b_i - \mathbf{w} \cdot \mathbf{x}_i)^2.
\tag{11.29}
$$

Equation 11.29 gives the total error for all the training patterns. The LMS algorithm, however, updates the weights in the nth iteration using only the nth pattern error,

$$
E_n = (b_n - \mathbf{w} \cdot \mathbf{x}_n)^2.
\tag{11.30}
$$

where the subscript, n, refers to the nth pattern presented to the discriminant. The LMS weight update rule is given by

$$\mathbf{w}^{n+1} = \mathbf{w}^n + \eta(b_n - \mathbf{w} \cdot \mathbf{x}_n)\mathbf{x}_n. \qquad (11.31)$$

where η is a positive step size, usually chosen between 0 and 1. It can be shown [15] that this algorithm will converge to the minimum squared error solution of (11.28). However, it must be realized that the positive constants, b_i, are arbitrarily selected. It is possible that the original problem could be linearly separable, but, due to a poor choice of the b_i, the minimum mean square solution not be a separating weight vector.

The following comment on the perceptron algorithm and the LMS algorithm is worth noting: In the linearly separable case, the perceptron algorithm finds a separating weight vector but does not converge in the non–linearly separable case. However, the LMS algorithm converges in both the linearly separable and nonseparable case but there is no guarantee that the solution vector is a separating vector. Clearly, more complex systems than the perceptron are needed for non–linearly separable problems. One such system that has been studied in great detail is the MLP. The MLP is described in Sections 11.5 and 11.6.

We have described how a single neuron is used for a two-class problem. In the case of multiclass problems, more that one neuron is used. In this case, there are different ways in which the different classes can be represented by the neurons. This is the subject of the next section.

11.4 MULTICLASS PROBLEMS

In the two-class case, the output of the neuron is required to be 0 or 1, depending on whether the input feature vector belongs to class 0 or class 1. If there are more than two classes, then the two possible states (0 or 1) of the neuron cannot represent all the classes. Thus, more than one neuron must be used. We now discuss some methods used to label classes when the number of classes, M, is greater than 2.

Extending the two-class case, we use binary vectors to label the classes in the M-class case. For example, if there are 8 classes, then all the binary vectors from $(0, 0, 0)$ to $(1, 1, 1)$ can be used to label the classes. Thus, 3 neurons will be used and the desired output of the neurons will be determined by the binary vector class label. In general, for M classes, p neurons are used where $\lceil \log_2 M \rceil \leq p \leq M$. $\lceil x \rceil$ denotes the smallest integer greater than or equal to x. Since each of the neurons has a desired output of either 0 or 1, each neuron is essentially implementing a two-class problem. All the classes for which the desired output of the neuron is 0 form one class and the rest of the classes form the other class. If $p < M$, then, for at least one neuron, both the classes in the two-class problem corresponding to that neuron are comprised of more than one class. Such class-labeling methods are called distributed encoding schemes. If, on the other hand, M neurons are used, then the M binary basis vectors can be used as class labels. Thus the ith class is labeled with a 1 in the ith component and a 0 for all the other $M-1$ components. In this case, the ith neuron separates the ith class from the rest of the $M-1$ classes. This labeling method is called the local encoding scheme.

If a distributed encoding scheme is used to label the classes, then a threshold of 0.5 is set for each neuron. If the output of the neuron is greater than 0.5, then it is treated as 1; if the output is less than 0.5, then it is taken to be 0. The outputs of all the neurons are then taken together to form a binary vector. If this corresponds to a class label, then the input feature vector is classified accordingly. If the binary vector does not correspond to a class label, then the feature vector can be rejected or a decision made by some other technique. In the local encoding scheme, since each class is represented by a single neuron, we can use the following decision rule:

$$\mathbf{x} \in \text{class } i \qquad \text{iff} \quad f(\mathbf{w}_i \cdot \mathbf{x}) > f(\mathbf{w}_j \cdot \mathbf{x}), \quad \forall \, j \neq i, \tag{11.32}$$

where $\mathbf{w}_i$ is the weight vector for the ith neuron and f is the activation function of the neuron. This is also called a *winner-take-all* or *competition* rule.

11.5 MULTILAYER PERCEPTRON ARCHITECTURE

As mentioned in Section 11.2, neurons can be arranged in layers to form feed-forward neural networks. The most commonly used feedforward neural network is the multilayer perceptron (MLP), shown in Figure 11.8. In this network, neurons are arranged in layers, with connections between adjacent layers. There are no connections between neurons in the same layer. The input to the network is fed into the first layer, called the input layer, and the output is derived from the last layer, called the output layer. The layers in between the input and output layers are called hidden layers. Such systems have been studied in Nilsson [8], but no efficient training algorithm existed until recently, when the backpropagation algorithm was discovered [1, 9, 10].

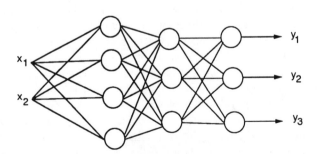

Figure 11.8 Multilayer perceptron (MLP).

In Section 11.3, linear discriminant functions were discussed. A linear discriminant can be implemented by a single neuron, like the perceptron, and suffices for linearly separable problems. However, in practice, most problems are not linearly separable and more complex structures are necessary. The MLP provides one such structure capable of solving non–linearly separable tasks. It can be shown [16] that an MLP with one hidden layer and sigmoidal output functions can arbitrarily approximate any mapping from input to output space. Thus, any arbitrary classification task can be represented by a single-hidden-layer network. However, there are still no results giving the capacity of an MLP like the results of Cover (see Section 11.13) for single neurons.

The most popular learning algorithm for MLPs is the backpropagation algorithm. This is, essentially, a gradient-descent algorithm. This algorithm is discussed in the next section.

11.6 THE BACKPROPAGATION ALGORITHM

The backpropagation algorithm uses a set of labeled training feature vectors. Thus, for each input feature vector the MLP is required to produce a particular output corresponding to the class label. The difference between the desired output and the actual output is called the error. This error is transformed by some distortion function and used to update the weights of the neurons in the MLP. This approach is called *learning by example*. Recently, another approach— called *learning by doing*—has been proposed [17]. In this approach, the network is capable of a set of actions. For each input, the network action is judged by the user as to its appropriateness. The user's evaluation of the network is represented by a single bit of information corresponding to whether the action of the network was appropriate or not. This information is used to update the network parameters. Learning by doing is more useful than learning by example when on-line learning is required. This is because learning by example requires that the correct network outputs be provided for learning. However, if the correct output is available, then there is no need for the network itself.

Backpropagation [1, 9, 10] is a gradient-descent algorithm that minimizes the distortion function of the error between the desired outputs and the actual outputs calculated by the MLP. The algorithm is an elegant extension of the LMS algorithm and uses the chain rule for calculating derivatives to propagate the errors at the output neurons of the MLP back through the hidden layers. The most commonly used distortion metric is the squared error, or L_2-norm, metric given by

$$E = \frac{1}{2} \sum_{i=1}^{N} (t_i - y_i)^2, \qquad (11.33)$$

where N is the number of output neurons in the MLP; t_i is the desired output, or target, for neuron i; and y_i is the output at neuron i calculated by the MLP.

The gradient-descent method updates an arbitrary weight, w, in the network by the following rule:

$$w^{n+1} = w^n - \eta \frac{\partial E}{\partial w^n},$$ (11.34)

where the superscript, n, denotes the iteration number. Thus, it is required to calculate the derivatives, $\partial E / \partial w$, for every weight, w, in the network.

Figure 11.9 shows an arbitrary hidden neuron, j, in the network. It derives its inputs from the l outputs of the neurons in the previous layer and feeds its output to the inputs of the m neurons in the next layer. Let f be the nonlinear function implemented by the neuron, y_j the output of the neuron j, and net_j the weighted sum of its inputs, so that

$$y_j = f(net_j).$$ (11.35)

The function, f, is also called the activation function. The most common function used for backpropagation is the sigmoid function, given by

$$f(x) = \frac{1}{1 + e^{-x}}.$$ (11.36)

The algorithm to update the weights of the hidden neuron, j, is now given. Let the weight connecting neuron i to neuron j be $w_{i,j}$. Using the chain rule, we write the derivative of E with respect to the weights $w_{i,j}$ by

$$\frac{\partial E}{\partial w_{i,j}} = \frac{\partial E}{\partial net_j} \frac{\partial net_j}{\partial w_{i,j}}.$$ (11.37)

Since

$$net_j = \sum_{i=1}^{l} w_{i,j} y_i$$ (11.38)

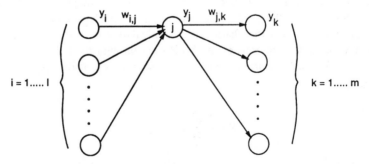

Figure 11.9 A hidden neuron in an MLP.

we have

$$\frac{\partial net_j}{\partial w_{i,j}} = y_i, \tag{11.39}$$

and, hence, (11.37) becomes

$$\frac{\partial E}{\partial w_{i,j}} = y_i \frac{\partial E}{\partial net_j}. \tag{11.40}$$

Again, using the chain rule, we get

$$\frac{\partial E}{\partial net_j} = \sum_{k=1}^{m} \frac{\partial E}{\partial net_k} \frac{\partial net_k}{\partial net_j} = \sum_{k=1}^{m} \frac{\partial E}{\partial net_k} \frac{\partial net_k}{\partial y_j} \frac{\partial y_j}{\partial net_j}. \tag{11.41}$$

Since

$$net_k = \sum_j w_{j,k} y_j, \tag{11.42}$$

we get

$$\frac{\partial net_k}{\partial y_j} = w_{j,k} \tag{11.43}$$

and, from (11.35), we get

$$\frac{\partial y_j}{\partial net_j} = f'(net_j). \tag{11.44}$$

Substituting (11.43) and (11.44) in (11.41), we get

$$\frac{\partial E}{\partial net_j} = f'(net_j) \sum_{k=1}^{m} \frac{\partial E}{\partial net_k} w_{j,k}. \tag{11.45}$$

If the sigmoid function of (11.36) is used for the activation function, then

$$f'(net_j) = y_j(1 - y_j). \tag{11.46}$$

Equation 11.45 is the core of backpropagation; it gives a recursive relation that calculates $\partial E/\partial net_j$ for a hidden neuron. What remains is to find $\partial E/\partial net_j$, when j is an output neuron, since then we can use (11.45) to recursively calculate $\partial E/\partial net_j$ for all neurons in the network and finally use (11.40) to calculate the required error derivatives for every neuron in the network. If j is an output neuron, we have

$$\frac{\partial E}{\text{net}_j} = \frac{\partial E}{\partial y_j} f'(\text{net}_j). \tag{11.47}$$

If the squared error distortion metric is used, then, from (11.33), we have

$$\frac{\partial E}{\partial y_j} = -(t_j - y_j). \tag{11.48}$$

The backpropagation algorithm using the L_2-norm distortion metric and sigmoid activation functions can now be summarized. First, the outputs, y_j, for all the neurons in the network are calculated for a particular input vector, x. The error derivative needed for the gradient-descent update rule of (11.34) is calculated using (11.40) and (11.47), if j is an output neuron. Substituting (11.46) and (11.48) into (11.47), we get

$$\frac{\partial E}{\partial \text{net}_j} = -(t_j - y_j)y_j(1 - y_j). \tag{11.49}$$

If j is a hidden neuron, then the error derivative is calculated by using (11.40) and (11.45). Substituting (11.46) into (11.45), we get

$$\frac{\partial E}{\partial \text{net}_j} = y_j(1 - y_j) \sum_{k=1}^{m} \frac{\partial E}{\partial \text{net}_k} w_{j,k}. \tag{11.50}$$

Finally, the weights are updated using (11.34). Many modifications of this algorithm have been proposed to increase its learning speed. One method is to use an acceleration, or *momentum*, term [1] in the update equation so that

$$w^{n+1} = w^n - \eta \frac{\partial E}{\partial w^n} + \alpha \Delta w^n, \tag{11.51}$$

where Δw^n is the amount by which the weight w is updated at the nth iteration. Both η and the momentum term, α, are typically chosen to be between 0 and 1. The actual choice of these parameters has a strong effect on the learning rate. Another modification to the backpropagation algorithm is to use an extension of Newton's algorithm [18, 19]. Yet another approach that has been studied is to change the distortion metric [20, 21].

We shall now present a well-known example to illustrate the backpropagation algorithm. The problem considered here is the exclusive OR (XOR) problem. The XOR problem is a commonly used test for neural networks. The XOR function of two variables is written as $a \oplus b$; its truth table is given in Table 11.51. The boolean vectors $(0, 0)$, and $(1, 1)$ belong to one class and the vectors $(0, 1)$, and

TABLE 11.2
Truth Table XOR
Problem.

a	b	$(a \oplus b)$
0	0	0
0	1	1
1	0	1
1	1	0

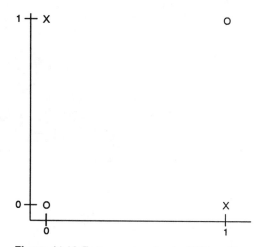

Figure 11.10 Feature vectors for the XOR problem.

$(1, 0)$ belong to the second class. Figure 11.10 shows the four feature vectors in feature space, marking one class by a "X" and the other by a "O." Clearly, this problem is not linearly separable, i.e., it is not possible to draw a line that separates the "X" class from the "O" class. Thus, an MLP with hidden neurons is needed to solve the problem. Figure 11.11 shows the MLP used for this problem. Two hidden neurons are used. In Figure 11.12, the decision regions of the MLP are shown. The figure shows two diagonal lines. By examining the weights of the hidden neurons, it is clear that the two lines are implemented by the hidden neurons. The output neuron then combines the outputs of the hidden neurons to make the class decision.

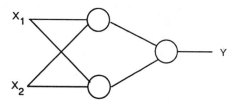

Figure 11.11 MLP for the XOR problem.

Figure 11.12 MLP decision regions for XOR.

Backpropagation has achieved widespread popularity as a training algorithm for neural networks. It has been successfully used in applications such as speech recognition [22], mapping text to phonemes [23], deducing the structure of proteins from amino acid sequences [24], playing backgammon [25], and performing nonlinear signal-processing functions [26, 27]. However, the algorithm is not without its problems. One of the main disadvantages of backpropagation is its slow convergence. Furthermore, since it is a gradient-descent algorithm, it has the problems associated with local minima. Finally, it is necessary to provide the number of neurons and their interconnections before learning can begin. Since the algorithm can get stuck in local minima, there is no guarantee that backpropagation will provide a solution for any particular network configuration. Hence the number of neurons is usually chosen by trial and error.

11.7 SUMMARY

In this chapter neural networks were introduced, with an emphasis on perceptrons and MLPs. The perceptron learning algorithm was described in Section 11.3. The backpropagation algorithm for training MLPs was discussed in Section 11.6. MLPs have been widely used for various problems such as speech recognition, game playing, and character recognition. The ability of MLPs to generalize well on test data makes them an attractive approach to pattern classification. However, the backpropagation algorithm is slow to converge to a solution. Also, since the MLP configuration has to be chosen by trial and error, the learning time can be very large. Recently, some algorithms have been proposed that grow neural networks as they learn. One such algorithm is the cascade correlation algorithm [28]. A method based on a combination of neural networks and decision trees is described in Chapter 12.

Although this chapter stressed the most widely used neural-network architecture and training algorithm—i.e., MLPs and backpropagation, there are many others that may be well suited to a particular application (see, for example, Mammone and Zeevi [29]).

ACKNOWLEDGMENTS

This research reported here was made possible through the support of the New Jersey Commission on Science and Technology and the Center for Computer Aids for Industrial Productivity at Rutgers University.

REFERENCES

1. D. E. Rumelhart and J. L. McClelland. *Parallel Distributed Processing.* Cambridge, Mass.: MIT Press, 1986.

2. A. Waibel. Modular Construction of Time Delay Neural Networks for Speech Recognition. *Neural Comput.*, 1, March 1989.

3. K. J. Lang, A. H. Waibel, and G. E. Hinton. A Time Delay Neural Network Architecture for Isolated Word Recognition. *Neural Networks*, 3(1):23–43, 1990.

4. A. Rajavelu, M. T. Musavi, and M. V. Shirvaikar. A Neural Network Approach to Character Recognition. *Neural Networks*, 2(5):387–394, 1989.

5. Y. Le Cun et al. Handwritten Zip Code Recognition with Multilayer Networks. In *Proceedings of the 10th International Conference on Pattern Recognition*, vol. 2, pp. 35–40, 1990.

6. F. Rosenblatt. *Principles of Neurodynamics.* New York: Spartan, 1962.

7. M. Minsky and S. Papert. *Perceptrons.* Cambridge, Mass.: MIT Press, 1969.

8. N. J. Nilsson. *Learning Machines.* New York: McGraw-Hill, 1965.

9. P. Werbos. *Beyond Regression: New Tools for Prediction and Analysis in the Behavioral Sciences.* PhD thesis, Harvard University, 1974.

10. D. B. Parker. Learning Logic. Technical Report TR-47, Center for Computational Research in Economics and Management Science, MIT, 1985.

11. T. M. Cover. Geometrical and statistical properties of systems of linear inequalities with applications in pattern recognition. *IEEE Trans. Elect. Comput.*, 1965.

12. R. O. Duda and P. E. Hart. *Pattern Classification and Scene Analysis.* New York: Wiley, 1973.

13. S. Agmon. The Relaxation Method for Linear Inequalities. *Canad. J. Math.*, 6(3):382–392, 1954.

14. T. S. Motzkin and I. J. Schoenberg. The Relaxation Method for Linear Inequalities. *Canad. J. Math.*, 6(3):393–404, 1954.

15. B. Widrow and S. D. Stearns. *Adaptive Signal Processing.* Englewood Cliffs, N.J.: Prentice-Hall, 1985.

16. G. Cybenko. Approximation by Superposition of a Sigmoidal Function. *Math. Control Signals Syst.*, **2**:303–314, 1989.

17. A. L. Gorin, S. E. Levinson, A. N. Gertner, and E. Goldman. Adaptive Acquisition of Language. In *Neural Networks: Theory and Applications*, R. Mammone and Y. Y. Zeevi, eds., pp. 125–167. San Diego, Calif.: Academic Press, 1991.

18. D. B. Parker. Optimal Algorithms for Adaptive Networks: Second Order Backpropagation, Second Order Direct Propagation, and Second Order Hebbian Learning. In *Proceedings of the IEEE International Conference on Neural Networks*, pp. 593–600, San Diego, Calif., 1987.

19. R. L. Watrous. Learning Algorithms for Connectionist Networks: Applied Gradient Methods for Non-Linear Optimization. In *Proceedings of the IEEE International Conference on Neural Networks*, pp. 619–627, San Diego, Calif., 1987.

20. S. J. Hanson and D. J. Burr. Minkowski-r backpropagation: Learning in connectionist models with non-euclidean error signals. New York: American Institute of Physics, 1988.

21. A. Sankar and R. J. Mammone. A New Fast Learning Algorithm for Feedforward Neural Networks using the L1 Norm of the Error. Technical Report TR-115, Rutgers University, 1990.

22. R. P. Lippmann. Review of Neural Networks for Speech Recognition. *Neural Comput.*, **1**(1):1–38, 1989.

23. T. J. Sejnowski and C. M. Rosenberg. Parallel Systems that Learn to Pronounce English Text. *Complex Syst.*, **1**:145–168, 1987.

24. N. Qian and T. J. Sejnowski. Predicting the Secondary Structure of Globular Proteins using Neural Network Models. *J. Molec. Biol.*, **202**:865–884, 1988.

25. G. Tesauro and T. J. Sejnowski. A Neural Network that Learns to Play Backgammon. In *Neural Information Processing Systems*, D. Anderson, ed., pp. 794–803. New York: American Institute of Physics, 1988.

26. R. P. Lippmann and P. E. Beckman. Adaptive Neural-Net Processing for Signal Detection in Non-Gaussian Noise. In *Advances in Neural Information Processing Systems 1*, D. S. Touretzky, ed. San Mateo, Calif.: Morgan Kauffman, 1989.

27. S. Tamura and A. Waibel. Noise Reduction using Connectionist Models. In *Proceedings of the IEEE International Conference on Acoustics, Speech and Signal Processing*, pp. 553–556, April 1988.

28. S. E. Fahlman and C. Lebiere. The Cascade-Correlation Learning Architecture. Technical Report CMU-CS-90-100, Carnegie-Mellon University, 1990.

29. R. J. Mammone and Y. Y. Zeevi, eds. *Neural Networks: Theory and Applications.* San Diego, Calif.: Academic Press, 1991.

12

NEURAL TREE NETWORKS

Ananth Sankar[1]

AT&T Bell Laboratories
600 Mountain Avenue
Murray Hill, New Jersey 07947
email: sankar@research.att.com

12.1 INTRODUCTION

In this chapter, a hybrid pattern classifier that combines neural networks and decision trees is discussed. The new classifier is called the Neural Tree Network (NTN). Before entering into the details of the NTN, the concepts of neural networks and decision trees will be reviewed. Chapter 11 has already provided a discussion of the subject of neural networks. This chapter begins with a discussion of decision trees in Section 12.2; the NTN is then described in Section 12.3. The ideas presented in this chapter are summarized in Section 12.4 with some comments on future research.

12.2 DECISION TREES

In Chapter 11, MLPs and the backpropagation algorithm were discussed. Another approach to pattern recognition is decision trees [1, 2]. Decision trees classify a feature vector, x, as follows:

Associated with each internal node of the decision tree is a test function that operates on the vector x. The test function has M_i possible outcomes. Each outcome corresponds to a different child node. When an input vector, x, enters the root of the decision tree, the test function at the root routes the vector to

---

[1]This research was done when Dr. Sankar was a Ph.D. candidate at Rutgers University.

one of the child nodes. This process is repeated until the feature vector ends up in a leaf node. The leaf nodes are used to classify the feature vector. When $M_i = 2$ for all the internal tree nodes, the decision tree is a binary decision tree. Figure 12.1 shows a binary decision tree and the decision regions it implements for a two-class problem in a two-dimensional feature space. At each internal node of the tree, a hyperplane test is made to find out on which side of the hyperplane the feature vector, x, lies. The most common decision-tree algorithms use hyperplanes that are perpendicular to the feature space axes, as shown in Figure 12.1. CART [1], a widely tested decision-tree method, does allow for nonperpendicular hyperplanes, but this option is computationally expensive.

Since decision trees trace a path from the root to a leaf node during classification, the number of tests that must be evaluated is, typically, not very large (the maximum being the length of the tree). However, MLPs evaluate the outputs of every single neuron during classification. Decision trees are also easier to analyze, since the functions implemented at the internal nodes are easy to interpret. On the other hand, the role of hidden neurons in MLPs is much more difficult to interpret. One problem with decision trees is the computationally inefficient training algorithms that are used, such as ID3 [2] and CART [1]. These algorithms use an exhaustive search through an arbitrarily generated list of hyperplanes to find the best hyperplane at each internal decision-tree node [1, 2].

Decision trees are trained by using a set of labeled training vectors. Thus, the training is supervised, as is the case in MLPs. The parameters that have to be learned in a decision tree are the test functions at each internal node. Most decision-tree methods use hyperplane tests. However, the training does not proceed by a gradient-descent or relaxation method, as in the training of linear discriminants [3]. Rather, a set of possible hyperplanes is generated and a

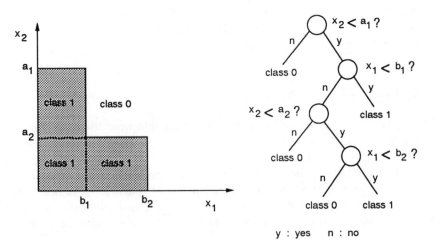

Figure 12.1 A binary decision tree and the corresponding decision regions.

distortion metric is used to evaluate each hyperplane. For example, consider the CART algorithm [1], as used for a binary decision tree.

First, it is assumed that the training vectors are drawn from an M-dimensional feature space, so that a vector can be represented as

$$\mathbf{x} = (x_1, x_2, \ldots, x_M). \tag{12.1}$$

The set of all possible hyperplanes to be evaluated is given by

$$\mathcal{D} = \{d_i: \ x_i = c_i\}, \tag{12.2}$$

where c_i ranges over all the possible values of the component, x_i. These hyperplanes use only a single feature variable and, hence, are perpendicular to the feature-space axes. In [1], a method is given that includes more than one feature variable; thus, the set of possible hyperplanes is given by

$$\mathcal{D} = \left\{ d_i: \ \sum_{i=1}^{M} a_i x_i = c \right\}, \tag{12.3}$$

where c ranges over all possible values.

The best hyperplane is found by exhaustively searching the set $\mathcal{D}$ for the hyperplane which results in the maximum decrease in some distortion metric, $\Phi(t)$. $\Phi(t)$ measures the class homogeneity in the tree node, t. Thus, $\Phi(t)$ is largest when the classes are equally mixed in the node, t, and is smallest when the node, t, contains only one class [1]. One function that satisfies these requirements is the entropy function, given by

$$\Phi(t) = -\sum_{i=1}^{C} p(i \mid t) \log p(i \mid t), \tag{12.4}$$

where $p(i \mid t)$ is the probability that a feature vector in node t belongs to class i and C is the number of classes.

The hyperplane, $d_i : x_i = c_i$, splits the data into two sets,

$$\begin{aligned} \mathbf{x} \in \mathcal{X}_1, & \quad \text{if } x_i < c_i, \\ \mathbf{x} \in \mathcal{X}_2, & \quad \text{if } x_i \geq c_i, \end{aligned} \tag{12.5}$$

and the decrease in the distortion is measured by

$$\Delta_{\Phi(t)}(d_i, t) = \Phi(t) - p_L \Phi(t_L) - p_R \Phi(t_R), \tag{12.6}$$

where p_L and p_R are the probabilities that the hyperplane, d_i, sends a feature vector from t to the left child, t_L, or the right child, t_R, respectively. An exhaustive search is used to evaluate every discriminant, d_i, with respect to the distortion, $\Phi(t)$, and the one that gives the largest decrease in the distortion is chosen. The data is then partitioned into two sets, $\mathcal{X}_L$ and $\mathcal{X}_R$, and the algorithm is repeated for each child node corresponding to $\mathcal{X}_L$ and $\mathcal{X}_R$.

The algorithm can be stopped when the number of data points in the node, t, is below some threshold, n_t. Breiman et al. [1] report that this method does not work well in practice. If n_t is picked to be too large, then the tree-growing process is prematurely stopped; if n_t is picked to be too small, then a very large tree is grown. It has been suggested in [1] that the best strategy is to grow a large tree and then use a pruning algorithm to prune off unnecessary nodes. It is clear that a very large tree will perfectly classify all the training data. Pruning is done so that good performance might be obtained on an independent test set of data. An optimal pruning algorithm is given in [1, 4] for binary decision trees. An extension of this algorithm to arbitrary trees is discussed in Section 12.3.4.

12.3 THE NEURAL TREE NETWORK

The NTN is a hierarchical pattern classifier that combines neural networks and decision trees in a very effective way. The basic idea of the NTN is to classify a pattern in a sequential fashion, similar to decision trees. However, each internal node of the decision tree is replaced with a neural network whose training is more efficient than the standard decision-tree learning algorithms. The NTN has been shown to be superior in performance to both MLPs and decision trees [5, 6]. The NTN also offers an advantageous trade-off of classification speed for hardware implementation [5].

As mentioned in Chapter 11, the backpropagation algorithm used to train MLPs has problems associated with local minima. However, in Section 12.3.1, it is shown that the NTN algorithm has a lower probability of being trapped in local minima. In an MLP, one may view each hidden neuron as implementing a hyperplane in feature space. The training algorithm simultaneously moves all the hyperplanes so as to achieve the best partitioning of the feature space. This task can be very difficult when the decision regions are very complex. Another problem with backpropagation is that the number of neurons and their interconnections must be specified before learning can begin. Since there is no guarantee that the algorithm will converge to a solution for a given network configuration, the number of neurons and their interconnections are usually chosen by trial and error. This can result in very long training times. The NTN, on the other hand, provides a method that grows the network as it learns. Thus, the algorithm provides the correct number of neurons. Also there are no problems associated with local minima—the NTN always converges to a solution.

Decision trees use a sequential decision-making strategy to solve pattern-classification problems. The tree is grown during learning, rather than being specified *a priori*, as are MLPs. However, the standard decision-tree training algorithms are extremely complex. As mentioned in Section 12.2, decision-tree training is the process of finding the optimum hyperplane at each node. The most common method used is to exhaustively search an arbitrarily generated search space of hyperplanes [1, 2]. This process is very inefficient. In most decision-tree algorithms, the hyperplanes are constrained to be perpendicular to the feature-space axes. This is very restrictive and can result in a large number of hyperplane tests, even in a linearly separable problem, if the separating hyperplane is not perpendicular to a feature-space axis. Figure 12.2 illustrates this point. The NTN uses a gradient-descent algorithm, which is more efficient than the exhaustive search techniques used in decision trees. The NTN also allows for hyperplanes that are not perpendicular to the feature-space axes. This characteristic can be obtained with the CART decision tree [1], but the algorithm is computationally expensive.

The architecture for the NTN is described in Section 12.3.1 and the new learning algorithm is discussed in Section 12.3.2. The treatment of the architecture and learning algorithm is carried out for the two-class case. In Section 12.3.3, these ideas are generalized to an arbitrary M-class problem. Different possible generalizations are discussed and arguments are provided to select one. In Section 12.3.4, an optimal pruning algorithm that improves the performance on an independent test set of data is described. This pruning algorithm is a generalization of the algorithm given by Breiman et al. [1] for binary decision trees. In Section 12.3.5, computer simulation results are presented for some boolean function learning tasks, a "mesh" problem, and a speaker-independent vowel-recognition task. It is shown that the NTN is superior in performance to both decision trees and MLPs.

12.3.1 The Architecture

In this section, the new architecture is described for the two-class case. This leads to a binary tree structure. The more general multiclass case is discussed in Section 12.3.3. The NTN uses a tree architecture that implements a sequential, linear decision-making strategy. Figure 12.3 shows a binary NTN and the corresponding decision regions in feature space. A single neuron is used at each internal node of the NTN. This neuron provides an elegant way to implement hyperplanes that are not constrained to be perpendicular to the feature-space axes, as they are in the case of decision trees. Consider the feature vector, x, shown in Figure 12.3. The feature vector enters the NTN at the root node. If the output of the neuron at the root node is less than 0.5, then the left branch is taken; if the output is greater than 0.5, then the right branch is taken. In this case, the left branch is taken and the neuron at the left child node is used to route the feature vector further down the NTN. Figure 12.3 shows the path taken by the feature vector, x, from the root to a leaf of the NTN. Each leaf node corresponds to a class and the feature vector is classified appropriately.

During the training phase, the neuron in the root node operates on the entire training set, $\mathcal{X}$. This neuron splits the training data into two subsets according

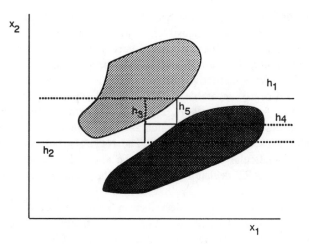

Linearly separable problem

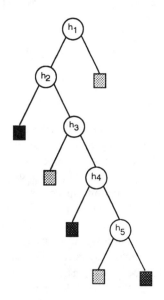

Decision Tree to solve above problem

Figure 12.2 Many hyperplanes are required by decision trees to solve linearly separable problems.

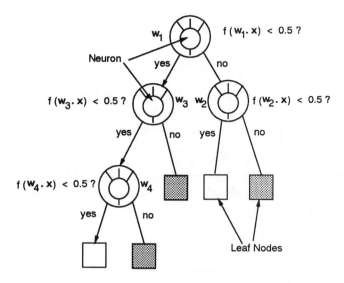

Binary NTN

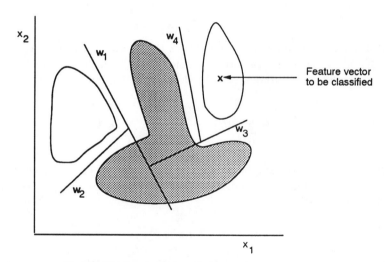

Decision Regions in Feature Space

Figure 12.3 Binary NTN and the corresponding decision regions in feature space.

to the following rule:

$$
\begin{aligned}
\mathbf{x} \in \mathcal{X}_0, & \quad \text{if } f(\mathbf{w}_1 \cdot \mathbf{x}) > 0.5, \\
\mathbf{x} \in \mathcal{X}_1, & \quad \text{if } f(\mathbf{w}_1 \cdot \mathbf{x}) \leq 0.5,
\end{aligned}
\tag{12.7}
$$

where $\mathbf{w}_1$ is the weight vector of the neuron at the root and $f(\cdot)$ is the sigmoid activation function of the neuron. The subsets, $\mathcal{X}_0$ and $\mathcal{X}_1$, are further divided by the child nodes. The subsets of training data corresponding to the leaf nodes partition the training set, $\mathcal{X}$.

When compared with MLPs, the single neuron used in the internal NTN nodes results in a lower probability of getting trapped in local minima. For example, if there are n hidden layers in an MLP with one neuron in each layer (Figure 12.4), then the output y can be written as

$$
y = f(w_n f(w_{n-1} \cdots f(w_1 x) \cdots)),
\tag{12.8}
$$

where x is the input to the network. If there are N labeled training patterns, then the L_2 distortion can be written as

$$
E = \sum_{j=1}^{N} (t_j - y_j)^2,
\tag{12.9}
$$

where t_j and y_j are the target and network output for the jth input vector. From (12.8) and (12.9), it is clear that the distortion is a function of the weights, w_i. All minima of this distortion, whether local or global, are characterized by

$$
\frac{\partial E}{\partial w_i} = -\sum_{j=1}^{N} 2(t_j - y_j) \frac{\partial y_j}{\partial w_i} = 0, \qquad i = 1, \ldots, n.
\tag{12.10}
$$

Solving this equation for the weights w_i gives the positions, in weight space, of the minima. Let the sigmoid function $f(x)$ be expressed by a Taylor series expansion as

$$
f(x) = 0.5 + a_1 x^3 + a_2 x^5 + \cdots,
\tag{12.11}
$$

with only odd powers of x, since the sigmoid is an odd function. If all the terms in the expansion of higher order than x^m are ignored, then the highest power of, say, w_1, in the left-hand side of (12.10) is w_1^{2mn-1}. Keeping all other weights constant, there are $2mn - 1$ possible solutions for w_1 and correspondingly that

Figure 12.4 MLP with a single neuron in each layer.

many minima. However, in the NTN, since only one neuron is trained at a time, $n = 1$, and the number of possible minima is $2m - 1$. Thus it is clear that the probability of getting trapped in a local minima for the NTN is proportional to m, whereas for the MLP it is proportional to mn, where n is the number of layers in the MLP. Also, since the NTN successively divides the training set until each leaf node contains only one class, there are no problems with local minima.

Recently, Baum [7] has shown that an arbitrary dichotomy of N points in general position in d-dimensional space can be implemented by an MLP with a single hidden layer containing $\lceil N/d \rceil$ neurons. The notation $\lceil x \rceil$ denotes the smallest integer greater than or equal to x. Sankar and Mammone [5] showed that, for large d, an NTN with $\lceil N/d \rceil$ internal nodes can implement any arbitrary dichotomy of N points in general position in d-dimensional space. This result can be improved by removing the requirement of large d, as in the following theorem.

Theorem 12.1 *A binary NTN with $\lceil N/d \rceil$ internal nodes can implement any arbitrary dichotomy of N points in general position in d-dimensional space.*

Proof The proof of this theorem is constructive. Let the set, $\mathcal{X}$, of N points be arbitrarily divided into two sets, $\mathcal{X}^+$ and $\mathcal{X}^-$, containing N^+ and N^- points, respectively. Without loss of generality, assume that $N^+ \geq N/2 \geq N^-$. Let H^- be the convex hull of the points in $\mathcal{X}^-$. Choose any face, F, of H^-. Since the points are in general position, there are d points on the face, F, and all of them belong to $\mathcal{X}^-$. Define the outside of F to be the side away from the convex hull and the inside to be the side towards the convex hull (see Figure 12.5). Move F infinitesimally parallel to itself towards its outside to get the hyperplane, F_{out}, and towards its inside to get the hyperplane, F_{in}, so that no points in $\mathcal{X}^+$ lie between F_{out} and F_{in}. Define the outside and inside of F_{out} and F_{in} to be in the same directions as the outside and inside of F. Let $\mathcal{X}_{\text{out}}^{\text{out}}$ be the set of all points on the outside of F_{out}. Clearly, $\mathcal{X}_{\text{out}}^{\text{out}} \subseteq \mathcal{X}^+$. Let the neuron at the root of the NTN implement F_{out}. Denote the two child nodes as C_1 and C_2 so that C_1 corresponds to the set of points $\mathcal{X}_{\text{out}}^{\text{out}}$ and C_2 corresponds to the set $\mathcal{X} - \mathcal{X}_{\text{out}}^{\text{out}}$. Since $\mathcal{X}_{\text{out}}^{\text{out}} \subseteq \mathcal{X}^+$, C_1 is a child node. Let the neuron in C_2 implement the hyperplane, F_{in}. The set of points in $\mathcal{X} - \mathcal{X}_{\text{out}}^{\text{out}}$ on the outside of F_{in} belong to $\mathcal{X}^-$. Let this set be denoted by $\mathcal{X}_{\text{in}}^{\text{out}}$. Thus, the hyperplane F_{in} divides $\mathcal{X} - \mathcal{X}_{\text{out}}^{\text{out}}$ into the set $\mathcal{X}_{2,1} = \mathcal{X}_{\text{in}}^{\text{out}}$, corresponding to the child node $C_{2,1}$, and the set $\mathcal{X}_{2,2} = \mathcal{X} - \mathcal{X}_{\text{out}}^{\text{out}} - \mathcal{X}_{\text{in}}^{\text{out}}$, corresponding to the child node $C_{2,2}$. Since $\mathcal{X}_{\text{in}}^{\text{out}} \subseteq \mathcal{X}^-$, $C_{2,1}$ is a leaf node. The NTN is grown again under the child $C_{2,2}$ by using the same construction as above: dividing $\mathcal{X}_{2,2}$ into $\mathcal{X}_{2,2}^+$ and $\mathcal{X}_{2,2}^-$ and finding the convex hull of $\mathcal{X}_{2,2}^-$. The number of points in $\mathcal{X}_{2,2}^-$ is $N^- - d$, since the d points on the face F of the convex hull H^- have already been removed. Since d points are removed from $\mathcal{X}^-$ at each step of the construction, the process stops after $\lceil N^-/d \rceil$ steps. At each step, two internal nodes are added to the NTN. Thus the total number of internal nodes is $2\lceil N^-/d \rceil \leq \lceil N/d \rceil$. This proves Theorem 12.1.

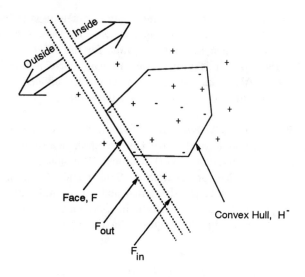

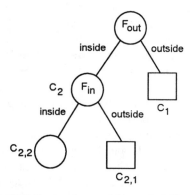

Figure 12.5 Construction used in proof of Theorem 12.1.

We can relax the general-position requirement a little, since the above construction only requires that no $d+1$ points containing points from both $\mathcal{X}^+$ and $\mathcal{X}^-$ lie on a $(d-1)$-dimensional hyperplane. If this condition is satisfied, it does not matter whether or not the points in each set, $\mathcal{X}^+$ and $\mathcal{X}^-$, are in general position or not. The above proof shows that MLPs and NTNs require the same number of neurons to implement an arbitrary dichotomy. The construction given only guarantees perfect classification of a training set of points. There is no guarantee of good generalization. Thus, Theorem 12.1 only deals with a representational issue. This is also true of Baum's proof for MLPs [7].

The construction given in the proof of Theorem 12.1 grows an NTN with a maximum path length from root to leaf of $\lceil N/d \rceil$. Thus, at most $\lceil N/d \rceil$ neurons are fired in sequence to classify a feature vector. This number can be reduced by the following construction. We know from Cover [8] that, if the number of patterns in an internal NTN node is $d + 1$ or less, then any arbitrary dichotomy of these points can be implemented with a single neuron implementing a $(d-1)$-dimensional hyperplane. If each of the last but one level nodes in the NTN contain $d + 1$ points, then there are $\lceil N/(d + 1) \rceil$ such nodes. Since the NTN is a binary tree, there are $2\lceil N/(d+1) \rceil - 1$ internal nodes. Furthermore, if a hyperplane is used at each stage to divide the number of patterns in half, then the maximum path length from the root to a leaf node is

$$\text{path length} = \log_2 2 \left\lceil \frac{N}{d+1} \right\rceil = 1 + \log_2 \left\lceil \frac{N}{d+1} \right\rceil .$$

Thus, the number of time units required to classify a feature vector is $1 + \log_2 \lceil N/(d+1) \rceil$. An MLP with one hidden layer requires two time units to classify a feature vector—the first time unit to fire the hidden-layer neurons and the second to fire the output neuron. This is true only if all the neurons are implemented in parallel so that the hidden layer neurons can be fired together. The NTN, however, needs only one neuron to be implemented in hardware since, at each stage, only one neuron is fired. A look-up table is used to store the weights of all the neurons. The output of the neuron controls which weights are loaded onto it for the next step. This implementation is shown in Figure 12.6. Thus, there is a trade-off between the speed of classification and the number of neurons that must be implemented. If the implementation complexity is defined as the product of the classification time and the number of neurons implemented in hardware, then the complexity of the NTN is

$$C_{\text{NTN}} = 1 + \log_2 \left\lceil \frac{N}{d+1} \right\rceil \tag{12.12}$$

and the complexity of the MLP is

$$C_{\text{MLP}} = 2 \left\lceil \frac{N}{d} \right\rceil . \tag{12.13}$$

From these expressions, it is clear that there is a logarithmic relation between C_{NTN} and C_{MLP}:

$$C_{\text{NTN}} \approx \log_2 C_{\text{MLP}}. \tag{12.14}$$

Clearly, the complexity of the NTN is less; hence, the NTN offers an attractive implementation advantage over the MLP.

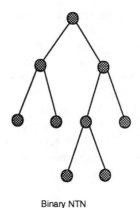

Binary NTN

Look-up
Table →

x_1

weights

x_2

x_3

x_4

Figure 12.6 Implementation of NTN by a single processor.

12.3.2 The Learning Algorithm

As described in the previous section, the NTN successively divides the training set into subsets, assigning each subset to a different child node. Each division adds a level to the NTN. It is always possible to grow a very large NTN that correctly classifies all the training data. However, for good generalization, the number of nodes should not be very large. Thus, it is desired to grow the smallest possible NTN that correctly classifies the training data. This problem has been shown to be NP-complete [9]. Thus, the best way to solve the problem is to look for good heuristic solutions.

The heuristic used for the NTN is to minimize the number of classification errors at each internal NTN node. In other words, the neuron at node t of the NTN should be trained so as to make minimum classification errors on the training subset $\mathcal{X}_t$ corresponding to node t. To date, no algorithms have minimized the number of misclassifications, except in the linearly separable case. All techniques that address this problem are heuristic methods. For examples of such methods, see references [10–12]. A heuristic approach based on minimizing the L_1 norm of the error is used at each step for the NTN. Another method that has been considered is the thermal-perceptron algorithm [11].

The learning algorithm is recursive and so we consider its application to a particular node, t, of the NTN that processes a subset, $\mathcal{X}_t$, of the training data. If $\mathcal{X}_t$ contains only one class, then the node becomes a leaf node and is labeled by the appropriate class label. If $\mathcal{X}_t$ contains two classes, then a single neuron is used to find a hyperplane that separates the feature vectors into two subsets. These subsets of training data are processed by each of two child nodes using the same algorithm.

Since the case where $\mathcal{X}_t$ has only one class is trivial, we shall go to the case where the number of classes is two. In this case, a neuron is used to process $\mathcal{X}_t$. The neuron computes a nonlinear function of the dot product of its inputs and its weights. The output of the neuron is given by

$$y = \frac{1}{1 + \exp\left(\sum_i w_i x_i\right)}, \tag{12.15}$$

where x_i is the ith input to the neuron and w_i is the weight from input i to the neuron.

As mentioned earlier, the criterion used in the NTN is to minimize the number of misclassifications in the set $\mathcal{X}_t$. The standard gradient-descent techniques used to train MLPs minimize the L_2 norm of the errors; this is, however, sensitive to a few statistical outliers, i.e., a few misclassifications. This results in poor solutions if the problem is not linearly separable [13, 14]. The L_1 norm is more robust than the L_2 norm, in that the effect of outliers is diminished. Therefore, minimizing the L_1 norm results in a better partition of a non–linearly separable training set. Minimizing the L_1 norm also results in faster training [15]. In the new algorithm, each neuron is trained by using a gradient-descent technique to minimize the L_1 norm of the errors. The weights of the neuron are updated using a gradient-descent algorithm.

The L_1 error when pattern j is presented to the network is defined by

$$E_j = |t_j - y_j|, \tag{12.16}$$

where t_j is the required output for the neuron when training pattern j is presented and y_j is the actual output of the neuron. The derivative of E_j with respect to the weight w_i from the ith input is given by

$$\frac{\partial E_j}{\partial w_i} = -\operatorname{sgn}(t_j - y_j) f'(\mathbf{w} \cdot \mathbf{x}) x_i, \tag{12.17}$$

where the sgn function gives the sign of its argument and x_i is the ith input to the neuron.

The weights of the neuron are updated only if its output and target do not agree. The update rule for the neuron when pattern j is presented to the network is given by

$$w_k^{n+1} = w_k^n - \eta \frac{\partial E_j}{\partial w_k}, \tag{12.18}$$

where η is a gain term that is taken to lie between 0 and 1 and w_k is the weight from the kth input to the neuron.

Substituting the derivative of the L_1 error from (12.17) into (12.18), we get

$$w_k^{n+1} = w_k^n - \eta y_j (1 - y_j) \operatorname{sgn}(t_j - y_j) x_k. \tag{12.19}$$

The gradient-descent algorithm requires that the error function E_j be differentiable at all points. Note that $\operatorname{sgn}(t_j - y_j)$ is not defined when $t_j = y_j$. However, this does not affect the algorithm, since no updates are made when the output agrees with the target.

After the error has converged, a hyperplane is formed by the above algorithm, which partitions the training set into two subsets. Each subset is assigned to a child node and the algorithm is then used recursively on each child node. If the hyperplane does not split the training data, then the neuron is retrained with the two feature vectors from opposite classes that produce the maximum error in each class. This procedure guarantees that the neuron will split the training set. In practice, it was found that this situation rarely occurs. The training algorithm is summarized in the following pseudocode:

Input An NTN node, t; training examples $\{x_k, y_k\}$; and the number of classes, $N_{C,t}$. x_k is a real-valued feature vector of dimension m. y_k is a binary-valued class label.

Output An NTN with a single neuron at each internal NTN node. The weights for the neurons at these NTN nodes are calculated by the learning algorithm and the leaf nodes are labeled by the appropriate class.

Algorithm

If $N_{C,t} = 1$
Then

1. Label the node by the appropriate class.
2. STOP.

Else

1. Set up a neuron at the NTN node.
2. Use the L_1 norm update rule to train the neuron until the error has converged.

3. **If** the neuron splits the training data
   Then

   1. Grow 2 child nodes.
   2. Use the neuron to split the training data into the two child nodes.
   3. Calculate the number of classes, $N_{C,i}$, for each child node i.
   4. Repeat the algorithm for each child node.

   Else

   1. Find the two training examples of opposite output value that result in maximum error.
   2. Retrain the neuron with these two training examples.
   3. Repeat the same process as above for growing child nodes and repeat the algorithm for each child node.

The L_1-norm algorithm described above is one approach to minimizing the number of classification errors. We now consider another approach called the thermal perceptron algorithm [11]. Recall from Chapter 11 that the perceptron algorithm updates the weights for an incorrectly classified pattern by

$$
\begin{aligned}
\mathbf{w}^{n+1} &= \mathbf{w}^n + \eta\mathbf{x}, && \text{if } \mathbf{x} \in \text{class 1 and } \mathbf{w}\cdot\mathbf{x} \le 0, \\
\mathbf{w}^{n+1} &= \mathbf{w}^n - \eta\mathbf{x}, && \text{if } \mathbf{x} \in \text{class 0 and } \mathbf{w}\cdot\mathbf{x} > 0, \\
\mathbf{w}^{n+1} &= \mathbf{w}^n && \text{otherwise.}
\end{aligned}
\tag{12.20}
$$

If the input pattern is far away from the hyperplane, then the above update rule will probably make too large a weight correction, thereby misclassifying many patterns that were previously correctly classified. The thermal perceptron algorithm decreases the amount of the correction when the distance of the pattern from the hyperplane is large. The thermal perceptron weight update rule is given by

$$
\begin{aligned}
\mathbf{w}^{n+1} &= \mathbf{w}^n + \eta(n)\exp\left(-\frac{|\mathbf{w}^n\cdot\mathbf{x}|}{T(n)}\right)\mathbf{x}, && \text{if } \mathbf{x} \in \text{class 1,} \\
\mathbf{w}^{n+1} &= \mathbf{w}^n - \eta(n)\exp\left(-\frac{|\mathbf{w}^n\cdot\mathbf{x}|}{T(n)}\right)\mathbf{x}, && \text{if } \mathbf{x} \in \text{class 0.}
\end{aligned}
\tag{12.21}
$$

where $\eta(n)$ and $T(n)$ are time-varying step size and temperature functions, respectively. Initially, T is chosen to be large, so that the algorithm behaves like the perceptron algorithm and corrects large and small errors equally. However, the temperature is annealed as the iterations increase so that, as the algorithm proceeds, larger errors are corrected less than smaller errors, the assumption being that when low temperatures are reached, we are close to a solution. This idea is very similar to simulated annealing [16, 17]. A similar idea has also been used for depth restoration from a monocular 2-dimensional image (see Chapter 5). The step size can also be annealed over time.

We now describe how the neural tree network classifies a test feature vector. The feature vector enters the NTN at its root. The neuron at each node is used to route the feature vector to one of the child nodes. Classification is done at the leaf nodes of the NTN. The pseudocode for the classifying algorithm is given below:

Input A feature vector, x_k, and a node t in the NTN.

Output The class label y_k classifying the feature vector.

Algorithm

> **If** the node is a leaf
>
> **Then** Classify the feature vector by the label of the leaf node.
>
> **Else** Use the neuron at the NTN node to route the feature vector to a child node. Repeat the above algorithm for the child node.

12.3.3 Generalization to Multiclass Problems

In the two-class case described in the previous section, a single neuron was used in each internal NTN node. If there are M classes, where $M > 2$, then p neurons are necessary, where $\lceil \log_2 M \rceil \leq p \leq M$. In the case where $\lceil \log_2 M \rceil$ neurons are used, the M classes are represented by M different bit strings. If $\log_2 M$ is an integer, then all possible bit strings of the $\log_2 M$ bits are used as class labels. If $\log_2 M$ is not an integer, then M out of the possible $2^{\lceil \log_2 M \rceil}$ bit combinations are used as class labels. These methods are called distributed encoding schemes. For example, 8 classes are represented by the 3 bit binary numbers from 000 to 111. If there are 9 classes, 4 bits would be required but only 9 of the 16 possible bit strings are used as labels. Another technique is to use M neurons for M classes. In this case, the M binary basis vectors can be used to label the classes. Thus, class i is represented with a 1 in the ith bit and 0 in all other bits. This is called the local encoding scheme. Using different labeling methods results in different ways to grow the NTN. It is argued below that the local encoding scheme is superior to distributed encoding schemes for the NTN.

Labeling schemes are essentially different methods of representing the classes. Suppose that some scheme is agreed upon and p neurons are used to label the M classes. It is required to use these p neurons to partition the training set, $\mathcal{X}$, into subsets so that each subset can be processed by a child node. In the two-class case, a single neuron was used and the training set was partitioned into two subsets corresponding to two child nodes. For the M-class case, two different cases exist—the first when $p < M$ neurons are used and the second when M neurons are used.

If $p < M$ neurons are used, then classification cannot be done using a winner-take-all rule. This is because each neuron may be labeled by a "1" for more than one class. A threshold must, therefore, be set for each neuron. Thus, if the output of a neuron is greater than 0.5, then it is treated as 1 and if the output is less than 0.5, then it is treated as 0. Since the neuron uses a sigmoid activation function, whose value is 0.5 when the weighted sum of its inputs is 0, this thresholding implies that the output of the neuron is 0 or 1 depending on whether the feature vector is on one or the other side of the hyperplane implemented by the neuron. The p hyperplanes form regions in feature space, each corresponding to a different p-bit string. Thus, the thresholded outputs of the p neurons indicate which region of feature space the feature vector belongs to. Some of these regions correspond to an actual class decision (if the p-bit string corresponding to that region is a class label), and other regions may not correspond to any class.

It can be shown, using a recursion given in Muroga [18], that the number of regions formed by p hyperplanes in d-dimensional space is

$$
R_{p,d} = \begin{cases} \displaystyle\sum_{i=0}^{d} \binom{p}{i}, & \text{if } p > d, \\ 2^p, & \text{if } p \leq d. \end{cases} \tag{12.22}
$$

Of these $R_{p,d}$ regions, some may not contain training data. Thus, the training set is partitioned by the p hyperplanes into at most $R_{p,d}$ subsets and each subset is sent to a different child node for processing. For example, if $d = 10$ and the number of classes is 17, then 5 neurons can be used to label all classes, since $17 < 2^5$. The number of regions formed by the 5 neurons in a 10-dimensional feature space is $2^5 = 32$. Thus, 32 child nodes must be grown. This results in a very wide NTN. Simulation results also show that good classification performance was not achieved when using $p < M$ neurons. This point can be understood in the following way.

Suppose that the ith bit in the class labels is 1 for class A and class B and 0 for all other classes. Since the labels are assigned arbitrarily, the case may be such that the training data corresponding to class A and class B are not be close to each other in feature space. In fact, it is possible that class A and B form widely separated clusters. This will obviously result in poor performance, since the similarity in the ith bit of the class labels is not reflected in closeness in feature space. Figure 12.7 shows a 4-class problem in 2-dimensional feature space with two different schemes to label the classes. One labeling scheme groups classes that are clustered together and the other does not. In this case, the first labeling scheme can be implemented by two neurons and the second

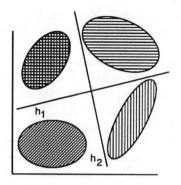

Linearly Separable Labeling	Non-linearly separable Labeling

Linearly Separable Labeling

0 0

0 1

1 1

1 0

Non-linearly separable Labeling

0 0

1 1

0 1

1 0

Figure 12.7 Linearly separable and non–linearly separable labelings of the same problem.

cannot. We call the first a linearly separable labeling, since it can be implemented by an MLP without hidden neurons. The second labeling is called a non–linearly separable labeling, since an MLP without hidden neurons cannot be used to solve this formulation. In general, there is no known optimal method to choose the labels when $p < M$. For these reasons, M neurons are used for the NTN and a local encoding scheme is used to label the M classes.

Suppose that node t of the NTN corresponds to a subregion R_t of the feature space. Let the subset of training data in R_t be denoted by $\mathcal{X}_t$. Also, let the number of classes in $\mathcal{X}_t$ be M_t. Thus M_t neurons are used at node t. Since a local encoding scheme is used to label the classes, each neuron corresponds to a different class. After training, feature vectors are classified according to a winner-take-all:

$$\mathbf{x} \in \text{class } i \qquad \text{iff} \quad f(\mathbf{w}_i \cdot \mathbf{x}) \geq f(\mathbf{w}_j \cdot \mathbf{x}) \quad \forall \, j \neq i, \qquad (12.23)$$

where f is the sigmoid activation function and $\mathbf{w}_i$ is the weight vector of the ith neuron.

The L_1-norm algorithm and the thermal perceptron algorithm described in Section 12.3.2 can be extended to the multiclass case as follows. For the L_1-norm algorithm, each of the M_t neurons is treated separately. For example, the ith neuron separates class i from the rest of the classes. Thus, this neuron is trained by the L_1-norm algorithm, as in the two-class case described in Section 12.3.2.

In the case of the thermal perceptron algorithm, if the feature vector $\mathbf{x}$ belongs to class i, but the network classifies $\mathbf{x}$ into class j, then the multiclass thermal perceptron algorithm updates $\mathbf{w}_i$ and $\mathbf{w}_j$, the weight vectors of the ith and jth neurons, by the following rule:

$$\mathbf{w}_i^{n+1} = \mathbf{w}_i^n + \eta(n) \exp\left(-\frac{|\mathbf{w}_i^n \cdot \mathbf{x}|}{T(n)}\right) \mathbf{x},$$

$$\mathbf{w}_j^{n+1} = \mathbf{w}_j^n - \eta(n) \exp\left(-\frac{|\mathbf{w}_j^n \cdot \mathbf{x}|}{T(n)}\right) \mathbf{x}. \tag{12.24}$$

Note that the thermal perceptron algorithm only makes weight updates for misclassified patterns and, even then, only two weight vectors are updated. The L_1-norm algorithm, on the other hand, updates all the weight vectors for every pattern presented. Thus, if similar performance can be achieved, the thermal perceptron algorithm will result in faster training than the L_1-norm algorithm.

The winner-take-all rule of (12.23) partitions the feature space into M_t regions. The training data in each region is sent to a different child node; thus, there are at most M_t child nodes. These M_t regions are convex, as can be shown by the following argument. Let the region R_t be partitioned by the neurons—using the winner-take-all rule of (12.23)—into M_t regions, $R_{t,i}$, $i = 1, \ldots, M_t$. Suppose that $\mathbf{x}_1$ and $\mathbf{x}_2$ belong to $R_{t,i}$. From (12.23),

$$f(\mathbf{w}_i \cdot \mathbf{x}_1) > f(\mathbf{w}_j \cdot \mathbf{x}_1), \qquad \forall \, j \neq i, \tag{12.25}$$

$$f(\mathbf{w}_i \cdot \mathbf{x}_2) > f(\mathbf{w}_j \cdot \mathbf{x}_2), \qquad \forall \, j \neq i. \tag{12.26}$$

Since the sigmoid function, f, is monotonically increasing, this gives

$$\mathbf{w}_i \cdot \mathbf{x}_1 > \mathbf{w}_j \cdot \mathbf{x}_1, \qquad \forall \, j \neq i, \tag{12.27}$$

$$\mathbf{w}_i \cdot \mathbf{x}_2 > \mathbf{w}_j \cdot \mathbf{x}_2, \qquad \forall \, j \neq i. \tag{12.28}$$

Consider the point $\mathbf{x}$ such that

$$\mathbf{x} = \lambda \mathbf{x}_1 + (1 - \lambda)\mathbf{x}_2, \qquad 0 \leq \lambda \leq 1.$$

From (12.27) and (12.28),

$$\lambda \mathbf{w}_i \cdot \mathbf{x}_1 > \lambda \mathbf{w}_j \cdot \mathbf{x}_1, \qquad \forall \, j \neq i, \tag{12.29}$$

$$(1 - \lambda)\mathbf{w}_i \cdot \mathbf{x}_2 > (1 - \lambda)\mathbf{w}_j \cdot \mathbf{x}_2, \qquad \forall \, j \neq i. \tag{12.30}$$

Adding (12.29) and (12.30), we get

$$\mathbf{w}_i \cdot \mathbf{x} > \mathbf{w}_j \cdot \mathbf{x} \qquad \forall \, j \neq i.$$

From this and the fact that f is monotonically increasing,

$$f(\mathbf{w}_i \cdot \mathbf{x}) > f(\mathbf{w}_j \cdot \mathbf{x}) \qquad \forall \ j \neq i.$$

Thus, if $\mathbf{x}_1$ and $\mathbf{x}_2$ belong to $R_{t,i}$, then $\mathbf{x} = \lambda \mathbf{x}_1 + (1 - \lambda)\mathbf{x}_2$ also belongs to $R_{t,i}$; hence, $R_{t,i}$ is convex.

To summarize, the M_t neurons at each node t of the NTN are trained using the L_1-norm update rule given by (12.18) using the training set $\mathcal{X}_t$. The winner-take-all rule of (12.23) is then used to partition $\mathcal{X}_t$ into M_t subsets. Each subset is assigned to a different child node. The number of classes in each child node is calculated and the algorithm repeated for each child node. The branch-growing algorithm stops if there is only one class in the node t. Figure 12.8 shows an NTN and the corresponding division of feature space for a 3-class pattern-recognition problem. The neural network at the root divides the feature space into 3 regions corresponding to 3 child nodes. One of these regions has only one class in it and, thus, the corresponding child node is a leaf node. The other two regions are further divided into subregions by the neural networks at the child nodes.

We have discussed the NTN architecture and the learning algorithm used to grow an NTN. However, in order to get a good generalization on an independent test set of data, it is necessary to prune the NTN. The pruning algorithm is discussed in the next section.

12.3.4 Pruning

The NTN grown by the learning algorithm described in Section 12.3.2 will perform well on the training data. However, the NTN may be too large for a

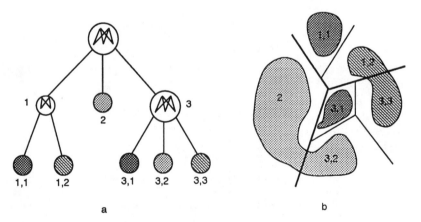

a   b

Figure 12.8 (a) Two-level NTN with neural nets in internal nodes. The leaf nodes are shaded according to their appropriate class. (b) Feature-space regions: the first number refers to the first-level split of the feature space and the second number, to a second-level split. First-level region boundaries are shown by thick lines.

good generalization on an independent test set. Since finding the shortest NTN that correctly classifies all the training data is an NP-complete problem [9], heuristic techniques must be used to grow the NTN. The use of heuristics may result in an NTN that is not optimum. The solution that has been suggested in the case of decision trees is to grow a large tree and then prune the tree [1]. This approach can also be used for the NTN [19]. Thus, the NTN grown by the learning algorithm is pruned to get the best performance on the test set.

Pruning can also be carried out for MLPs. A relation has recently been given relating the size of an MLP, the training set size, and the degree of generalization required [20]. Thus, it is possible to find the appropriate number of neurons for an MLP that will generalize well. However, there is no guarantee that learning algorithms such as backpropagation [21] will find the correct weight values for a given number of neurons. A standard approach to get around this problem is to start with a large network and then remove neurons that have little effect on the classification error [22, 23]. The neurons are ordered according to the amount by which the classification error is decreased when the neuron is removed. The neurons with the smallest effect are removed, or pruned. These methods are not optimal in that, for a given classification error, a smaller network than that found by these algorithms may exist.

In this section, an optimal pruning algorithm for the NTN is discussed. This is an extension of an algorithm given in [1] for binary decision trees. The pruning algorithm generates a sequence of pruned subtrees. Each of these subtrees is the smallest optimally pruned subtree with respect to a Lagrangian cost function. This cost has two terms. The first term is a measure of the number of misclassifications and the second term is a penalty attached to each leaf. The sequence of subtrees is generated by increasing the penalty factor from 0 to ∞. For each value of the penalty factor, a different subtree in the sequence is the optimally pruned subtree. After generating a sequence of subtrees, each subtree is evaluated on an independent test set of data.

The Pruning Algorithm   Let the NTN grown by the learning algorithm given in Section 12.3.2 be denoted by T. The classification performance of T on the training data will be good. However, the training set performance will not give a good measure of the performance of the algorithm on an independent test set. Pruning is a technique that generates a smaller NTN which will have better performance on the test data.

Before proceeding further, some notations and definitions should be established. Figure 12.9 illustrates these definitions. As already mentioned, the NTN grown by the learning algorithm is denoted by T.

Definition 12.1   *A trivial tree consists of just a single node.*

Definition 12.2   *A subtree T' of T has the same root node as T and if $t \in T'$, then $t \in T$. This relation is written as $T' \leq T$. If $T' \neq T$, then $T' < T$.*

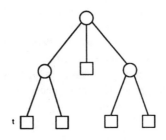

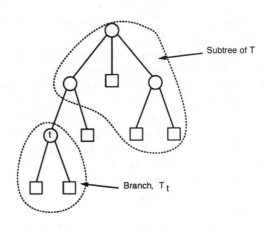

Trivial Tree

Subtree of T

Branch, T$_t$

Pruned subtree, T - T$_t$

Figure 12.9 Tree definitions.

Definition 12.3 *A branch T_t of T is defined as the union of the node t and all the descendants of t.*

Definition 12.4 *Pruning a branch T_t from T consists of deleting from T all the descendants of t. The new tree is now $T - T_t$. The node, t, now becomes a leaf node and is labeled by the class that has maximum membership in X_t, the subset of training data corresponding to t. The pruned tree is called a pruned subtree of T.*

One obvious way to generate the set of pruned subtrees is simply to list all subtrees of T. However, this is infeasible as the number of subtrees is extremely large even for small NTNs. For example, consider a binary NTN, T, with root node t_1. Let the two trees rooted at the child nodes of t_1 be denoted by T_L and T_R. Then the number of subtrees of T is

$$N(T) = 1 + N(T_L)N(T_R). \tag{12.31}$$

Let T_n denote the n-level binary balanced NTN with 2^n leaf nodes. Then $N(T_0) = 1$, $N(T_1) = 1$, $N(T_2) = 5$, $N(T_3) = 26$, and $N(T_4) = 677$. Thus, even for a very small NTN, the number of subtrees is prohibitively large. A procedure is required to generate a sequence of subtrees, decreasing in size, such that each subtree is the "best" subtree in its size. The "best" subtree is one that minimizes some cost function. We may wish to minimize the number of misclassifications on the training set. In this case, we get the following pruning strategy. Start with a T having L leaf nodes. Now, for each H, $1 \le H < L$, find the subtree, T_H, with $L - H$ leaf nodes that minimizes the number of misclassifications on the training set. This algorithm has the drawback that each pruned subtree, T_H, is not necessarily a subtree of T_{H+1}. Thus, some nodes that have been pruned may appear later as the algorithm progresses.

The algorithm given next generates a nested sequence of pruned subtrees such that each pruned subtree can be found from the previous pruned subtree. The pruning algorithm makes use of a Lagrangian cost function given by

$$C_\alpha(T) = D_T + \alpha|\tilde{T}|, \tag{12.32}$$

where D_T is a measure of the number of misclassifications of the NTN T, $|\tilde{T}|$ is the number of leaf nodes of T, and α is a penalty associated with each leaf node. The cost function thus takes into consideration the performance of the NTN and also its complexity, measured by the number of leaf nodes. If $\alpha = 0$, then the best pruned NTN is T. On the other hand, if $\alpha = \infty$, then the best pruned NTN is the root of T denoted by $\{t_1\}$. We now define an optimally pruned subtree with respect to the cost function of (12.32).

Definition 12.5 *A pruned subtree, T_1, of T is an optimally pruned subtree with respect to α if $C_\alpha(T_1) = \min_{T' \le T} C_\alpha(T')$. T_1 is the smallest optimally pruned subtree of T with respect to α if $T_1 \le T'$ for all the optimally pruned subtrees, T', of T. There can be at most one smallest optimally pruned subtree of T. This subtree is denoted $T(\alpha)$.*

The pruning algorithm works by finding a sequence of smallest optimally pruned subtrees of T as α ranges from 0 to ∞. This sequence starts with T and ends with the root node $\{t_1\}$. Even though α takes on a continuum of values, there are only a finite number of subtrees of T. Thus, T remains the smallest optimally pruned subtree for a range of values of $\alpha, 0 \le \alpha < \alpha_1$. At this point a new subtree, $T(\alpha_1)$, becomes the smallest optimally pruned subtree and continues to remain so until $\alpha = \alpha_2$. Now $T(\alpha_2)$ becomes the smallest optimally pruned subtree. This process continues until, at some value of $\alpha = \alpha_L$, the root $\{t_1\}$ remains; $\{t_1\}$ then continues to be the smallest optimally pruned subtree for $\alpha \geq \alpha_L$. Pruning consists of finding the sequence of subtrees T, $T(\alpha_1)$, $T(\alpha_2)$, ..., $T(\alpha_{L-1})$, $\{t_1\}$, for $0 < \alpha_1 < \alpha_2 < \cdots < \alpha_{L-1} < \alpha_L$.

An algorithm is now given and in Section 12.35 it is proved that this algorithm generates the sequence of smallest optimally pruned subtrees, T, $T(\alpha_1)$, $T(\alpha_2)$, ..., $\{t_1\}$.

Consider a node $t \in T$ and the branch T_t. Using the cost function of (12.32), the cost of node t can be written as

$$C_\alpha(t) = D_t + \alpha \tag{12.33}$$

and the cost of the branch T_t is

$$C_\alpha(T_t) = D_{T_t} + \alpha |\tilde{T}_t|. \tag{12.34}$$

As long as $C_\alpha(t) > C_\alpha(T_t)$, the branch T_t is not pruned off. As α is increased, there is a point when $C_\alpha(t) = C_\alpha(T_t)$. At this point, T_t is pruned off, since the resulting NTN, $T - T_t$, is less complex but has the same cost as T. This value of α is given by equating the right-hand sides of (12.33) and (12.34). Solving for α, we get

$$\alpha = g(t, T) = \frac{D_{T_t} - D_t}{1 - |\tilde{T}_t|}. \tag{12.35}$$

The pruning algorithm works by finding this value of α for all the nodes $t \in T - \tilde{T}$ and then pruning off the branch T_{t^*} corresponding to the minimum value of $\alpha_t = \alpha_1$. This process is repeated starting with the NTN, $T_1 = T - T_{t^*}$ and so on until only the root $\{t_1\}$ is left. Thus, the algorithm produces a sequence of NTNs, T, T_1, T_2, ..., $\{t_1\}$. It is shown in the next section that $T_i = T(\alpha_i)$. In other words, T_i is the smallest optimally pruned subtree with respect with α_i.

After generating the sequence of smallest optimally pruned subtrees, an independent test set of data is used to evaluate each pruned subtree. The best subtree is picked and used for classification purposes.

Properties of the Algorithm All trees rooted at the child nodes of the root of T are called the primary branches of T. This is stated in the following definition and illustrated in Figure 12.10.

Definition 12.6 *Let the root of T be denoted t_1. The primary branches of T are the set of all branches rooted at the child nodes of t_1 and are denoted by T_i, $1 \leq i \leq N$, where N is the number of child nodes of t_1.*

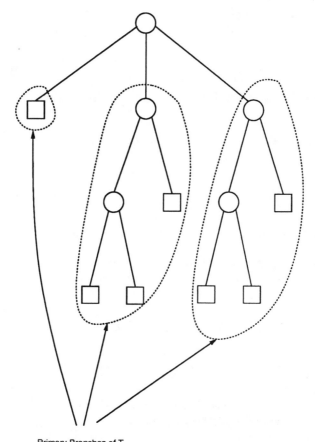

Primary Branches of T

Figure 12.10 Primary branches.

The following theorem states that the sequence of smallest optimally pruned subtrees is a nested sequence, i.e., $T < T(\alpha_1) < \cdots < \{t_1\}$. In fact, the algorithm given in Section 12.3.4 uses this fact to find $T(\alpha_j)$ from $T(\alpha_{j-1})$.

Theorem 12.2 *If* $\alpha_2 \geq \alpha_1$, *then* $T(\alpha_2) \leq T(\alpha_1)$.

Proof The theorem is clearly true if T_1 is a trivial tree, i.e., if $T_1 = T(\alpha_1)$, then $T_1 = T(\alpha_2)$ for $\alpha_2 \geq \alpha_1$. Consider the tree T rooted at t_1 and assume the theorem is true for all primary branches T_i, $1 \leq i \leq N$, where N is the number of child nodes of t_1; i.e.,

$$\text{if } \alpha_2 \geq \alpha_1, \quad \text{then} \quad T_i(\alpha_2) \leq T_i(\alpha_1), \quad 1 \leq i \leq N. \tag{12.36}$$

The cost of T can be written in terms of the costs of T_i as

$$C_\alpha(T) = \sum_{i=1}^{N} C_\alpha(T_i).$$ (12.37)

Let $T_1 = T(\alpha)$. From (12.37) it can be shown by mathematical induction that

$$T_1 = \{t_1\}, \quad \text{if} \quad C_\alpha(t_1) \le \sum_{i=1}^{N} C_\alpha(T_i(\alpha)),$$ (12.38)

$$T_1 = \{t_1\} \cup (\cup_{i=1}^{N} T_i(\alpha)), \quad \text{if} \quad C_\alpha(t_1) > \sum_{i=1}^{N} C_\alpha(T_i(\alpha)).$$ (12.39)

From this and the assumption of (12.36) above, Theorem 12.2 follows by mathematical induction.

This theorem paves the way for the next theorem, which states that the algorithm given in Section 12.3.4 generates the sequence of smallest optimally pruned subtrees.

Theorem 12.3 *The algorithm given in Section 12.3.4 generates the sequence of smallest optimally pruned subtrees $T > T(\alpha_1) > T(\alpha_2) > \cdots > T(\alpha_{L-1}) > \{t_1\}$.*

Proof It is clear that T is the smallest optimally pruned subtree of itself with respect to α, for $0 \le \alpha < \alpha_1$, and that $T_1 = T(\alpha_1)$ is the smallest optimally pruned subtree of T with respect to $\alpha_1 = \min_{t \in T - \tilde{T}} g(t, T)$. Also, T_1 is the smallest optimally pruned subtree of itself with respect to α, for $\alpha_1 \le \alpha < \alpha_2$, and T_2 is the smallest optimally pruned subtree of T_1 with respect to $\alpha_2 = \min_{t \in T_1 - \tilde{T}_1} g(t, T_1)$. We shall show that $\alpha_2 > \alpha_1$, $T(\alpha) = T_1$ for $\alpha_1 \le \alpha < \alpha_2$, and $T(\alpha_2) = T_2$.

From Theorem 10.11 of [1], we know that if $t \in T_1 - \tilde{T}_1$, then

$$g(t, T_1) > g(t, T) \quad \text{if} \quad T_{1_t} < T_t$$

$$g(t, T_1) = g(t, T) \quad \text{if} \quad T_{1_t} = T_t$$

But $\alpha_2 = \min_{t \in T_1 - \tilde{T}_1} g(t, T_1)$ and $\alpha_1 = \min_{t \in T - \tilde{T}} g(t, T)$; thus, $\alpha_2 > \alpha_1$.

If $\alpha_1 \le \alpha < \alpha_2$, then, from Theorem 12.2, $T(\alpha) \le T(\alpha_1) = T_1 < T$. From this and the fact that the relation $\le$ is transitive, we get $T(\alpha) = T_1(\alpha) = T_1$. Again from Theorem 12.2, $T(\alpha_2) \le T(\alpha_1) = T_1 < T$ and, from the transitivity of $\le$, we get $T(\alpha_2) = T_1(\alpha_2) = T_2$.

If T_2 is trivial, then $T(\alpha) = T_2$ for $\alpha \ge \alpha_2$; otherwise, the process can be repeated until only the root $\{t_1\}$ remains. This proves Theorem 12.3.

An Example In this section, an example of the working of the pruning algorithm is given. The speaker-independent vowel-recognition problem described in the next section is used in this example.

The NTN learning algorithm was first used to grow the NTN. The pruning algorithm was then used to generate a sequence of smallest optimally pruned subtrees. Finally, the test set was used to find the classification error for each of the pruned subtrees. In the graph of Figure 12.11, we show a plot of the classification success percentage for the test set for each of the pruned subtrees. The success rate is plotted against the number of neurons used in the pruned subtrees. Since the number of neurons decreases while pruning, the X-axis of the graph should be read from right to left. In this particular run of the algorithm there are 30 optimally pruned subtrees. The graph shows the success rate for the first 27 optimally pruned subtrees. As can be seen from the graph, the classification success rate increases to a maximum and then decreases again as

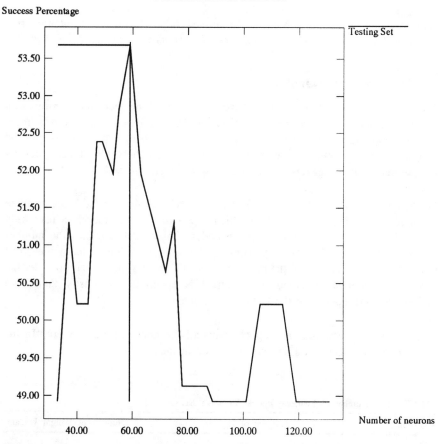

Figure 12.11 Classification error for a sequence of optimally pruned subtrees. X axis: Number of neurons, pruning proceeds from right to left; Y axis: classification success percentage.

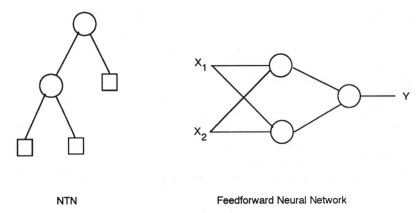

NTN   Feedforward Neural Network

Figure 12.12 NTN and MLP for XOR.

pruning is continued. In this case, the pruned subtree with the minimum error on the testing set is $T(\alpha_{17})$. The number of neurons used in this NTN was 59.

12.3.5 Numerical Results

The performance of the NTN is first demonstrated on two boolean-function learning problems. These are the exclusive OR (XOR) problem and a four-variable non–linearly separable function given by $(a \vee b) \oplus (c \wedge d)$. A "mesh" problem with real-valued features is then used to test the NTN. Finally, results are presented on a speaker-independent vowel-recognition task.

Boolean Functions Figure 12.12 shows the NTN grown for the XOR problem and also the MLP that solves the problem. As can be seen, the NTN uses only two nodes, while the MLP needs three nodes. Table 12.1 tabulates the number of epochs taken by the NTN algorithm and backpropagation to learn the XOR problem. One epoch consists of presenting all the training data once to the network. It should be noted that each NTN epoch is less costly than each backpropagation epoch for the following reasons:

1. In backpropagation, each epoch trains all the nodes in the network. In the NTN, each epoch trains only one node in the network.

TABLE 12.1 Number of Learning Epochs to Solve XOR.

Algorithm	Number of Epochs	Number of Weight Updates
NTN	30	315
Backpropagation	256	9216

2. In the NTN, as we traverse the tree from the root to the leaf nodes, the number of training patterns becomes less; hence, each epoch of training has fewer training patterns presented to the network.

For the above reasons, we also show the number of weight updates for both algorithms. From the table, it is clear that the NTN is about 30 times faster than backpropagation.

Figure 12.13 shows the decision regions formed for the XOR problem by the NTN and Figure 12.14 shows the decision regions formed by an MLP with two hidden neurons trained using the backpropagation algorithm. The vertices of the square in the figures represent the four training data. As can be seen, the decision regions for both algorithms are formed by two lines. These lines are implemented by the two hidden neurons in the MLP and are learned simultaneously. However, in the NTN, each line is learned separately. The root node of the NTN implements one of the lines and the other line is implemented by one of its child nodes. From the figures, it can also be seen that the NTN places the lines more symmetrically than the MLP.

The four-variable boolean function given by $(a \lor b) \oplus (c \land d)$ is considered next. The truth table for this function is given in Table 12.14. For this problem, we compare the performance of the NTN with backpropagation and, also, two decision-tree approaches. The first, ID3 [2], is a popular decision-tree method used by the artificial intelligence community. ID3 splits the training data at each internal node on the basis of the value of a feature variable selected by an exhaustive search over all variables. The second, perceptron trees [24], uses ID3 nodes for the internal tree nodes and perceptrons at the leaf nodes. In this

Figure 12.13 NTN decision regions for XOR.

Figure 12.14 MLP decision regions for XOR.

method, the perceptron algorithm is used to train a neuron at each tree node. If the perceptron algorithm converges to a solution, then the node becomes a leaf node. If the algorithm does not converge after a predetermined number of iterations, the problem is assumed to be non–linearly separable and the ID3 algorithm is used. Since the NTN uses a neural net at all internal nodes, its performance is better than these methods.

TABLE 12.2 Truth Table for $(a \vee b) \oplus (c \wedge d)$.

a	b	c	d	$(a \vee b) \oplus (c \wedge d)$
0	0	0	0	0
0	0	0	1	0
0	0	1	0	0
0	0	1	1	1
0	1	0	0	1
0	1	0	1	1
0	1	1	0	1
0	1	1	1	0
1	0	0	0	1
1	0	0	1	1
1	0	1	0	1
1	0	1	1	0
1	1	0	0	1
1	1	0	1	1
1	1	1	0	1
1	1	1	1	0

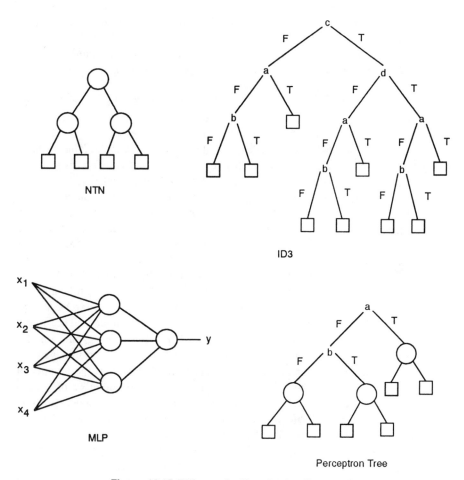

Figure 12.15 Different classifiers for $(a \lor b) \oplus (c \land d)$.

Figure 12.15 shows the MLP that solves the problem given in Table 12.14 and also the trees which were grown using the NTN, ID3, and perceptron trees. The NTN requires only 3 internal nodes, as compared to 8 nodes for ID3 and 5 nodes for the perceptron trees. The MLP required 3 hidden neurons for this problem.

Table 12.3 compares the NTN learning algorithm with backpropagation in terms of the number of epochs and the number of weight updates needed to

TABLE 12.3 Number of Learning Epochs to Solve $(a \lor b) \oplus (c \land d)$.

Algorithm	Number of Epochs	Number of Weight Updates
NTN	77	4350
Backpropagation	715	217360

learn the problem. As can be seen from the table, the new algorithm trained about 50 times faster than backpropagation for this problem. Recall that, for the XOR problem, the NTN trained 30 times faster than backpropagation. It has been found that the improvement in training over backpropagation increases as the complexity of the problem increases.

It is interesting to note that an NTN can be mapped onto an mLP in the following way. Consider the two-class case corresponding to a binary NTN structure. The leaf nodes are labeled "1" and "0" corresponding to the two classes. Each internal NTN node corresponds to a boolean variable. This variable corresponds to the hyperplane test which is evaluated at that node and is a "1" or a "0," depending on which child node the feature vector is routed to. An NTN essentially implements a disjunct of conjuncts of these boolean variables, where each path from the root of the NTN to a leaf node labeled by "1" corresponds to a conjunct. Thus, the NTN implements the disjunct of the paths from the root to the leaves that are labeled "1." If this boolean function (the disjunct of conjuncts implemented by the NTN) is linearly separable, then the NTN can be mapped onto an MLP with one hidden layer in the following way. Each NTN node corresponds to a hidden neuron in the MLP and the output node of the MLP implements the boolean function (which is linearly separable). The weights of the hidden neurons are initialized by the NTN nodes. The output node weights are initialized to small random numbers. The training time for this neural network is greatly decreased due to the fact that the hidden unit weights have been set correctly by using the NTN. This method also gives the number of hidden nodes required in the MLP as opposed to finding the number of hidden nodes by trial and error. Table 12.4 shows the average number of epochs to train a neural network for the XOR problem and the boolean problem, $(a \lor b) \oplus (c \land d)$, using different initializations. The first row shows the number of epochs when all the weights were randomly initialized and the second row shows the number of epochs when the hidden unit weights were initialized using the NTN. We must also include the number of epochs which the NTN took to train. The table takes this into consideration. Again, note that each NTN epoch is less costly than each neural net epoch, so the increase in learning rate is actually better than the table indicates.

Mesh Problem   The problems described in the previous section were boolean function learning tasks. In this section, a more complicated problem with real-valued features is presented. The problem is to separate the two classes shown in Figure 12.16. Clearly, this is a non–linearly separable problem and, thus, hidden neurons will be necessary in the case of MLPs.

TABLE 12.4   Effect of Different Initializations.

Initialization Type	XOR	$(a \lor b) \oplus (c \land d)$
Random initialization	256	715
Initialization using NTN weights	79	91

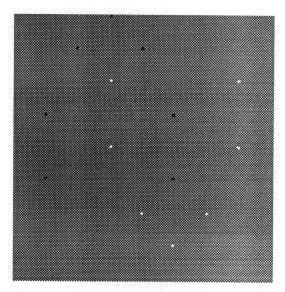

Figure 12.16 "Mesh" problem.

Figure 12.17 NTN decision regions for "mesh."

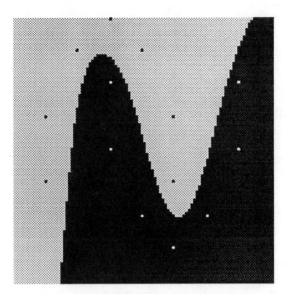

Figure 12.18 MLP decision regions for "mesh."

Figure 12.17 shows the decision regions formed by the NTN for this problem; the decision regions formed by the MLP are shown in Figure 12.18. From these figures, it can be seen that the decision boundaries for the MLP are smoother than those of the NTN. This is because the NTN sequentially uses hyperplane tests to classify a feature vector while the MLP's hidden neurons allow for a nonlinear decision boundary. The NTN and MLP architectures for this problem are shown in Figure 12.19. As can be seen, the MLP could solve the problem with 5 hidden neurons and the NTN also used 5 internal nodes. However, the training times were very different. The number of epochs and weight updates for the NTN and MLP are shown in Table 12.5. In this case, the NTN learns around 20 times faster than backpropagation.

The decision boundaries in the NTN are piecewise linear. However, it is possible to have higher-order boundaries by including higher-order terms in the input vector. By including all second-order terms, the input vector (x_1, x_2) can be transformed to the vector $(x_1^2, x_2^2, x_1 x_2, x_1, x_2)$. Figure 12.20 shows the decision regions corresponding to an NTN solution with all second-order terms included in the input. Thus, the NTN allows for higher-order nodes. This is an area for future research.

Speaker-Independent Vowel Recognition Speech recognition is a problem that has received considerable attention in the past [25, 26]. There has recently been much work on the use of neural networks for speech recognition [27–30]. Comparative studies of neural networks and decision trees for speech recognition have also been performed [31, 32].

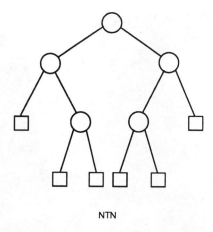

NTN

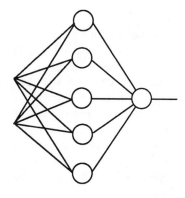

MLP

Figure 12.19 NTN and MLP architectures for "mesh."

TABLE 12.5 Number of learning epochs to solve "mesh."

Algorithm	Number of epochs	Number of weight updates
NTN	750	15300
Backpropagation	953	280182

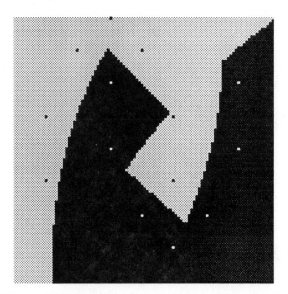

Figure 12.20 Higher-order NTN decision regions for "mesh."

In this section, we present simulation results on a speaker-independent vowel-recognition task. The training set consists of 528 feature vectors representing 11 vowel sounds obtained from 4 male speakers and 4 female speakers. The problem is to train the system on this data and then test it on an independent set of 462 feature vectors obtained from a different set of 4 male speakers and 3 female speakers. For both the training and testing set, each vowel is repeated 6 times by each speaker. The features are 10 log area parameters calculated using six 51-sample Hamming-windowed segments from the steady part of the vowel. Details of this data set can be obtained from [33]. This particular database represents a difficult problem and the best performance of any classifier was about 55%.

Experiments have been performed on this data set using various MLP neural networks with different numbers of hidden nodes and different types of nodes, such as radial basis function nodes and Gaussian nodes [33]. The performance of CART, one of the most popular decision-tree methods, on this data set was reported in [31]. These results are summarized in Table 12.6. The table also shows the performance of the best pruned NTN on this data set. These results have been averaged over 20 runs of the NTN algorithm.

We see that the NTN's performance is much better than CART's, but is similar in classification accuracy to MLP's with radial basis function (RBF) nodes, Gaussian nodes (GN), or square nodes (SN). As shown in the table, if the number of hidden nodes is not chosen correctly, MLP's performance is quite poor. The NTN algorithm, however, grows the correct number of neurons while training.

TABLE 12.6 Results on Speaker-Independent Vowel Recognition.

Classifier	Number of Hidden Units	Percent Correct
Single-layer perceptron	—	33
MLP	88	51
MLP	22	45
MLP	11	44
RBF	528	53
RBF	88	48
GN	528	55
GN	88	53
GN	22	54
GN	11	47
SN	88	55
SN	22	51
SN	11	50
Nearest Neighbor	—	56
CART	—	44
NTN	59[a]	54

[a]The neurons in the NTN are not hidden neurons. Rather, this number refers to the total number of neurons used in the NTN.

Table 12.6 also shows the number of neurons used in the various methods. The number of neurons used in the NTN is less than that of an MLP with similar performance. The average depth of the NTN was only 3.1, which is quite a shallow tree. The average retrieval time for a test pattern is thus about 3.1 time units. An MLP with one hidden layer implemented in parallel requires two time units—one to fire the hidden layer nodes and one to fire the output nodes. The NTN is, therefore, only very slightly slower than the MLP for retrieval. However, as described in Section 12.3.1, only one processor needs to be implemented for each clock cycle of the NTN [5], whereas all the processors must be implemented in the case of the MLP. Thus, the NTN provides an attractive trade-off of speed for hardware implementation, as explained in Section 12.3.1.

We also tested the performance of the NTN trained by the thermal perceptron algorithm. As mentioned before, the step size, $\eta(n)$, and the temperature, $T(n)$, decrease with time. This corresponds to the annealing schedule of simulated annealing [16, 17]. In this section, a Gaussian annealing function is used for the step size and the temperature. Thus, the step size and temperature are given by

$$\eta(n) = \eta_0 \exp\left(-\frac{an^2}{n_{\max}^2}\right),$$

$$T(n) = T_0 \exp\left(-\frac{an^2}{n_{\max}^2}\right), \qquad (12.40)$$

where η_0 is the initial step size, T_0 is the initial temperature, a is a constant that controls the rate at which η and T decrease, and $n_{\max}$ is the maximum number

TABLE 12.7   Speaker Independent Vowel Recognition Using Thermal Perceptron for the NTN.

anneal η?	yes	no	yes	no
anneal T?	yes	yes	no	no
percent correct	52	52	48	48

of epochs of training. It is possible to anneal just the temperature or the step size, or both the temperature and the step size together.

The thermal perceptron algorithm was used to grow the NTN for the speaker-independent vowel-recognition problem. Many different annealing schemes were tried, both annealing the temperature and step size independently and annealing them together. Table 12.7 shows the classification performance of the NTN for different annealing schemes. For all the experiments, we used $\eta_0 = 0.1$ and $T_0 = 2.0$. The values of a and n_{max} were 7 and 1500, respectively. The results are averaged over 20 runs of the algorithm. From Table 12.7 it can be seen that the best performance is 52%. This is slightly worse than the L_1-norm NTN algorithm, which had a 54% classification accuracy (see Section 12.20). However, it is possible that a better choice of initial values for the step size and the temperature or a better annealing schedule may improve the performance of the thermal perceptron NTN algorithm.

It was also observed that the thermal perceptron NTN did not need much pruning. In most runs of the algorithm, the best performance was achieved by the NTN grown by the learning algorithm with no pruning. The thermal perceptron algorithm also grows the NTN faster than the L_1-norm algorithm. This is because at each internal node of the NTN, the thermal perceptron only updates two neurons for every misclassified pattern. However, the L_1-norm algorithm updates all the neurons for every pattern presented.

12.4   SUMMARY AND CONCLUSIONS

The NTN classifier described in this chapter combines the concepts of neural networks and decision trees in an effective way. The NTN uses a decision-tree structure with each internal node being replaced by an MLP with no hidden neurons. The neural network at the internal NTN nodes provides a natural way to implement hyperplanes that are not restricted as in the case of decision trees. A gradient-descent learning rule that is more efficient than the exhaustive search techniques of decision trees is used. The NTN uses a minimum misclassification technique based on robust surface fitting to train the neural network at each internal tree node. The NTN learning algorithm grows the network during training, unlike MLPs. Thus, the designer need not specify the network configuration for the NTN. The NTN can also be implemented with a single processor and a look-up table. This provides an attractive trade-off of classification speed against hardware implementation cost as compared to MLPs. The NTN can be pruned using an optimal pruning algorithm. Pruning will provide the NTN with better performance on an independent test set. Simulation results have shown that the NTN learns faster than MLPs. The NTN's superior classification performance is superior to that of decision trees on the speaker-independent vowel-recognition task. The classification performance of the NTN was found to be similar to MLPs. However, the NTN has other advantages over MLPs, as discussed above. Thus, the NTN provides an attractive hybrid approach to pattern classification.

There are many avenues for future work. The NTN can be applied to a variety of problems, such as word and character recognition. The NTN can be used as a training system for intelligent machines such as LESTRADE [34].

The NTN can also be modified to apply it to signal and image restoration. For this problem, the blurred image and the original image are used to train the NTN to learn the inverse of the blurring operation. The gray-level value of each pixel is represented as an m-bit binary number. An MLP is used at each internal NTN node to learn the bit corresponding to the level of the NTN node, the most significant bit being learned by the root node and the least significant bit being learned by the leaf nodes. Preliminary results on this problem have been encouraging [35].

Another modification to the NTN is to use more powerful neurons at each internal node. An example of this idea using polynomial inputs was given in Section 12.4.

The NTN can also be used to sequentially form subgroups of the training data. For example, in phoneme recognition, the root node of the NTN can be used to distinguish between consonant and vowel sounds. Further subgroups could be formed at each child node.

Finally, different learning rules can be used to grow the NTN. A possible approach would be to define a Lagrangian cost function, such as that given in (12.32). This cost is a sum of a misclassification measure and the complexity of the NTN. The NTN can then be grown by minimizing this cost function. This idea is similar to the entropy-constrained vector quantizer (ECVQ) used in source coding [36]. The classification problem can also be viewed as one of entropy minimization, where the entropy of the problem measures the class homogeneity in the training set. This entropy should be minimized. An area which should be looked into is the derivation of an appropriate entropy measure that can be expressed in terms of the parameters of the classifier. This would allow for a gradient-descent technique to minimize the entropy.

ACKNOWLEDGMENTS

This research was made possible through the support of the New Jersey Commission on Science and Technology and the Center for Computer Aids for Industrial Productivity at Rutgers University.

REFERENCES

1. L. Breiman, J. H. Friedman, R. A. Olshen, and C. J. Stone. *Classification and Regression Trees*. Belmont, Calif.: Wadsworth International group, 1984.

2. J. R. Quinlan. Induction of Decision Trees. *Machine Learning*, 1:81–106, 1986.

3. R. O. Duda and P. E. Hart. *Pattern Classification and Scene Analysis*. New York: Wiley, 1973.

4. P. A. Chou, T. Lookabaugh, and R. M. Gray. Optimal Pruning with Applications to Tree-Structured Source Coding and Modeling. *IEEE Trans. Inform. Theory*, IT-35:2999–315, 1989.

5. A. Sankar and R. J. Mammone. Neural Tree Networks. In R. J. Mammone and Y. Y. Zeevi, editors, *Neural Networks: Theory and Applications*. San Diego, Calif.: Academic Press, 1991.

6. A. Sankar and R. Mammone. Speaker Independent Vowel Recognition using Neural Tree Networks. In *Proc. IJCNN*, July 1991.

7. E. B. Baum. On the capabilities of multilayer perceptrons. *J. Complexity*, 4(3):193–215, 1988.

8. T. M. Cover. Geometrical and statistical properties of systems of linear inequalities with applications in pattern recognition. *IEEE Trans. Electron. Comput.*, 1965.

9. L. Hyafil and R. L. Rivest. Constructing Optimal Decision Trees is NP-Complete. *Inform. Process. Lett.*, 5(1):15–17, 1976.

10. S. Amari. A Theory of Adaptive Pattern Classifiers. *IEEE Trans. Elect. Comput.*, EC-16(3):299–307, 1967.

11. M. Frean. *Small Nets and Short Paths: Optimising Neural Computation*. PhD thesis, University of Edinburgh, 1990.

12. A. Sankar. *Neural Tree Networks: A Hybrid Approach to Pattern Classification*. PhD thesis, Rutgers—The State University of New Jersey, 1991.

13. A. Sankar and R. J. Mammone. A Fast Learning Algorithm for Tree Neural Networks. In *Proceedings of the 1990 Conference on Information Sciences and Systems*, pp. 638–642, March 1990.

14. A. Sankar and R. J. Mammone. Tree Structured Neural Networks. Technical Report TR-122, Rutgers University, 1990.

15. A. Sankar and R. J. Mammone. A New Fast Learning Algorithm for Feedforward Neural Networks using the L1 Norm of the Error. Technical Report TR-115, Rutgers University, 1990.

16. S. Kirkpatrick, C. D. Gelatt, Jr., and M. P. Vecchi. Optimization by simulated annealing. *Science*, 220(4598), 1983.

17. S. Kirkpatrick. Optimization by simulated annealing: Quantitative studies. *J. Stat. Phys.*, 34(5&6):975–986, 1984.

18. S. Muroga. *Threshold Logic and Its Applications*, pp. 271–272. New York: Wiley, 1971.

19. A. Sankar and R. Mammone. Optimal Pruning of Neural Tree Networks for Improved Generalization. In *Proc. IJCNN*, July 1991.

20. E. B. Baum. What Size Net Gives Valid Generalization? *Neural Comput.*, 1(1):151–160, 1989.

21. D. E. Rumelhart and J. L. McClelland. *Parallel Distributed Processing*. Cambridge, Mass.: MIT Press, 1986.

22. M. C. Mozer and P. Smolensky. Skeletonization: A Technique for Trimming the Fat from a Network via Relevance Assessment. In *Advances in Neural Information Processing I*, D. S. Touretzky, ed., pp. 107–115. San Mateo, Calif.: Morgan-Kaufmann, 1989.

23. E. D. Karnin. A Simple Procedure for Pruning Back-Propagation Trained Neural Networks. *IEEE Trans. Neural Networks*, 1(2):239–242, 1990.

24. P. E. Utgoff. Perceptron Trees: A Case Study in Hybrid Concept Representation. In *Proceedings of the Seventh National Conference on Artificial Intelligence*, St. Paul, MN, 1988. Morgan-Kaufman.

25. J. L. Flanagan. *Speech Analysis Synthesis and Perception*. New York: Springer-Verlag, 1972.

26. L. R. Rabiner and R. W. Schafer. *Digital Processing of Speech Signals*. Englewood Cliffs, N.J.: Prentice Hall, 1978.

27. A. Waibel. Modular Construction of Time Delay Neural Networks for Speech Recognition. *Neural Comput.*, 1, March 1989.

28. K. J. Lang, A. H. Waibel, and G. E. Hinton. A Time Delay Neural Network Architecture for Isolated Word Recognition. *Neural Networks*, 3(1):23–43, 1990.

29. R. P. Lippmann. Review of Neural Networks for Speech Recognition. *Neural Comput.*, 1(1):1–38, 1989.

30. K. K. Paliwal. Neural Net Classifiers For Robust Speech Recognition Under Noisy Environments. In *Proceedings of the IEEE International Conference on Acoustics, Speech, and Signal Processing*, 1990.

31. Ah Chung Tsoi and R. A. Pearson. Comparison of Three Classification Techniques, CART, C4.5, and Multi-Layer Perceptrons. Proceedings of the 1990 IEEE Conference on Neural Information Processing Systems, Denver.

32. L. Atlas, R. Cole, Y. Muthusamy, A. Lippman, J. Connor, D. Park, M. El-Sharkawi, and R. Marks II. A Performance Comparison of Trained Multi-Layer Perceptrons and Trained Classification Trees. *Proc. IEEE*, 78(10), 1990.

33. A. J. Robinson. *Dynamic Error Propagation Networks*. PhD thesis, Cambridge University Engineering Department, 1989.

34. H. Freeman, G. Hung, R. Mammone, and J. Wilder. Project LESTRADE: The Design of a Trainable Machine Vision Inspection System. In *Machine Vision for Three Dimensional Scenes*, H. Freeman, ed., pp. 219–242. San Diego, Calif.: Academic Press, 1990.

35. A. C. Surendran. Personal Communication.

36. P. A. Chou, T. Lookabaugh, and R. M. Gray. Entropy-Constrained Vector Quantization. *IEEE Trans. Acoust. Speech Signal Process.*, 37(1):31–42, 1989.

13

THE RECOGNITION OF MULTICOMPONENT SIGNALS

Adam B. Fineberg

CAIP Center
Rutgers University
Piscataway, New Jersey 08855–1390
email: fineberg@caip.rutgers.edu

13.1 INTRODUCTION

When studying the time-frequency distributions of many naturally occurring signals, such as speech and music, it is often visually evident that the signal can be decomposed into two or more structural components. These components are readily identified by human inspection as being areas of interest. There is, however, much controversy surrounding the development of an algorithmic identification procedure because there are an infinite number of ways to represent a signal $s(t)$ as a sum of components of various kinds, say $a(t)$ and $b(t)$, such as

$$S(t) = a(t) + b(t), \qquad (13.1)$$

where

$$a(t) = 0, \qquad (13.2)$$
$$b(t) = S(t).$$

Another representation may be given by

$$a(t) = S(t)/2, \qquad (13.3)$$
$$b(t) = S(t)/2.$$

Therefore some definition must be constructed that, when applied to a component of a multicomponent signal, describes the essence of what is seen when a human inspector views the spectrum of the signal. Additionally, a method of finding a unique representation of these components must be determined.

A multicomponent signal representation should have a similar structure for all of its components. This redundancy of structure allows for the elimination of any spectral distortion of one component by calibration against the others. If this structure can be extracted from naturally occurring signals such as speech and music, or generated in man-made signals such as communications, radar, and sonar signals, then recognition and classification algorithms can be designed to take advantage of this redundancy and thereby improve performance.

13.2 MULTICOMPONENT SIGNALS

The essence of a multicomponent signal can be found in the signal spectrum. The components which are visualized by human inspection form groups which can be segmented from the rest of the spectrum. These groups are loosely defined as being spectrally compact with respect to the extent of the entire signal spectrum. This local compactness may be witnessed when the dominant frequency (in some sense) of each group is separated by more than the sum of the apparent local half-bandwidth of the groups, which corresponds to the condition that

$$\omega_2(t) - \omega_1(t) > \frac{bw_1(t)}{2} + \frac{bw_2(t)}{2}, \tag{13.4}$$

where $\omega_i(t)$ represents the local dominant frequency of the ith component and $bw_i(t)$ represents the respective local bandwidth of the ith component. Since these are all functions of time, a signal may be considered to be multicomponent at some time, and not at others.

This definition is similar to that given by Cohen and Lee [1] when the local dominant frequency is the mean or instantaneous frequency and the respective local bandwidth is given by the variance about the mean. For a signal that can be represented by

$$S(t) = A(t)e^{j\varphi(t)}, \tag{13.5}$$

the instantaneous frequency, $\omega_i(t)$, is the average value of all frequencies which exist in the signal at a particular instant in time and is given by the derivative of the phase to be

$$\omega_i(t) = \frac{d\varphi(t)}{dt}. \tag{13.6}$$

The global bandwidth is defined as the variance about the instantaneous frequency and is given by [1, 2]

$$\sigma^2 = \int \left(\frac{A'(t)}{A(t)}\right)^2 A^2(t)dt + \int (\varphi'(t) - \langle\omega\rangle)^2 A^2(t)\,dt, \tag{13.7}$$

where the prime denotes differentiation, and $\langle\cdot\rangle$ is the expected value operator. It is therefore clear that the bandwidth is determined by both the amplitude and frequency modulation functions of the signal. This gives an intuitive definition that will be shown to be very useful for multicomponent signal analysis.

Analysis of the individual components of a multicomponent signal will require that they be extracted from the composite signal. This is accomplished by adaptively separating each component mode from the composite spectrum. This filtering operation must be performed adaptively in order to accurately track the instantaneous frequency of the component and the instantaneous bandwidth variations. The filters should be constructed with smooth transition regions to minimize the spectral distortion imposed on each component. A new technique for very fast, computationally efficient frequency tracking has been developed [3] that allows the component locations to be tracked sample by sample. This technique is described in Chapter 8. The bandwidth adaptation can then be performed by determining the bandwidth bounds that maximize the ratio between the energy captured by the adaptive filter and the total energy.

For analysis of signals from an ideal time-frequency distribution, this definition of multicomponent signals seems appropriate. It is interesting to note that several naturally occurring signals are of this form, such as human speech signals [4], animal calls [5], and music [6]. It is suggested that this form is desirable in nature since it includes inherent redundancy in the signal which can be favorably used for recognition in the presence of unknown distortions in the transmission medium.

13.3 COMPONENT REPRESENTATION

As discussed above, the multicomponent representation of the signal, and therefore of the component, is not unique. If, however, we decide that a signal meets the criteria established for a multicomponent signal, then we need a representation of each component that can adequately describe the time-varying nature of the component. Since each component will be spectrally compact, some form of adaptive filtering can be used to extract each component from the composite signal. Each component $s_i(t)$ will then be described by

$$s_i(t) = A_i(t) \cos(\omega_i t + \phi_i(t)), \tag{13.8}$$

where ω_i is the mean frequency of the ith component; $A_i(t)$ is the ith component's envelope or AM modulation function; and $\phi_i(t)$ is the ith component's instantaneous phase or FM modulation function. Therefore, in order to completely describe the character of the signal, it is necessary to know not only the mean frequencies of each component, but also the local characteristics which modulate the components.

One method of finding a representation based on the modulation functions $A_i(t)$ and $\phi_i(t)$ can be determined from the analytic representation given by

$$S_i(t) = s_i(t) + j\hat{s}_i(t), \tag{13.9}$$

where $\hat{s}_i(t)$ is the Hilbert transform of $s_i(t)$ [7], whose frequency components are in phase quadrature with those of $s_i(t)$. The analytic signal $S_i(t)$ for each component i is given by

$$S_i(t) = A_i(t)e^{j(\omega_i t + \phi_i(t))}. \tag{13.10}$$

This complex representation yields the modulation functions:

$$A_i(t) = \sqrt{\left(\Re(S_i(t)^2) + \Im(S_i(t)^2)\right)}, \tag{13.11}$$

$$\phi_i(t) = \arctan\left(\frac{\Im(S_i(t))}{\Re(S_i(t))}\right), \tag{13.12}$$

where $\Re$ represents the real portion and $\Im$, the imaginary portion of the complex signal.

Other representations are possible, since the representation given by (13.8) is not unique. Another possibility that may be appropriate is

$$\text{minimize} \quad \lambda_1 \int_{-\infty}^{\infty} \omega^2 \left[\int_{-\infty}^{\infty} A_i(t)e^{-j\omega t}\, dt\right] d\omega$$

$$+ \lambda_2 \int_{-\infty}^{\infty} \omega^2 \left[\int_{-\infty}^{\infty} \phi_i(t)e^{-j\omega t}\, dt\right] d\omega \tag{13.13}$$

$$\text{subject to} \quad s_i(t) = A_i(t)\cos(\omega_i t + \phi_i(t)),$$

which attempts to find the representation for which the combined bandwidth of $A_i(t)$ and $\phi_i(t)$ is minimum. This criterion seems appropriate to a multicomponent definition that requires a compactness of each component.

Since this representation is presented as an optimization problem with Lagrange multipliers, a simple example can be shown if $\lambda_1 = 0$. In this case,

$$\phi_i(t) = 0 \quad \text{and} \quad A_i(t) = \frac{s_i(t)}{\cos(\omega_i t)}. \tag{13.14}$$

If $\lambda_2 = 0$, then

$$A_i(t) = 1 \qquad \text{and} \qquad \phi_i(t) = \arccos(s_i(t)) - \omega_i t. \qquad (13.15)$$

This example shows one method of solving for the modulation functions and also illustrates that the criteria for an optimum decomposition must be carefully chosen. Several optimality constraints and methods for calculating the modulating functions can be developed for specific tasks; however, it will be shown that the analytic signal representation is a useful one.

13.4 COMPONENT MATCHING

Many distortions experienced by signals traveling through an uncontrollable medium are manifested as modulations of the underlying signal. It is easily shown [8] that phase and frequency modulation are related and indistinguishable from the waveform or from an analytical expression for the waveform. It can also be shown that time modulation produces the same effect on the resulting waveform as the substitution of variables given by

$$t' = t + \frac{1}{\omega}\beta(t) \qquad (13.16)$$

in a representation of a signal, $f(t) = A\cos(\omega t)$. This yields

$$f(t') = A\cos(\omega t') = A\cos(\omega t + \beta(t)). \qquad (13.17)$$

Since these types of distortion are not component selective when applied to a multicomponent signal, the modulation from the distortion will be imposed on each component. Therefore, if the modulation function corresponding to the first component can be estimated, it would have the same value as the estimates from each other component in the signal. This consistency across the components will be exploited as a means of classifying the underlying signal.

13.4.1 Relative Modulation Estimation

It is usually impossible to estimate the absolute modulation functions without any *a priori* knowledge of the undistorted signal. However, if the problem is to classify a distorted signal as being a modulated version of one signal of a class of signals, then only the relative modulation functions need be estimated. This can be seen from the following example, which shows the formulation for the analytic signal representations $S_{\text{ref}}(t)$ and $S_{\text{test}}(t)$ for some known multicomponent reference and test signals, $s_{\text{ref}}(t)$ and $s_{\text{test}}(t)$, respectively. The test signal is a distorted version of the reference signal, where t was replaced by $t + \cos \Omega t$.

The known signals are therefore given by

$$S_{\text{ref}}(t) = \sum_{i=1}^{n} A_i \cos(\omega_i t + \phi) \tag{13.18}$$

and

$$S_{\text{test}}(t) = \sum_{i=1}^{n} A_i \cos(\omega_i(t + \cos \Omega t) + \phi), \tag{13.19}$$

where n represents the number of components in the signal, ω_i is the dominant frequency of the component, ϕ is the unknown initial phase offset, and Ω determines the bandwidth.

By taking the analytic signal representation of each signal, the relative modulation functions for each component can be estimated:

$$S_{\text{ref}}(t) = \sum_{i=1}^{n} A_i e^{j(\omega_i t + \phi)} \tag{13.20}$$

and

$$S_{\text{test}}(t) = \sum_{i=1}^{n} A_i e^{j(\omega_i t + \omega_i \cos \Omega t + \phi)}. \tag{13.21}$$

By separating the terms, (13.20) and (13.21) become

$$S_{\text{ref}}(t) = \sum_{i=1}^{n} A_i e^{j\omega_i t} e^{j\phi} \tag{13.22}$$

and

$$S_{\text{test}}(t) = \sum_{i=1}^{n} A_i e^{j\omega_i t} e^{j\omega_i \cos \Omega t} e^{j\phi}. \tag{13.23}$$

The relative modulation functions for each component can then be found by dividing the equation for a component of the test signal by the equation for the same component of the reference signal. This is given by

$$\frac{S_{\text{test}}(t)}{S_{\text{ref}}(t)} = \frac{A_i e^{j\omega_i t} e^{j\omega_i \cos \Omega t} e^{j\phi}}{A_i e^{j\omega_i t} e^{j\phi}} \tag{13.24}$$

$$= e^{j\omega_i \cos \Omega t}.$$

The modulation function of the ith component relative to the reference (ref), $\beta_{i,\text{ref}}(t)$, can now be determined from (13.25) by

$$\beta_{i,\text{ref}}(t) = \frac{1}{\omega_i} \arctan \left(\frac{\Im(e^{j\omega_i \cos \Omega t})}{\Re(e^{j\omega_i \cos \Omega t}))} \right).$$ (13.25)

The relative modulation function can then be calculated for each component of the signal. Since they should all be equal if the test signal was a distorted version of the reference, a consistency classification technique can be developed to verify the similarity of the modulation functions for each component.

13.4.2 Consistency Classification

If it is known that the test signal will be a distorted version of one of several reference signals, then the relative modulation functions can be determined for each component relative to each reference signal. If there are 3 components and 5 reference signals, then there will be a 3×5 array of relative modulation functions. The correct reference can then be determined by finding the distance between the entries in each column. Since the relative modulation functions should be highly similar along the column corresponding to the correct reference, the column which yields the smallest distance will have the highest probability of corresponding to the correct reference.

Due to measurement noise and limited computational precision, only an estimate of $\beta_{ij}(t)$ is observable. It is therefore necessary to define a distance measure that quantifies the similarity among the estimates of the relative modulation functions on each component. It has been found that the L_1-error norm provides a robust distance estimate [9]. Therefore, the template k, which yields the minimum value of

$$d_j = \sum_{i=1}^{n} \left\| \beta_{ij}(t) - \frac{1}{n} \sum_{k=1}^{n} \beta_{kj}(t) \right\|$$ (13.26)

would be selected. Alternatively, an L_2 error norm could be used to determine the distance:

$$d_j = \sqrt{\sum_{i=1}^{n} \left(\beta_{ij}(t) - \frac{1}{n} \sum_{k=1}^{n} \beta_{kj}(t) \right)^2}.$$ (13.27)

This consistency classification technique will determine the closest reference signal without any *a priori* knowledge of the statistics of the distortion. This is a very powerful tool that can be applied to many pattern classification tasks. An example of a speech-recognition system is shown in the next section.

13.5 SPEECH RECOGNITION BY THE MODULATION OF FORMANTS

Speech parameters vary widely from speaker to speaker and may even vary significantly for utterances by the same speaker. Therefore methods to recognize speech sounds must take into account both spectral and temporal variations in the speech signal. These variations from one utterance to the next may be due to changes in loudness (amplitude variations) and varying stress levels of the speaker. Temporal variations are a result of the speaker modifying the rate of speech production (articulation rate). If these variations can be removed from the composite speech signal, a reasonable attempt can be made to classify the sound by its spectral characteristics. Currently, hidden Markov models (HMM) and neural networks appear to be the most commonly used methods for speech recognition. [10, 11]

Hidden-Markov-model techniques are based on the assumption that speech can be modeled as a Markov process [10, 12–14]. Training data are used to create probabilistic ensemble models, where each model represents a word or some part of a word. The recognition task is then divided into state determination and classification. The state determination is accomplished by calculating the most probable state to be occupied at each time t. Once the state transition sequence is found, it can be decided which model gave rise to that sequence. It can then be said that the utterance has the highest probability of being represented by this model. HMM technology has certain inherent limitations, however [10]. The major limitations are statistical in nature and include the incorrect assumption that successive observations are statistically independent, the assumption that the distribution of observations is Gaussian, and the Markov assumption that that the probability of being at any state depends only on the previous state. HMM-based systems have been shown to correctly identify phonemes for large-vocabulary unrestricted grammar-recognition systems with an accuracy of about 32% [15]. The accuracy is a measure of the total number of correct decisions. Additional processing of the phonemes allows for contextual and grammatical rules [15, 16] to increase the correct word classification to about 95%.

Many pattern-classification techniques attempt to exploit features of the speech signal rather than the signal itself. Often, a linear predictive coding (LPC) analysis [17] is performed and the LPC coefficients are used as features [18], or the cepstral LPC coefficients may be used [15]. Another technique is to perform a vector quantization (VQ) and use the VQ codes as features [19]. These techniques tend to improve noise immunity and reduce dependence on speaker-dependent attributes. Once these features are identified, a pattern-classification algorithm is used to identify the speech pattern.

Neural networks can also be trained to classify the given phoneme pattern [11]. A multilayer perceptron has been shown to yield an accuracy of about 54% correct context-independent recognition of 16 vowels and about 67% with contextual information [20]. Many new learning algorithms [21] and architectures have been developed, but there is still much to be understood about selecting the correct features, training sets, and training data.

The approach presented here yields a method of speech recognition based on a decomposition of the speech signal into components and then studying the modulation functions found by analytic signal techniques. These modulation functions are then tested by a consistency check to ensure each component is distorted in a similar fashion. The speech-recognition task under investigation is a speaker-independent phoneme-recognition task similar to those undertaken by neural network and HMM systems.

13.5.1 Analysis of Speech Components

For the case of interest here, the FM modulation function can be considered bandlimited without adversely effecting its usefulness as a time-varying model of speech. It has been shown [22] that—although the formants can vary over a wide range of frequencies—for an individual speaker under normal voicing conditions, the formants will vary within a relatively small bandwidth.

Some reasonable assumptions about the maximum frequency deviations about the formants due to articulation variations can be made. Since all the formant frequencies experience identical deviations, it is only necessary to determine the deviation about one formant in order to determine the maximum frequency of deviation. It has been shown [4, 22, 23] that the deviation about the first formant is approximately ±200 Hz for a male speaker. This matches very well with the maximum frequency deviation allowed by the generalized Carson's rule [8]. For a male speaker, the first formant is located at approximately 500 Hz; therefore, the maximum frequency deviation is

$$\Omega_{\max} < \frac{1}{2\sqrt{2}}\omega_1 \approx \pm 200 \text{ Hz}, \tag{13.28}$$

which is Carson's rule for small carrier frequencies [24], as is the case for speech formants. A female speaker can have a larger maximum frequency deviation but, since the first formant is at a higher frequency, Carson's rule still holds. This ensures that there is no spectral ambiguity due to the FM modulation function.

13.5.2 Speech-Component Classification

The adaptive time-varying spectral estimation method described in Chapter 8 has been found [3] to provide good estimates of the formant frequencies and their bandwidths. The resulting estimates are used to decompose the speech signal into the formant components. The distortions on any particular speech utterance due to interspeaker or intraspeaker differences is estimated relative to a reference utterance and will be similar for each formant only if the correct reference is used.

An estimate of the relative modulation function, $\beta_{ij}(t)$, which relates the ith component of the test signal to the ith component of the jth reference can then be found by the methods described in Section 13.4.1. Since the maximum frequency contained in $\beta_{ij}(t)$ was limited by the variability of the human voice, it is possible to low-pass filter $\beta_{ij}(t)$ at Ω_{max} to increase the immunity to noise. As stated earlier, for a male speaker, a low-pass cutoff frequency of approximately 200 Hz should be used. For a female speaker, a low-pass cutoff frequency of approximately 300 Hz is appropriate. The error distance, d_j, between the estimates of the modulation functions corresponding to each formant relative to the jth reference can now be computed. If the test signal is a distorted version of the jth reference, then the error distance will be very small since

$$\beta_{1j}(t) \approx \beta_{2j}(t) \approx \cdots \approx \beta_{nj}(t), \tag{13.29}$$

where n represents the number of formants studied.

The consistency check given by (13.26) is then used to verify that two or more formants have similar modulation functions for a given reference and test signal. This matching technique is more robust to nonlinear distortions due to its dependence on an FM modulation function as opposed to standard template matching techniques, which essentially perform an AM modulation matching.

13.6 NUMERICAL RESULTS

The data used to determine the performance of phonemic recognition was taken from the TIMIT database [25, 26]. This database is comprised of both male and female speakers from eight dialectical regions. Each speaker recites several sentences, each of which is rich in phonemic content. Each sentence is provided with a transcription and alignment table, which identifies a set of 62 phonemes. Samples of the vowels were extracted from the database for both male and female speakers to serve as reference templates. No effort was made to ensure any special relationship between the template and test utterances (i.e., there is no guarantee as to the form of the relative time warping).

The algorithm as applied here was limited to processing only the first two formants. This was done to reduce the computational overhead and is justified because most of the information required for speech understanding can be obtained from the first two formants [4]. It is expected that processing and checking more formants would improve the performance of the algorithm, but not significantly.

The test signal was processed by the algorithm with comparisons to each of the templates. The results for $\beta_{ij}(t)$ are shown for several vowels in Figures 13.1 to 13.6. Each figure shows the warping of a single formant and can be compared to the warping of the second formant for the same template comparison. It can be seen from Figures 13.1 and 13.2 that when the vowels are compared against the proper template, the warping is nearly identical for both formants. It can also be seen in Figures 13.3 through 13.6 that the warping is very different when an incorrect template is used.

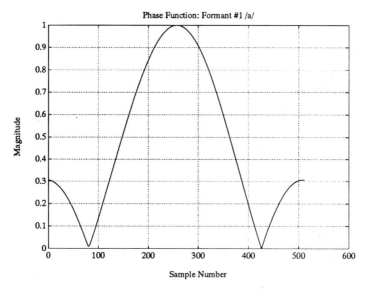

Figure 13.1 Relative modulation function of formant 1. Test signal /a/ compared to reference /a/.

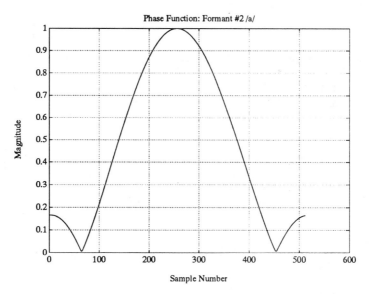

Figure 13.2 Relative modulation function of formant 2. Test signal /a/ compared to reference /a/.

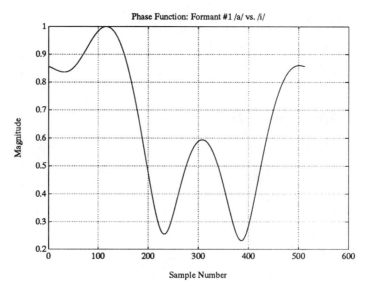

Figure 13.3 Relative modulation function of formant 1. Test signal /a/ compared to reference /i/.

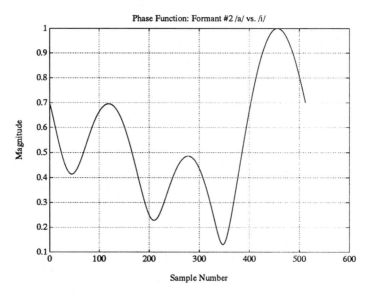

Figure 13.4 Relative modulation function of formant 2. Test signal /a/ compared to reference /i/.

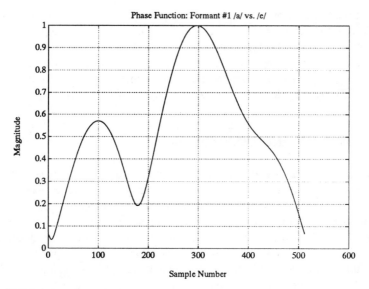

Figure 13.5 Relative modulation function of formant 1. Test signal /a/ compared to reference /e/.

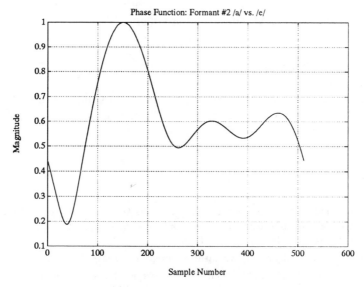

Figure 13.6 Relative modulation function of formant 2. Test signal /a/ compared to reference /e/.

The overall performance for vowel recognition using this classification algorithm was found to be 84% correct classification. Other techniques applied to this problem, such as hidden Markov models and neural networks, have yielded performance levels of approximately 64% [27, 20] correct classification. Therefore, it is clear that exploiting the inherent redundancy in the multicomponent nature of the vowel allows for increased recognition performance [28].

13.7 SUMMARY AND CONCLUSIONS

A new approach to multicomponent signal classification is presented that is based on the utilization of the spectral redundancy contained in many naturally occurring signals. This redundancy between components can be engineered into many man-made signals, such as the waveforms for radar, sonar and communications. An application to vowel recognition is shown here where decomposing the spectrogram into individual components allowed for the spectral variations due to time warping to be eliminated referentially. The technique for decomposing the spectrogram into its components is an *ad hoc* approach that is based on the quasistationary nature of the speech signal during vowel phonation and the multicomponent nature of the formants. Numerical results demonstrate the performance of the signal-classification algorithm on vowels and shows that it compares very favorably to more traditional approaches.

ACKNOWLEDGMENTS

The research reported here was made possible, in part, through the support of the New Jersey Commission on Science and Technology and the Computer Aids for Industrial Productivity (CAIP) Center at Rutgers University.

REFERENCES

1. L. Cohen and C. Lee. Instantaneous frequency, its standard deviation and multicomponent signals. In *Advanced Algorithms and Architectures for Signal Processing III*, F. T. Luk, ed., pp. 186–208. SPIE, 1988.

2. L. Mandel. Interpretation of instantaneous frequency. *Amer. J. Phys.*, **42**:840, 1974.

3. A. B. Fineberg and R. J. Mammone. An adaptive technique for high resolution time-varying spectral estimation. In *Proceedings of the IEEE International Conference on Acoustics, Speech, and Signal Processing 1991*, Toronto, Ontario, May 1991.

4. J. L. Flanagan. *Speech Analysis, Synthesis, and Perception*, 2nd ed. New York: Springer-Verlag, 1972.

5. J. C. Lilly. *Man and Dolphin*. New York: Pyramid Books, 1969.

6. R. Cann. An analysis/synthesis tutorial. In C. Roads and J. Strawn, eds., *Foundations of Computer Music*, chapter 9, pp. 114–144. Cambridge, Mass.: MIT Press, 1988.

7. R. J. Mammone, R. J. Rothacker, C. I. Podilchuk, S. Davidovici, and D. L. Schilling. Estimation of carrier frequency, modulation type and bit rate of an unknown modulated signal. In *Proceedings of IEEE International Conference on C-87*, Seattle, Wash., June 1987.

8. H. Taub and D. Schilling. *Principles of Communication Systems*. New York: McGraw Hill, 1986.

9. R. J. Mammone. Spectral extrapolation of constrained signals. *J. Opt. Soc. Am.*, **73**(11), 1983.

10. L. R. Rabiner. A tutorial on HMMs and selected applications in speech recognition. *Proc. IEEE*, **77**(2), 1989.

11. K. J. Lang, A. H. Waibel, and G. E. Hinton. A Time Delay Neural Network Architecture for Isolated Word Recognition. *Neural Networks*, **3**(1):23–43, 1990.

12. S. E. Levinson, L. R. Rabiner, and M. M. Sondhi. An introduction to the application of the theory of probabilistic functions of a Markov process to automatic speech recognition. *Bell Syst. Tech. J.*, **62**(4):1035–1074, 1983.

13. L. R. Rabiner and B. H. Juang. An introduction to Hidden Markov Models. *IEEE Acoust. Speech Signal Process. Mag.*, **3**(1), 1986.

14. L. R. Rabiner, B. H. Juang, S. E. Levinson, and M. M. Sondhi. Some properties of continuous Hidden Markov Model representations. *AT&T Tech. J.*, **64**(6), 1985.

15. K. F. Lee, H. W. Hon, and R. Reddy. An overview of the SPHINX speech recognition system. *IEEE Trans. Acoust. Speech Signal Process.*, **38**(1), 1990.

16. L. R. Rabiner and S. E. Levinson. A speaker-independent, syntax-directed, connected word recognition system based on Hidden Markov Models and level building. *IEEE Trans. Acoust. Speech Signal Process.*, **33**(3), 1985.

17. J. D. Markel and A. H. Gray. *Linear Prediction of Speech*. New York: Springer-Verlag, 1976.

18. F. Itakura. Minimum prediction residual principle applied to speech recognition. *IEEE Trans. Acoust. Speech Signal Process.*, **23**(1), 1975.

19. A. A. Buzo and et al. Speech coding based on vector quantization. *IEEE Trans. Acoust. Speech Signal Process.*, **28**(5), 1980.

20. H. C. Leung and V. W. Zue. Some phonetic recognition experiments using artificial neural nets. In *Proceedings of IEEE International Conference on Acoustics, Speech, and Signal Processing-88*, volume S-1, New York: April 1988.

21. A. Sankar and R. J. Mammone. A fast learning algorithm for tree neural networks. In *Conference on Information Sciences and Systems*, Princeton, N.J., March 1990.

22. W. Hess. *Pitch Determination of Speech Signals*. New York: Springer-Verlag, 1983.

23. L. R. Rabiner and R. W. Schafer. *Digital Processing of Speech Signals*. Englewood Cliffs, N.J.: Prentice Hall, 1978.

24. H. L. Van Trees. *Detection, Estimation, and Modulation Theory, II*. New York: Wiley 1971.

25. W. M. Fisher, G. R. Doddington, and K. M. Goudie-Marshall. The DARPA speech recognition research database: Specifications and status. In *DARPA Speech Recognition Workshop*, 1986.

26. L. F. Lamel, R. H. Kassel, and S. Seneff. Speech database development: Design and analysis of the acoustic-phonetic corpus. In *DARPA Speech Recognition Workshop*, 1986.

27. N. Hataoka and A. H. Waibel. Speaker-independent phoneme recognition on TIMIT database using integrated time-delay neural networks (TDNNs). Technical Report CMU-CMT-89-115, Center for Machine Translation, Carnegie-Mellon University, Nov. 1989.

28. A. B. Fineberg. *Phoneme Recognition: An Articulation Rate Independent Classification Technique*. PhD thesis, Rutgers University, October 1990.

REVIEW OF LINEAR ALGEBRA

Christine I. Podilchuk

AT&T Bell Laboratories
600 Mountain Avenue
Murray Hill, New Jersey 07947
email: chrisp@research.att.com

An $M \times N$ matrix $\mathbf{A}$ is defined as

$$\mathbf{A} = [a_{ij}] = \begin{pmatrix} a_{11} & a_{12} & \cdots & a_{1N} \\ a_{21} & a_{22} & \cdots & a_{2N} \\ \vdots & \vdots & \ddots & \vdots \\ a_{M1} & a_{M2} & \cdots & a_{MN} \end{pmatrix}. \tag{A.1}$$

Similarly, a vector $\mathbf{x}$ of dimension N is described as

$$\mathbf{x} = \begin{pmatrix} x_1 \\ x_2 \\ \vdots \\ x_N \end{pmatrix}. \tag{A.2}$$

The transpose of a matrix $\mathbf{A}$ is defined as

$$\mathbf{A}^T = [a_{ij}^T] = [a_{ji}] = \begin{pmatrix} a_{11} & a_{21} & \cdots & a_{M1} \\ a_{12} & a_{22} & \cdots & a_{M2} \\ \vdots & \vdots & \ddots & \vdots \\ a_{1N} & a_{2N} & \cdots & a_{MN} \end{pmatrix} \tag{A.3}$$

Additional properties of the transpose operator include

$$(A + B)^T = A^T + B^T$$
$$(A^T)^T = A$$
$$(AB)^T = B^T A^T$$

The matrix A is said to be symmetric when $A = A^T$. Note that when dealing with complex matrices, the Hermitian transpose is defined by taking the complex conjugate of the ordinary transpose; that is,

$$B^H = \bar{B}^T, \tag{A.4}$$

where the bar denotes the complex conjugate. A matrix B is said to be *Hermitian* when $B^H = B$. A diagonal matrix is a square matrix in which all the terms off the principal diagonal are zero, i.e., $a_{ij} = 0$, $i \neq j$. The identity matrix I is a diagonal matrix whose diagonal terms are all equal to one:

$$I_N = \begin{pmatrix} 1 & 0 & \cdots & 0 \\ 0 & \cdots & \cdots & 0 \\ \vdots & \vdots & \ddots & \vdots \\ 0 & \cdots & \cdots & 1 \end{pmatrix} \tag{A.5}$$

A linear vector space $\mathcal{V}$ satisfies the following conditions for all $v, w, x \in \mathcal{V}$:

$$v + w \in \mathcal{V}$$
$$v + w = w + v$$
$$v + (w + x) = (v + w) + x$$
$$v + 0 = v$$
$$v + -v = 0$$
$$\alpha v \in \mathcal{V} \qquad \text{for any scalar } \alpha$$
$$\alpha(v + w) = \alpha v + \alpha w$$
$$(\alpha + \beta)v = \alpha v + \beta v$$
$$(\alpha \beta)v = \alpha(\beta v)$$
$$1 \cdot v = v.$$

An inner product is defined as

$$\langle v, v \rangle \geq 0 \qquad \text{with equality iff } v = 0. \tag{A.6}$$

If a vector space with an inner product has a norm defined by

$$\|\mathbf{v}\| = \langle \mathbf{v}, \mathbf{v} \rangle^{1/2} \tag{A.7}$$

and is complete with respect to this norm, it is a *Hilbert space*.

A vector norm of $\mathbf{x}$ is a nonnegative number denoted by $\|\mathbf{x}\|$ that satisfies

$$\|\mathbf{x}\| > 0 \quad \text{for} \quad \mathbf{x} \neq 0 \text{ and } \|\mathbf{x}\| = 0 \text{ when } \mathbf{x} = 0,$$

$$\|k\mathbf{x}\| = |k|\|\mathbf{x}\| \quad \text{for any scalar} \quad k,$$

$$\|\mathbf{x} + \mathbf{y}\| \leq \|\mathbf{x}\| + \|\mathbf{y}\| \quad \text{(triangle inequality)}.$$

The following are common vector norms:

$$\mathcal{L}_1 \text{ norm} = \|\mathbf{x}\|_1 = |x_1| + |x_2| + \cdots + |x_N|,$$

$$\mathcal{L}_2 \text{ norm (Euclidean norm)} = \|\mathbf{x}\|_2 = \left\{x_1^2 + x_2^2 + \cdots + x_N^2\right\}^{1/2},$$

$$\mathcal{L}_\infty \text{ norm} = \|\mathbf{x}\|_\infty = \max |x_i|.$$

Two vectors $\mathbf{x}$ and $\mathbf{y}$ are *orthogonal* if their inner product is equal to zero; that is,

$$\langle \mathbf{x}, \mathbf{y} \rangle = 0. \tag{A.8}$$

A square matrix that has an inverse is called a nonsingular matrix. A square matrix that does not have an inverse is said to be singular. If the matrices $\mathbf{A}$ and $\mathbf{B}$ are nonsingular, the following is true:

$$\mathbf{A}\mathbf{A}^{-1} = \mathbf{A}^{-1}\mathbf{A} = \mathbf{I},$$

$$(\mathbf{A}^{-1})^{-1} = \mathbf{A},$$

$$(\mathbf{A}\mathbf{B})^{-1} = \mathbf{B}^{-1}\mathbf{A}^{-1}.$$

The rank of an $M \times M$ matrix $\mathbf{A}$ that is nonsingular and therefore has an inverse is M.

Consider the set of linear equations

$$\mathbf{A}\mathbf{x} = \mathbf{b}, \tag{A.9}$$

where $\mathbf{A}$ is an $M \times N$ matrix, $\mathbf{x}$ is an unknown vector of length N and b is an observed vector of length M. Exactly one of the following conditions must be true:

- If the rank of the augmented matrix $[\mathbf{A}, \mathbf{b}]$ is greater than the rank of $\mathbf{A}$, the system of equations is inconsistent.

- If the rank of $[\mathbf{A}, \mathbf{b}]$ is equal to the rank of $\mathbf{A}$, which is equal to the number of unknowns, there exists a unique solution.

- If the rank of $[\mathbf{A}, \mathbf{b}]$ is equal to the rank of $\mathbf{A}$, which is less than the number of unknowns, there exists an infinite number of solutions.

The Fredholm alternative states that $\mathbf{Ax} = \mathbf{b}$ has a unique solution for every $\mathbf{b}$ only if $\mathbf{x} = \mathbf{0}$ is the unique solution to the homogeneous equation $\mathbf{Ax} = \mathbf{0}$. This is true only when the rank of $\mathbf{A}$ is equal to the number of unknowns. A square $M \times M$ matrix $\mathbf{A}$ is nonsingular; that is, $\mathbf{A}$ has an inverse if and only if the rank of $\mathbf{A}$ is equal to M.

A set of vectors $\mathbf{v}_1$, $\mathbf{v}_2$, ..., $\mathbf{v}_S$ is linearly dependent if there exist numbers α_1, α_2, ..., α_S, not all zero, such that

$$\alpha_1 \mathbf{v}_1 + \alpha_2 \mathbf{v}_2 + \cdots + \alpha_S \mathbf{v}_S = 0. \tag{A.10}$$

Otherwise, the set of vectors are linearly independent. The rank r of a matrix $\mathbf{A}$ is the number of linearly independent rows or columns in the matrix. Therefore, a square matrix is nonsingular if and only if its rows (or columns) form an independent set. When every vector in a set S is orthogonal to every other vector in S and every vector satisfies $\|\mathbf{v}\| = 1$, the set is said to be *orthonormal*. Gram–Schmidt orthogonalization states that if vectors $\mathbf{v}_1$, ..., $\mathbf{v}_M$ form a linearly independent set in a vector space $\mathcal{V}$, then an orthonormal set of vectors $\mathbf{u}_1, \ldots, \mathbf{u}_i$ can be constructed with $1 \le i \le M$ so that the set of vectors $\mathbf{u}_1$, ..., $\mathbf{u}_i$ spans the same space as $\mathbf{v}_1$, ..., $\mathbf{v}_i$. Gram–Schmidt Orthogonalization is simply stated as

$$\mathbf{x}_i = \mathbf{v}_i - \langle \mathbf{u}_{i-1}, \mathbf{v}_i \rangle \mathbf{u}_{i-1} - \langle \mathbf{u}_{i-2}, \mathbf{v}_i \rangle \mathbf{u}_{i-2} - \cdots - \langle \mathbf{u}_1, \mathbf{v}_i \rangle \mathbf{u}_1 \tag{A.11}$$

and

$$\mathbf{u}_i = \frac{\mathbf{x}_i}{\|\mathbf{x}_i\|}. \tag{A.12}$$

A *basis* for a vector space is a linearly independent set of vectors that spans the space. The number of vectors in a basis is known as the dimension of the vector space.

If $\mathbf{A}$ is a linear transformation from a vector space $\mathcal{X}$ to a vector space $\mathcal{B}$, then the range space of $\mathbf{A}$ is defined as the linear subspace of $\mathcal{B}$ of all vectors of the form $\mathbf{Ax}$ as $\mathbf{x}$ varies over $\mathcal{X}$:

$$\mathcal{R}(\mathbf{A}) = \{\mathbf{b}: \ \mathbf{Ax} = \mathbf{b}\}. \tag{A.13}$$

The null space (or kernel) of $\mathbf{A}$ is the linear subspace of $\mathcal{X}$ consisting of all solutions to the equation $\mathbf{Ax} = \mathbf{0}$:

$$\mathcal{N}(\mathbf{A}) = \{\mathbf{x}: \ \mathbf{Ax} = \mathbf{0}\}. \tag{A.14}$$

The set of linear equations in (A.9) has at most one unique solution if the null space of $\mathbf{A}$ equals $\{\mathbf{0}\}$.

An $M \times M$ matrix $\mathbf{A}$ has an eigenvector $\mathbf{x}$ and associated eigenvalue λ that satisfies the relationship

$$\mathbf{Ax} = \lambda \mathbf{x}. \tag{A.15}$$

The Gershgorin circle theorem provides an estimate of the eigenvalues of a matrix $\mathbf{A}$. According to this theorem, every eigenvalue λ of an $M \times M$ matrix $\mathbf{A}$ satisfies at least one of the inequalities

$$|\lambda - a_{ii}| \leq r_i, \tag{A.16}$$

where

$$r_i = \sum_{j-1, j \neq i}^{M} |a_{ij}|, \qquad i = 1, 2, \ldots, M. \tag{A.17}$$

Each eigenvalue lies in at least one of the discs with center a_{ii} and radius r_i in the complex plane. If the union of N of the discs is disjoint from the other discs, then exactly N eigenvalues reside in the union of the N discs.

An orthogonal (unitary) matrix is a matrix whose transpose (Hermitian transpose) is equal to its inverse, that is, the columns of the matrix are orthonormal to each other:

$$\mathbf{AA}^T = \mathbf{A}^T\mathbf{A} = \mathbf{I} \qquad (\mathbf{AA}^H = \mathbf{A}^H\mathbf{A} = \mathbf{I}). \tag{A.18}$$

The Schur decomposition for an $M \times M$ matrix $\mathbf{A}$ is defined as

$$\mathbf{A} = \mathbf{PTP}^H, \tag{A.19}$$

where $\mathbf{P}$ is a unitary matrix and $\mathbf{T}$ is an upper triangular matrix. An important consequence of this transformation is that when $\mathbf{A}$ is a symmetric (Hermitian) matrix, the matrix $\mathbf{T}$ becomes a diagonal matrix.

A Householder transformation performs a reflection of a vector, preserving the length and angle, and therefore is a unitary transformation. It is expressed as

$$\mathbf{H} = \mathbf{I} - 2\mathbf{ww}^T, \tag{A.20}$$

where

$$\|\mathbf{w}\|_2 = 1. \tag{A.21}$$

This transformation is useful in performing the QR decomposition of a matrix **A**, where we wish to find a unitary transformation **Q** such that

$$\mathbf{A} = \mathbf{QR} \tag{A.22}$$

and **R** is upper triangular; that is, only terms on or above the diagonal are allowed to be nonzero. This technique can be used to determine the eigenvalues of **A** or to solve the least-squares problem for a set of linear equations.

Any $M \times N$ matrix **A** can be diagonalized using singular-value decomposition; this yields

$$\mathbf{A} = \mathbf{U}\boldsymbol{\Sigma}\mathbf{V}^H. \tag{A.23}$$

In the case of a real matrix **A**, the Hermitian transpose H can be replaced by the transpose T:

$$\mathbf{A} = \mathbf{U}\boldsymbol{\Sigma}\mathbf{V}^T, \tag{A.24}$$

where **U** and **V** are $M \times M$ and $N \times N$ unitary matrices respectively and, in the case $M > N$,

$$\boldsymbol{\Sigma} = \begin{pmatrix} \sigma_1 & 0 & \cdots & 0 \\ 0 & \sigma_2 & 0 & \cdots \\ & & \ddots & \\ 0 & \cdots & & \sigma_N \\ 0 & & & 0 \\ \vdots & \vdots & & \end{pmatrix}. \tag{A.25}$$

From (A.23), it follows that

$$\mathbf{A}^H\mathbf{A} = \mathbf{V}\boldsymbol{\Sigma}^H\mathbf{U}^H\mathbf{U}\boldsymbol{\Sigma}\mathbf{V}^H = \mathbf{V}(\boldsymbol{\Sigma}^H\boldsymbol{\Sigma})\mathbf{V}^H = \mathbf{V}\boldsymbol{\Sigma}_N^2\mathbf{V}_H \tag{A.26}$$

and

$$\mathbf{A}\mathbf{A}^H = \mathbf{U}\boldsymbol{\Sigma}\mathbf{V}^H\mathbf{V}\boldsymbol{\Sigma}^H\mathbf{U}^H = \mathbf{U}(\boldsymbol{\Sigma}\boldsymbol{\Sigma}^H)\mathbf{U}^H = \mathbf{U}\boldsymbol{\Sigma}_M^2\mathbf{U}^H. \tag{A.27}$$

The values along the diagonals of $\boldsymbol{\Sigma}_N^2$ and $\boldsymbol{\Sigma}_M^2$ are the eigenvalues of $\mathbf{A}^H\mathbf{A}$ and $\mathbf{A}\mathbf{A}^H$, respectively. **U** and **V** are composed of the orthonormal eigenvectors associated with $\mathbf{A}\mathbf{A}^H$ and $\mathbf{A}^H\mathbf{A}$, respectively. The strictly positive square roots σ_i of the nonzero eigenvalues of $\mathbf{A}^H\mathbf{A}$ (or $\mathbf{A}\mathbf{A}^H$) are called the singular values of **A**. When the set of linear equations describing the problem is inconsistent, (A.9) can be replaced with

$$\mathbf{Ax} + \mathbf{n} = \mathbf{b}, \tag{A.28}$$

where **n** is an error vector. The least-squares solution for (A.28) can be stated as

$$\text{Find } \mathbf{x} \text{ that minimizes} \quad \|\mathbf{n}\|_2 = \|\mathbf{Ax} - \mathbf{b}\|_2. \tag{A.29}$$

The solution for the least-squares problem is given by the solution to the normal equations:

$$\mathbf{A}^T\mathbf{Ax} = \mathbf{A}^T\mathbf{b}. \tag{A.30}$$

When the $M \times N$ matrix **A** is of rank r and its singular-value decomposition is described by (A.23) with $\sigma_1 \geq \sigma_2 \geq \cdots \geq \sigma_r > 0$ the Moore–Penrose generalized inverse (pseudo-inverse) $\mathbf{A}^\dagger$ of **A** is calculated as

$$\mathbf{A}^\dagger = \mathbf{V}\boldsymbol{\Sigma}^\dagger\mathbf{U}^H, \tag{A.31}$$

where

$$\boldsymbol{\Sigma}^\dagger = \begin{pmatrix} \mathbf{E} & 0 \\ 0 & 0 \end{pmatrix} \tag{A.32}$$

and

$$\mathbf{E} = \begin{pmatrix} \sigma_1^{-1} & 0 & \cdots & \\ 0 & \sigma_2^{-1} & 0 & \cdots \\ \vdots & & \ddots & \\ & & & \sigma_r^{-1} \end{pmatrix}. \tag{A.33}$$

Note that $\mathbf{A}^\dagger$ is unique, although **U** and **V** do not necessarily have to be unique. The Moore–Penrose generalized inverse must satisfy the following:

1. $\mathbf{AA}^\dagger\mathbf{A} = \mathbf{A}$.
2. $\mathbf{A}^\dagger\mathbf{AA}^\dagger = \mathbf{A}^\dagger$.
3. $\mathbf{AA}^\dagger$ and $\mathbf{A}^\dagger\mathbf{A}$ are Hermitian.
4. $\mathbf{A}^\dagger$ is unique.

From (A.31), it follows that

$$\mathbf{A}^\dagger\mathbf{A} = \mathbf{V}\boldsymbol{\Sigma}^\dagger\mathbf{U}^H\mathbf{U}\boldsymbol{\Sigma}\mathbf{V}^H = \mathbf{V}\boldsymbol{\Sigma}^\dagger\boldsymbol{\Sigma}\mathbf{V}^H, \tag{A.34}$$

where

$$\Sigma^\dagger \Sigma = \begin{pmatrix} \mathbf{I}_r & 0 \\ 0 & 0 \end{pmatrix}. \tag{A.35}$$

Proofs for the first three conditions are obtained from (A.34). For item 1,

$$\mathbf{A}\mathbf{A}^\dagger\mathbf{A} = \mathbf{U}\Sigma\mathbf{V}^H\mathbf{V}\Sigma^\dagger\mathbf{U}^H\mathbf{U}\Sigma\mathbf{V}_H = \mathbf{U}\Sigma\Sigma^\dagger\Sigma\mathbf{V}^H = \mathbf{U}\Sigma\mathbf{V}^H = \mathbf{A}; \tag{A.36}$$

for item 2,

$$\mathbf{A}^\dagger\mathbf{A}\mathbf{A}^\dagger = \mathbf{V}\Sigma^\dagger\mathbf{U}^H\mathbf{U}\Sigma\mathbf{V}^H\mathbf{V}\Sigma^\dagger\mathbf{U}^H = \mathbf{V}\Sigma^\dagger\Sigma\Sigma^\dagger\mathbf{U}^H = \mathbf{V}\Sigma^\dagger\mathbf{U}^H = \mathbf{A}^\dagger; \tag{A.37}$$

finally, for item 3, we get

$$\mathbf{A}\mathbf{A}^\dagger = \mathbf{U}\Sigma\mathbf{V}^H\mathbf{V}\Sigma^\dagger\mathbf{U}^H = \mathbf{U}\Lambda_M\mathbf{U}^H, \tag{A.38}$$

$$\mathbf{A}^\dagger\mathbf{A} = \mathbf{V}\Sigma^\dagger\mathbf{U}^H\mathbf{U}\Sigma\mathbf{V}^H = \mathbf{V}\Lambda_N\mathbf{V}_H, \tag{A.39}$$

where

$$\Lambda_M = \begin{pmatrix} I_r & 0 \\ 0 & 0 \end{pmatrix} \tag{A.40}$$

and

$$\Lambda_N = \begin{pmatrix} I_r & 0 \\ 0 & 0 \end{pmatrix}. \tag{A.41}$$

From (A.39) and (A.39) it is easy to see that $\mathbf{A}\mathbf{A}^\dagger$ and $\mathbf{A}^\dagger\mathbf{A}$ are Hermitian; that is,

$$(\mathbf{A}\mathbf{A}^\dagger)^H = (\mathbf{U}\Lambda_M\mathbf{U}^H)^H = \mathbf{U}\Lambda_M\mathbf{U}^H = \mathbf{A}\mathbf{A}^\dagger \tag{A.42}$$

and

$$(\mathbf{A}^\dagger\mathbf{A})^H = (\mathbf{V}\Lambda_N\mathbf{V}^H)^H = \mathbf{V}\Lambda_N\mathbf{V}^H = \mathbf{A}^\dagger\mathbf{A}. \tag{A.43}$$

The pseudo-inverse solution can be stated as

$$\mathbf{x}^\dagger = \mathbf{A}^\dagger\mathbf{A}\mathbf{x}. \tag{A.44}$$

The condition number of a matrix $\mathbf{A}$ is defined by

$$C = \frac{\sigma_{\max}}{\sigma_{\min}}, \tag{A.45}$$

where σ_{max} and σ_{min} denote the maximum and minimum singular values of the matrix A, respectively. Note that, for a rank-deficient operator, C is equal to infinity. In general, when C is a large number (> 100) for a noisy system of equations, the problem is said to be "ill conditioned"; a small difference in the input x can yield a large difference in the output b. For such cases, the pseudo-inverse solution expressed in (A.44) can lead to an unstable solution. Various techniques of smoothing and regularization have been developed to overcome the problems of ill conditioning.

COMPUTER PROGRAMS

Image Recovery The routine

$$\mathtt{rap2d(m,n,g,f,rows,cols,h,R,L,lamda)}$$

implements (4.25). Given the blurred image `g[1..rows][1..cols]`, the point-spread function (PSF) `h[-R..R][-L..L]` centered at pixel location (`m,n`), and relaxation factor `lamda`, this routine updates those pixel values of the estimated image `f[1..rows][1..cols]` which are covered by the support of PSF.

Simulated Annealing The main program is called `PlaceVoxel.c`. The executable program is run by specifying 6 arguments:

$$H,\ g,\ f,\ g_{\text{error}},\ f_{\text{error}},\ \text{and}\ d_{\text{error}}, \tag{B.1}$$

where H is an $mn \times mnd$ matrix, g and f are $mn \times 1$ and $mnd \times 1$ vectors that define the given $m \times n$ pixel values and $m \times n \times d$ voxel radiance values, respectively. H, g, and f are all specified in double-precision floating-point format. G_{error}, f_{error}, and d_{error} are output files created to save the error in the reconstructed image (g), voxel radiance (f), and voxel depth (d). Also contained in `PlaceVoxel.c` are the two subroutines `metrop` and `updategesto`. Metro uses the Metropolis criterion to decide whether to accept or reject a choice of voxel placement. Updategesto updates the estimate of the reconstructed image, g.

The file `s.c` containing the following subroutines, which perform all the constrained optimization strategies:

`votesmooth` Votes for the bestbest smooth surface.

`convertfd2tsti` Convert surface 2 tubes.

`ferrr` Computes error in depth.

`gerror` Computes L_2 norm of image error.

`gest` Computes *g*-estimate (image-estimate).

`newconf` Generates a new configuration.

`ran3` Random number generator.

`scalvect` Scalar vector.

`updateconf` Update an exiting configuration.

`stepudlig` Step up intensity by 1 grain.

`stepudlsg` Step up depth by 1 slice.

`irbit1` Random bit—to perturb.

`newconf1` Generates another new configuration.

`srtconf1` Generates starting configuration.

`voxsav` Save voxel configuration.

`smootherr` Smooth error.

`stepudlsgs` Step up depth by 1 slice.

`newconfs` Generates another new configuration.

`stepudlsgsf` Step up 1 slice.

`stepudligf` Step up or down 1 intensity grain.

`gderror` Compute depth error.

`gest2` Compute second form of image error.

`newconfs2` Step up or down 1 slice grain.

`voxsavf`

Utility The file `u.c` containing the matrix/vector memory allocation subroutines.

Two header files (`s.h`, `u.h`) are provided with definitions for `s.c` and `s.h`, respectively.

Spectral Estimation The program `ftvse` applies the row-action projection algorithm to the spectral estimation problem with a time varying signal. A low-resolution frequency-position estimation is used to initialize the frequency subspace partitioning. A setup file called `setup.fse` is also provided to initialize all parameters. The initialization sets the following parameters:

1. Seed for random number generator.
2. Acceleration (1 = no, 2 = yes).
3. Maximum number of times through equations.
4. Lower limit on variance (stopping criteria).
5. Step size.
6. Threshold value (for active signal detection).
7. Number of frequency bins to evaluate.
8. Maximum number of samples per iteration.
9. Total number of samples in input signal.

10. Segment size for frequency finding.

11. SNR (for noise added to input signal).

12. Number of tones (in input signal).

13. Real amplitude.

14. Imaginary amplitude.

15. Sample start number.

16. Sample stop number.

17. Frequency bin number.

18. Each of the last 5 values are repeated.

The following subroutines and functions are included in the program and perform the following actions:

RAP Perform the row-action projection algorithm.

DP Returns the dot product between a row of a matrix and a column vector.

NORM Returns the Euclidean norm of a row of a matrix.

SCALE Multiplies each element of a row of a matrix by a scalar.

ADDVEC Adds the elements of two vectors.

COPY Copies the contents of one vector to another vector.

VAR Calculates the variance between two vectors.

GASDEV Returns a Gaussian-distributed deviate with zero mean and unit variance.

RAN1 Returns a uniform random deviate between 0.0 and 1.0.

Channel Equalization This subroutine implements an adaptive linear or decision feedback equalizer. The input to the equalizer is specified in the complex vector X. The desired output of the equalizer should be supplied in the complex vector X D. The equalizer is run in reference-directed mode, i.e., the correct symbols X D are used for error estimation. To run in decision-directed mode, the user should insert the line $T(J)=Q(D)$ for $T(J)=XD(ISAMP)$, where $Q(D)$ implements the decision device. A linear equalizer can be implemented by setting the number of feedback coefficients to zero. The output of the dfe is stored in the input vector X. The equalizer coefficients are stored in the complex vector CX.

Subroutine RAP(K1,KTOT,X,XD,NSAM,NEQ,NIT,MU,CX)

K1 Number of coefficients in feedforward part of filter.

KTOT Total number of coefficients in filter. By default, this is equal to the number of feedback coefficients given by KTOT−K1.

X Complex input sample vector.

XD Complex desired response vector.

NSAM Total number of complex samples in input vectors X and XD.

NEQ Number of equations per processing block.

NIT Number of iterations per processing block.

MU Step size, 0<MU<2.0. Typically, MU=0.1-0.5.

CX Complex filter coefficients (output).

Adaptive Beamforming Two programs are provided for adaptive beamforming. These programs are for narrowband array processing and broadband array processing. Both programs are similar, except for the weight-updating subroutine and the scenario generator.

The number of sensors and iterations are defined as constants in the beginning of the program. The broadband-array program also includes a variable to control the number of taps. The scenario is generated in the scenario() subroutine. Note that if the desired signal generated by scenario is changed, then it must also be changed in out_stats() to give the proper performance outputs.

Subroutines included:

scenario Scenario generator.

beamform Execute beamforming procedure.

init_0 Initialize vector to zero.

next_x Input X vector.

calc_y Calculate y=W X.

update_w Update W vector via RAP.

out_stats Output data files for SNR, output error, etc.

ran Uniform random variable generator.

gran Gaussian random variable generator.

init_f Initialize F vector, broadband only.

shift_x Propagate X vector through array, broadband only.

Array constants:

TAPS Row length (always one for narrowband).

SENSORS Column length.

N Number of iterations.

Neural Networks This program allows the user to define a feedforward neural network with connections only between one layer and the next. The user may specify initial conditions in a file. The training data is also to be read from a file. The names of the two files are passed as command line arguments when running the program. The first file is the training file and the second contains initial values for the weights. If the second file name is not specified, then the program generates random initial weights. At any particular time, the user may test the network to see what outputs are generated for the various input data.

The training file contains the feature-vector elements and the network output. Therefore, if there is a three-dimensional feature vector and two output nodes, each line of the file would have five entries. An example is

```
0   1   1   0   1
9   8   1   1   0
8   5   1   1   1
```

If you want to test the network from a previously saved weight file, the weight file name must have been written by this program and be listed as the second file on the command line.

Neural Tree Networks The NTN program is made up of the following files:

```
ntn.c
ntn_func.c
func_def.h
incl.h
```

To run the program, three input files are needed:

1. Training file.
2. Class label file.
3. Testing file.

The formats for the above files are as follows:

1. Training file. Each row of this file corresponds to a feature vector followed by its class number. The class labels must be numbered from 0 to $M - 1$ if there are M classes. An example is:

```
0   1   1   0
9   8   1   1
8   5   1   0
```

The above example has three 3-dimensional feature vectors with two classes.

2. Class label file. This file specifies the class labels for the different classes. In this program, only local encoding is allowed for the class labels. Thus for an M-class problem, M bits are used and the labels are the binary basis vectors. For the above 2-class training file, the following is the class label file:

```
0.1   0.9   0
0.9   0.1   1
```

The first two entries in the rows are the bit values for the class labels and the third entry is simply the class number. Thus, the following is a three-class example:

```
0.1   0.1   0.9   0
0.1   0.9   0.1   1
0.9   0.1   0.1   2
```

3. Testing file. This file contains the data on which the NTN is to be tested. The format is the same as the training file, except that the first number gives the number of test data (the program prompts for number of training data). An example is:

```
4
0   1   0   0
1   1   1   0
2   2   1   1
0   1   1   1
```

The program is run as follows:

ntn file1 file2 file3 file4

where **file1** is the training file, **file2** is the class label file, **file3** is the testing file, and **file4** is the output file. The output file is created by the program and gives the performance for all the pruned NTNs.

The number of epochs for each NTN node is hardwired in the file ntn_func.c. It is currently 1500 but it can be changed to a user-determined value if necessary. Search for the string "1500" to find the appropriate line in ntn_func.c.

Multicomponent Matching A routine is provided to perform FM analysis of multicomponent signals. The subroutine has the following form

SUBROUTINE FM (in1, in2, num_samps, match, minsum)

and uses the following arguments and parameters:

in1 Contains the first component of input signal.

in2 Contains the second component of input signal.

num_samps Defines the number of samples in input signal.

match Returns template id.

minsum Returns matching error.

num_tmpls Defines the number of templates.

tmpl Is an array of template phase functions.

SELECTED EXAMPLES, PROBLEMS, AND COMPUTER PROJECTS

C.1 SELECTED NUMERICAL EXAMPLES

1. If

$$
\mathbf{H} = \frac{1}{4} \begin{pmatrix} 5 & 1 & -3 & 1 \\ 1 & 5 & 1 & -3 \\ -3 & 1 & 5 & 1 \\ 1 & 3 & 1 & 5 \end{pmatrix},
$$

find **U** and **D** such that

$$
\mathbf{H} = \mathbf{U}\mathbf{D}\mathbf{U}^T
$$

Answer We can see that **H** is a cyclic matrix. Hence, **U** is the 4-point DFT matrix,

$$
\mathbf{U} = \begin{pmatrix} 1 & 1 & 1 & 1 \\ 1 & -j & -1 & j \\ 1 & -1 & 1 & -1 \\ 1 & -j & -1 & j \end{pmatrix},
$$

and **D** is simply the FFT of the first column of **H**. Hence,

$$
\mathbf{D} = \begin{pmatrix} 1 & 1 & 1 & 1 \\ 1 & j & -1 & -j \\ 1 & -1 & 1 & -1 \\ 1 & -j & -1 & j \end{pmatrix} \begin{pmatrix} 5 \\ 1 \\ -3 \\ 1 \end{pmatrix} \tag{C.1}
$$

$$= \begin{pmatrix} 1 \\ 2 \\ 0 \\ 2 \end{pmatrix}. \tag{C.2}$$

$$\mathbf{D} = \begin{pmatrix} 1 & 0 & 0 & 0 \\ 0 & 2 & 0 & 0 \\ 0 & 0 & 0 & 0 \\ 0 & 0 & 0 & 2 \end{pmatrix}$$

2. Find $\mathbf{H}^\dagger$,

$$\mathbf{H}^\dagger = \frac{1}{4}\mathbf{U}^{-1}\mathbf{D}^\dagger(\mathbf{U}^{-1})^T.$$

Answer

$$\mathbf{H}^\dagger = \frac{1}{4} \begin{pmatrix} 2 & 1 & 0 & 1 \\ 1 & 2 & 1 & 0 \\ 0 & 1 & 2 & 1 \\ 1 & 0 & 1 & 2 \end{pmatrix} \tag{C.3}$$

$$= \frac{1}{4} \begin{pmatrix} 1 & 1 & 1 & 1 \\ 1 & j & -1 & -j \\ 1 & -1 & 1 & -1 \\ 1 & -j & -1 & j \end{pmatrix} \begin{pmatrix} 1 & 0 & 0 & 0 \\ 0 & \frac{1}{2} & 0 & 0 \\ 0 & 0 & 0 & 0 \\ 0 & 0 & 0 & \frac{1}{2} \end{pmatrix} \begin{pmatrix} 1 & 1 & 1 & 1 \\ 1 & -j & -1 & j \\ 1 & -1 & 1 & -1 \\ 1 & j & -1 & -j \end{pmatrix}. \tag{C.4}$$

3. If

$$\mathbf{y} = \frac{1}{4} \begin{pmatrix} -2 \\ 6 \\ 6 \\ -2 \end{pmatrix},$$

and $\mathbf{y} = \mathbf{Hx}$, find $\mathbf{x}^\dagger$.

Answer

$$\mathbf{x}^\dagger = \mathbf{H}^\dagger \mathbf{y} \tag{C.5}$$

$$= \begin{pmatrix} 2 & 1 & 0 & 1 \\ 1 & 2 & 1 & 0 \\ 0 & 1 & 2 & 1 \\ 1 & 0 & 1 & 2 \end{pmatrix} \frac{1}{4} \begin{pmatrix} -2 \\ 6 \\ 6 \\ -2 \end{pmatrix} \tag{C.6}$$

$$= \begin{pmatrix} 0 \\ 1 \\ 1 \\ 0 \end{pmatrix}. \tag{C.7}$$

4. What are the projection matrices **P** and **Q** and the projections **Py** and **Qy** if **H** and **y** are as above?

Answer

$$\mathbf{P} = \mathbf{H}^{\dagger}\mathbf{H} = \frac{1}{4} \begin{pmatrix} 2 & 1 & 0 & 1 \\ 1 & 2 & 1 & 0 \\ 0 & 1 & 2 & 1 \\ 1 & 0 & 1 & 2 \end{pmatrix} \frac{1}{4} \begin{pmatrix} 5 & 1 & -3 & 1 \\ 1 & 5 & 1 & -3 \\ -3 & 1 & 5 & 1 \\ 1 & 3 & 1 & 5 \end{pmatrix},$$

$$\mathbf{P} = \frac{1}{4} \begin{pmatrix} 3 & 1 & -1 & 1 \\ 1 & 3 & 1 & -1 \\ -1 & 1 & 3 & 1 \\ 1 & -1 & 1 & 3 \end{pmatrix},$$

$$\mathbf{Q} = \mathbf{I} - \mathbf{P},$$

$$\mathbf{Q} = \frac{1}{4} \begin{pmatrix} 1 & -1 & 1 & -1 \\ -1 & 1 & -1 & 1 \\ 1 & -1 & 1 & -1 \\ -1 & 1 & -1 & 1 \end{pmatrix},$$

$$\mathbf{Py} = \mathbf{y},$$

$$\mathbf{Qy} = \mathbf{0}.$$

5. Find $\mathbf{x}_{LS}$ if

$$H = \begin{pmatrix} 1 & -3 \\ 5 & 1 \\ 1 & 5 \\ -3 & 1 \end{pmatrix}$$

and **y** is as above.

Answer

$$\mathbf{x}_{LS} = (\mathbf{H}^{T}\mathbf{H})^{-1}\mathbf{H}^{T}\mathbf{y},$$

$$\mathbf{H}^{T}\mathbf{H} = 4 \begin{pmatrix} 9 & 1 \\ 1 & 9 \end{pmatrix},$$

$$(\mathbf{H}^{T}\mathbf{H})^{-1} = \frac{1}{320} \begin{pmatrix} 9 & -1 \\ -1 & 9 \end{pmatrix}.$$

Therefore

$$x_{LS} = \frac{1}{4} \begin{pmatrix} 1 \\ 1 \end{pmatrix}.$$

6. Find **x** using RAP with the constraints

$$x_1^2 + x_2^2 + x_3^2 + x_4^2 \leq 2 \qquad \text{and} \qquad 0 \leq x_i \leq 1.1.$$

(Do only the first six iterations.)
The RAP iteration can be written as

$$\mathbf{x}^{i+1} = \mathbf{x}^i + \mu e^i \frac{\mathbf{h}_j}{\|\mathbf{h}_j\|^2},$$

where $j = i \bmod n$ and n is the number of equations (here $n = 4$); i is the number of iterations done; e^{i+1} is the error for the ith iteration, or

$$e_{i+1} = y_j - \sum_{k=1}^{4} x_k^i h_{kj};$$

and μ is a scalar step size. Let $\mu = 1$ and the initial guess for **x** be

$$\mathbf{x}^0 = \begin{pmatrix} 0 \\ 0 \\ 0 \\ 0 \end{pmatrix}.$$

Substituting into the first equation, we get

$$\mathbf{e}_1 = -2 \qquad \text{and} \qquad \|h_0\|^2 = 25 + 1 + 9 + 1 = 36,$$

$$\mathbf{x}^1 = \mathbf{x}^0 + \frac{-2}{36} \begin{pmatrix} 5 \\ 1 \\ -3 \\ 1 \end{pmatrix} = \begin{pmatrix} -\frac{5}{18} \\ -\frac{1}{18} \\ \frac{3}{18} \\ -\frac{1}{18} \end{pmatrix}$$

Applying the constraint of positivity, we truncate all the negative numbers to zero; thus, Cx^0 is a convex operator that acts on x^0 to make it

$$Cx^0 = \begin{pmatrix} 0 \\ 0 \\ \frac{3}{18} \\ 0 \end{pmatrix}.$$

Now, substituting $\mathbf{Cx}^0$ into the second row,

$$e_2 = \frac{35}{6} \quad \text{and} \quad \|h_1\|^2 = 36,$$

and

$$\mathbf{x}^1 = \mathbf{Cx}^0 + \frac{35}{6}\frac{1}{36}\begin{pmatrix} 1 \\ 5 \\ 1 \\ -3 \end{pmatrix} = \begin{pmatrix} .162 \\ 0.81 \\ 0.162 \\ -0.486 \end{pmatrix}.$$

Applying positivity (the solution satisfies the energy criterion), we get

$$\mathbf{Cx}^1 = \begin{pmatrix} .162 \\ 0.81 \\ 0.162 \\ 0 \end{pmatrix}.$$

Similarly, using this as the input to row 3,

$$\mathbf{x}^2 = \begin{pmatrix} -0.2435 \\ 0.9454 \\ 0.8378 \\ 0.1352 \end{pmatrix}$$

and

$$\mathbf{Cx}^2 = \begin{pmatrix} 0 \\ 0.9454 \\ 0.8378 \\ 0.1352 \end{pmatrix};$$

after using this as an input to the last row and applying convex constraints,

$$\mathbf{Cx}^4 = \begin{pmatrix} 0 \\ 0.986 \\ 0.8676 \\ 0.0222 \end{pmatrix}.$$

It is clear that it is already close to the pseudo-inverse solution

$$\mathbf{x} = \begin{pmatrix} 0 \\ 1 \\ 1 \\ 0 \end{pmatrix}.$$

Try a few more iterations and check the convergence of the method.

7. Find **x** using the block projection method with the same constraints as above. (**Do only the first three iterations.**)

Answer Our initial guess is the pseudo-inverse solution. We see that this solution has no component in the null space, i.e., $Q\mathbf{x}^\dagger = \mathbf{0}$. Hence, the solution using the block method is the pseudo-inverse solution.

C.2 PROBLEMS IN IMAGE RECOVERY

1. Find an orthonormal set of vectors from

$$\mathbf{v}_1 = (1, 2, 3, 3),$$
$$\mathbf{v}_2 = (1, 1, 2, 2),$$

and

$$\mathbf{v}_3 = (-1, 2, 2, -1),$$

using Gram–Schmidt orthogonalization.

2. An $M \times M$ matrix **A** that is strictly diagonal is characterized by $|a_{ii}| > \sum_{j=1, j \neq i}^{M} |a_{ij}|$ for $i = 1, 2, \ldots, M$. Show that such a matrix is nonsingular using the Gershgorin theorem.

3. Compute the singular-value decomposition of the matrix

$$\mathbf{A} = \begin{pmatrix} 1 & 2 & 3 \\ 2 & 2 & 1 \end{pmatrix} \tag{C.8}$$

4. Use the singular-value decomposition from the previous problem to find **x** in

$$\begin{pmatrix} 2 \\ 3 \end{pmatrix} = \begin{pmatrix} 1 & 2 & 3 \\ 2 & 2 & 1 \end{pmatrix} \mathbf{x}. \tag{C.9}$$

5. Write a computer program to compute the singular-value decomposition of an arbitrary $M \times N$ matrix.

6. Convolve the following true image **f** with the PSF h to get the blurred image **g**, then study the following questions:

   a. Using the RAP algorithm (4.25 or the routine `rap2d( )` provided on the disk) to restore the blurred image **g** with the known PSF h, show how the result can be improved by using maximum and/or minimum bound constraint and by changing the relaxation factor λ. Compare the relative error and the speed of convergence.

b. Add 50 dB (signal-to-noise ratio) Gaussian noise to the blurred image **g**, then repeat the above question.

c. Repeat the above question with 25 dB Gaussian noise added to **g**.

$$f = \begin{pmatrix}
0 & 0 & 0 & 0 & 0 & 0 & 0 & 0 & 0 \\
0 & 60 & 60 & 60 & 60 & 60 & 60 & 60 & 0 \\
0 & 60 & 0 & 0 & 0 & 0 & 0 & 60 & 0 \\
0 & 60 & 160 & 160 & 160 & 160 & 0 & 60 & 0 \\
0 & 60 & 160 & 0 & 255 & 160 & 0 & 60 & 0 \\
0 & 60 & 160 & 255 & 0 & 160 & 0 & 60 & 0 \\
0 & 60 & 160 & 160 & 160 & 160 & 0 & 60 & 0 \\
0 & 60 & 0 & 0 & 0 & 0 & 0 & 60 & 0 \\
0 & 60 & 60 & 60 & 60 & 60 & 60 & 60 & 0 \\
0 & 100 & 100 & 100 & 255 & 180 & 120 & 80 & 0 \\
0 & 140 & 140 & 140 & 255 & 180 & 120 & 80 & 0 \\
0 & 180 & 180 & 180 & 255 & 180 & 120 & 80 & 0 \\
0 & 220 & 220 & 220 & 255 & 180 & 120 & 80 & 0 \\
0 & 255 & 255 & 255 & 0 & 0 & 0 & 0 & 0 \\
0 & 220 & 220 & 220 & 80 & 120 & 180 & 255 & 0 \\
0 & 140 & 140 & 140 & 80 & 120 & 180 & 255 & 0 \\
0 & 60 & 60 & 60 & 80 & 120 & 180 & 255 & 0 \\
0 & 0 & 0 & 0 & 0 & 0 & 0 & 0 & 0
\end{pmatrix}, \quad \text{(C.10)}$$

$$h = \begin{pmatrix}
0.05 & 0.1 & 0.05 \\
0.1 & 0.7 & 0.1 \\
0.05 & 0.1 & 0.05
\end{pmatrix}.$$

C.3 PROBLEMS IN 3-D IMAGE RESTORATION

1. A defocused imaging system, characterized by a PSF matrix

$$\mathbf{H} = \begin{bmatrix} 0.5 & 1.0 & 0.0 & 0.0 \\ 0.0 & 0.0 & 0.5 & 1.0 \end{bmatrix} \quad \text{(C.11)}$$

forms a two-pixel defocused image $\mathbf{g} = (50, 100)^T$, where the superscript T denotes the vector-transpose operation. Using the imaging relationship $\mathbf{g} = \mathbf{Hf}$, compute the four-voxel radiance vector **f** for the following cases:

a. No *a priori* information is available. Use the pseudo-inverse of **H**.

b. It is known that the object is opaque. Hence, in each of the sets (f_1, f_2) and (f_3, f_4) that correspond to voxels arranged along a line-of-sight, at most one element can have a nonzero value.

c. In addition to the opacity condition in (b), it is also known that the voxel radiances are quantized to one of three possible values, $\{0, 50, 100\}$.

2. Evaluate, for each of the three cases in problem 1, the image-irradiance error. Recall that this error is defined as $e_{image} = ||g - \hat{g}||^2$. Which case— a, b, or c—has the least error? Given that the lexicographic mapping of voxel-radiance indices i to 3-D coordinates (x, y, z) is $\{1, 2, 3, 4\} \rightarrow \{(0, 1, 0), (0, 1, 1), (0, 0, 0), (0, 0, 1)\}$, and assuming opacity, what are the voxel coordinates of the visible voxels associated with the least image-irradiance error?

3. An object space $\mathcal{O}$ is tessellated into two slices in depth (along the z axis), each slice being composed of eight cube-shaped voxels. An opaque staircase object is contained within $\mathcal{O}$. Its visible surface is orthogonal and forms a 2×2 pixel image g. A coarse depth map (D_{coarse}) reveals that the visible voxel coordinates have a constant depth (z-coordinate) along the horizontal (x) axis. That is,

$$(\mathbf{D}_{coarse}) = \begin{bmatrix} z(1) & z(1) \\ z(2) & z(2) \end{bmatrix}. \tag{C.12}$$

The defocused imaging system is characterized by a 8×16 matrix,

$$\mathbf{H} = \begin{bmatrix} A & 0 \\ 0 & A \end{bmatrix}, \tag{C.13}$$

where

$$\mathbf{A} = \begin{bmatrix} 1.0 & 0.5 & 0.0 & 0.0 \\ 0.0 & 0.0 & 1.0 & 0.5 \end{bmatrix} \tag{C.14}$$

and 0 is a null 2×4 matrix. The observed defocused image is given to be $\mathbf{g} = [64, 64, 16, 16]^T$. Using the imaging relationship $g = Hf$, estimate the optimal visible voxel coordinates, utilizing the coarse depth map, the opacity constraint, and the fact that the the voxel radiances are quantized to one of five possible levels, $\{0, 16, 32, 64, 128\}$.

C.4 PROBLEMS IN ADAPTIVE FILTERING

1. A signal undergoes amplitude scaling and is subsequently corrupted by additive white noise. We wish to construct a two-coefficient adaptive filter that undoes the amplitude distortion. The input and output of the adaptive filter are modeled as

$$x_i(k) = a\, d(k) + n(k) \tag{C.15}$$

and

$$x_o(k) = c_0 x_i(k) + c_1 x_i(k - 1), \tag{C.16}$$

respectively, where a is the fixed amplitude distortion and c_0 and c_1 are the filter coefficients. Calculate c_0 and c_1 by forming a 2×2 system of equations

$$\begin{bmatrix} d(k) \\ d(k-1) \end{bmatrix} = \begin{bmatrix} x_i(k) & x_i(k-1) \\ x_i(k-1) & x_i(k-2) \end{bmatrix} \begin{bmatrix} c_0 \\ c_1 \end{bmatrix} + \begin{bmatrix} e(k) \\ e(k-1) \end{bmatrix} \qquad \text{(C.17)}$$

and solving using the least-squares solution of (6.23). Show that it is possible for the least-squares estimate $c_1 \to \infty$ as $a \to 0$. Note that the correct values for the coefficients are $c_0 = 1/a$ and $c_1 = 0$ if $n(k) = 0$.

2. Using the same setup as in problem 1, we wish to regularize the solution. A simple form of regularization is to add a diagonal matrix to the data matrix of the form

$$\begin{bmatrix} x_i(k) & x_i(k-1) \\ x_i(k-1) & x_i(k-2) \end{bmatrix} + \begin{bmatrix} 0 & 0 \\ 0 & \gamma \end{bmatrix}, \qquad \text{(C.18)}$$

where $|\gamma| \gg 1$.

a. The eigenvalues of the original data matrix (C.17) are given as $\{\lambda_0, \lambda_1\}$, with $\lambda_0 \geq \lambda_1$. Show that the smallest eigenvalue of the augmented data matrix (C.18) is $\min\{\lambda_0, \lambda_1 + \gamma\}$.

b. Show that the regularized least-squares estimate of c_1 is proportional to $1/\gamma$ as $a \to 0$.

3. Using (6.55) and (6.56), derive the expression in (6.78).

4. Using (6.55) and (6.56), derive the expression in (6.81).

5. An inconsistent set of three hyperplanes is represented by Figure C.1.

Find the optimum coefficients, i.e., the point with the minimum sum of squared projection distances. Show that a sequence of orthogonal projections approaches the optimum point only as $\mu \to 0$. This condition is analogous to excess mean-square error in the LMS algorithm.

C.5 PROBLEMS IN SPECTRAL ESTIMATION

1. Solve the following system of equations to find the least-squares solution, X, which meets the given constraints. Use RAP, the positivity constraint, and an upper bound of 1.0:

$$\begin{pmatrix} -0.5 \\ -0.5 \\ -0.5 \\ -0.5 \\ -0.5 \end{pmatrix} = \begin{pmatrix} 0.625 & 0.223 & -0.223 & -0.625 & -0.9 \\ -0.223 & -0.9 & -0.9 & -0.223 & 0.625 \\ -0.9 & -0.625 & 0.625 & 0.9 & -0.223 \\ -0.9 & 0.625 & 0.625 & -0.9 & -0.223 \\ -0.223 & 0.9 & -0.9 & 0.223 & 0.625 \end{pmatrix} \cdot \begin{pmatrix} x_1 \\ x_2 \\ x_3 \\ x_4 \\ x_5 \end{pmatrix} . \qquad \text{(C.19)}$$

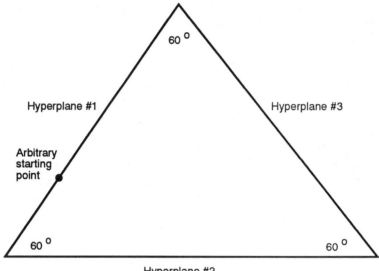

Figure C.1

2. For the chirp signal given by $s(t) = e^{j\beta t^2 + j0.5\pi t}$, find the time-versus-frequency plot for $\beta = 0.0, 0.5, 1.0$, and 2.0 and show that the slope of the frequency peak is equal to the derivative of the phase.

3. Calculate the Wigner distribution and the time-versus-frequency plot for the two-chirp signal given by $s(t) = e^{j0.5t^2 + j0.1\pi t} + e^{j-0.5t^2 + j0.9\pi t}$. Use the subspace partitioning constraint for the RAP method to constrain the signal within each chirp's bandwidth. Explain the decrease in performance as the chirps approach the crossing point.

C.6 PROBLEMS IN CHANNEL EQUALIZATION

1. Derive a finite-time-duration waveform that satisfies Nyquist's criterion. Show that the sampled waveform spectrum is constant.

2. Derive a first order approximation of the minimum eigenvalue of a correlation matrix for a raised cosine pulse transmitted over an ideal channel. Assume that you have a linear transversal equalizer of length N taps spaced at T seconds. Your answer should be a function of T and the roll-off parameter β. (Hint: Use (6.41) in Chapter 6.)

3. Derive the equation for the Euclidean norm of the DFE correlation matrix (9.34), assuming correct symbol feedback and an infinite length feedback section. The transmitted symbol sequence and the noise are both white and mutually uncorrelated.

4. Show that $\lim_{n \to \infty} X^n = 0$ only if the eigenvalues of X are all less than unity.

5. Show that the operation of the decision device in (9.21) can be interpreted as finding a hyperplane with minimum projection distance.

C.7   PROBLEMS IN ADAPTIVE BEAMFORMING

1. A 15-element sonar array ($c = 1.5 \times 10^3$) has its elements uniformly spaced 0.2 m apart:

   a. Find the maximum frequency that may be transmitted without creating grating lobes for a $-90°$ to $90°$ visible region.

   b. The angle of arrival for a 50 kHz wavefront is $30°$. Find the electrical angle.

   c. Find the spatial frequency of the wavefront in b in units of cycles/m.

2. Obtain an expression for the far-field pattern of a linear array whose elements are equispaced and have triangular aperture patterns. Follow the explanation given in Section 10.4 and assume that the triangles have a base width of $2M$ and a height of $1/M$.

3. A five-element narrowband array is focused on a target signal arriving normal to the array:

   a. What is the maximum number of signals (arriving from directions other than the target signal) that the array can cancel while remaining focused on the target signal?

   b. At a given time instant, it is known that the target signal has a value of 1.0. Find a vector that will allow the target signal to pass with unity gain while canceling a single noise source if the measured values at the array are

   $$X = [2.0, 1.707, 1.0, 0.293, 2.0]$$

4. Using (10.30), show that the optimum weight vector for the broadband array described in Section 10.12 is given by

   $$W_{opt} = R_{xx} C [C^T R_{xx} C]^{-1} F.$$

5. Using the program narrow.c with 5 sensors, obtain the output files for scenarios with 2, 3, and 4 interference sources and a boresight target signal. Using an FFT algorithm, obtain the output spectra for each scenario. Discuss the results of each scenario with respect to degrees of freedom.

6. Using the program `broad.c` with a scenario having one boresight signal and two wideband interference signals,

    a. Obtain the output files for an array with 5 sensors and 5 taps.

    b. Obtain the output files for an array with 3 sensors and 9 taps.

    c. Compare the results of a and b with respect to performance.

C.8 PROBLEMS IN NEURAL NETWORKS

1. The exclusive or problem is described in Chapter 11 ("Neural Networks"). Apply the neural tree network and the multilayer perceptron to this problem. Generate the neural tree network and compare the classification performance to a MLP with 2, 3, and 4 hidden nodes. Give some explanation for the differences in performance.

2. Apply the neural tree network and the multilayer perceptron programs to the four-variable boolean function problem described in Chapter 12 ("Neural Tree Networks") and compare the classification performance to an MLP with 4, 8, and 12 hidden nodes. Give some explanation for the differences in performance.

INDEX